Ecosystem Services Design from Single Space Solution to Landscape Vision

Ecosystem Services Design from Single Space Solution to Landscape Vision

Guest Editor

Teodoro Semeraro

MDPI

Basel • Beijing • Wuhan • Barcelona • Belgrade • Novi Sad • Cluj • Manchester

Guest Editor
Teodoro Semeraro
Research Institute on
Terrestrial Ecosystems
(IRET-URT Lecce)
National Research Council of
Italy (CNR)
Lecce
Italy

Editorial Office
MDPI AG
Grosspeteranlage 5
4052 Basel, Switzerland

This is a reprint of the Special Issue, published open access by the journal *Land* (ISSN 2073-445X), freely accessible at: https://www.mdpi.com/journal/land/special_issues/ecosystem_services_design.

For citation purposes, cite each article independently as indicated on the article page online and as indicated below:

Lastname, A.A.; Lastname, B.B. Article Title. *Journal Name* **Year**, *Volume Number*, Page Range.

ISBN 978-3-7258-5449-3 (Hbk)
ISBN 978-3-7258-5450-9 (PDF)
https://doi.org/10.3390/books978-3-7258-5450-9

Contents

About the Editor

Teodoro Semeraro

I am a PhD researcher and ecologist specializing in environmental science and landscape and urban sustainability. As the Principal Investigator (PI) of the Strategic Environmental Assessment (SEA, Directive 2001/42/CE) for landscape and urban planning, I have managed complex processes that integrate technical evaluation, stakeholder engagement, and interdisciplinary collaboration. My responsibilities included designing and assessing plans, contracting consultants, and managing client relationships. Through these experiences, I have developed a transdisciplinary approach that bridges business investment needs, societal requirements, and ecological priorities. My expertise spans Geographic Information Systems (GISs), biodiversity assessment, ecosystem service evaluation, nature-based solutions, green Infrastructure analysis, and environmental assessments. These skills enable me to connect diverse disciplines and contribute to innovative and sustainable planning strategies.

Preface

The aim of this Special Issue is to promote innovation in the design and management of green spaces as multifunctional infrastructures for regenerating ecosystems and delivering priority ecosystem services such as climate regulation, biodiversity conservation, and social well-being. The issue will highlight 'learning by doing' approaches, from practical case studies to policy visions.

Teodoro Semeraro
Guest Editor

land

MDPI

Article

Habitat Restoration: An Applicative Approach to "Biodiversity Heritage Relicts" in Social-Ecological Systems

Teodoro Semeraro [1,*], Alessio Turco [2], Stefano Arzeni [2], Giuseppe La Gioia [3], Roberta D'Armento [4], Riccardo Taurino [5] and Pietro Medagli [2]

1 Department of Biological and Environmental Sciences and Technologies (DiSTeBA), University of the Salento, 73100 Lecce, Italy
2 DHITECH S.c.a r.l. Prov.le Lecce Monteroni, 73100 Lecce, Italy; alessio.turco@unisalento.it (A.T.); stefano.arzeni@unisalento.it (S.A.); pietro.medagli@unisalento.it (P.M.)
3 Mediterranean Ornithology (Or.me): Ornithology in Puglia, Via Saponaro n 7, 73100 Lecce, Italy; lagioiagiu@gmail.com
4 Architecture and Urban Planning, V. Taranto 33, Campi Salentina, 73012 Lecce, Italy; robedarme@hotmail.com
5 Campi Salentina Amministration, Piazza Libertà, 33, Campi Salentina, 73012 Lecce, Italy; riccardotaurino@libero.it
* Correspondence: teodoro.semeraro@unisalento.it or teodorosemeraro@gmail.com

Abstract: Many landscapes are the result of interactions between ecological processes, economic activities, and the administrative and political organisation of society. Therefore, as a consequence of human transformations over time, some landscapes may contain residual damaged habitats hosting testimony of past biodiversity that can be called "biodiversity heritage relicts". From this perspective, the aim of the paper is to describe an applicative approach to habitat restoration in social-ecological landscapes. The approach entails the restoration of vegetation using GIS analysis integrated with field activities and a phytosociological method. The methodology includes expert and stakeholder involvement in order to increase the resilience of the measures over time, thereby consolidating landscape value. The approach was applied in the municipality of Campi Salentina, Province of Lecce, Italy, and the result was the restoration of an important riparian habitat classified under Directive 92/43/EEC as "*Salix alba and Populus alba galleries*" (code 92A0), which had not previously been recorded in the Province of Lecce. In this case, the project re-established a natural habitat that represented a "biodiversity heritage relict" in the landscape. The paper shows that direct knowledge of the landscape and the ability to identify "biodiversity heritage relicts", in combination with a phytosociological approach, can enhance the effectiveness of ecological restoration projects. Moreover, social and institutional integration in projects helps ensure the management of the measures over time.

Keywords: habitat damage; biodiversity conservation; phytosociology; landscape

1. Introduction

Many landscapes are the expression of the interactions between natural environments and human activities, which adapt ecosystems to human needs [1–4]. In these conditions, the landscape may be regarded as a Social-Ecological System (SES) characterised by the coevolution of ecological, economic, and social components, producing specific and distinctive features. Therefore, landscape evolution can be influenced by different values and importance that humans give to biodiversity at a certain time considering human needs [5–8]. For example, in Italy, in the past, landscape transformation was strongly influenced by Institutions that aimed to reinforce the social and sanitary aspects against the spread of malaria. Indeed, from 1882 up to the fascist era they started to develop a series of reclamation actions that promoted the conversion of wetlands, marshes, areas of temporary flooding or ponds into agricultural land [9–11]. In contrast with the past approach, today, the ecological role of wetlands, marshes, areas of temporary flooding and ponds is recognised to such an extent that there are some international strategies and policies to preserve and support them, such as the Ramsar Convention and European bio-diversity directive 92/43/EEC [12,13].

The technological and cultural evolution of society can influence landscape evolution. Indeed, in the past, forests had an important role in food and raw material production, e.g., chestnut forests were introduced by Romans to overcome periods of famine while other types of forests were used for ships' construction. The advance of forests in the first centuries of the Middle Ages coincides with their enhancement, not only economic but also cultural. They are perceived with a positive value that was not seen in the Roman world. In the modern era, wood has been replaced in many applications by new artificial materials and forests are protected and developed as important landscape elements for the biodiversity conservation and to fight climate change [14]. In addition, economic and social aspects can in time push landscape evolution, e.g., the growing needs of clean energies produced the development of wind and photovoltaic farms in the agricultural landscape in recent years [15]. Today, natural ecosystems, intended as potential vegetation that can grow in a place in harmony with the biotic and abiotic components that characterise a place and without human influence [16,17], are largely confined within protected areas. This represents the main international strategy to ensure their enhancement and conservation. By means of specific regulations and restrictions on land use, the protected areas play an important role in biodiversity conservation limiting the negative effects of human activities [12–18].

The interaction between nature and human needs represents the basis for the historical development of traditional agricultural landscapes [19–21] and is at the basis of the concept of the cultural landscape [22,23]. This holistic philosophy maintained the concept of connectivity and interdependence in the forefront of the community's consciousness, and it influenced the development of communities as well as the agroecology systems needed to sustainably support them [24]. For example, in the Puglia region, South Italy, the historical monumental olive groves are protected by Regional Law 14/2007 considering both data like shape and dimension of the trunk of the olive trees and the spiritual value that the olive groves could have [4]. In the SES there are still elements of naturalness that are not protected or enhanced. These areas represent the testimony of past landscape patches that have been heavily damaged, defined as acute and obvious changes [25], by anthropic activities such as agriculture and urbanisation. Therefore, some landscapes can be characterised by residual vegetation patches which testify to the presence of habitats that were strongly damaged by human activities or natural events of which sometimes the value or link with biodiversity and cultural value is not known or detected. Such vegetations can be described as "Biodiversity Heritage Relicts" (BHR) because they can represent an important testimony of past ecological diversity and socio-cultural and economic evolution of the landscape.

The problem is that BHR value in the landscape could be undetected or unknown and therefore not preserved, and their value goes unnoticed. Indeed, due to their state of degradation and limited size, their cultural and ecological value in a landscape is often not perceived or studied, so they are left to human pressures like agriculture and pasture, which can lead to their complete disappearance and hence the loss testimony of past landscapes [26–30]. Therefore, in order to fulfil human needs in synergy with landscape conservation and valorisation, habitat restoration in BHR is essential to avoid their total destruction and transformation over time [31].

While it is important to invest economic resources in protected areas to strengthen biodiversity, it is even more important to invest in habitat restoration projects outside them, identifying "biodiversity heritage relicts" and preserving them, at least where abiotic and biotic components characteristic of the natural habitat are still present [32,33].

Some applications of habitat restoration projects are based on landscape indices or surrogate variables such as "patch area" or "dominant vegetation type". Alternatively, it may focus on the arrangement and size of natural patches and the fragmentation of the landscape [34,35]. Community and vegetation indices and biodiversity indices (such as the Coefficient of Conservatism) are used to assess the level of biodiversity or the relations between the presence of native species and human disturbances levels, but they cannot give information about the cultural and ecological value of the vegetation present and mainly the knowledge that links the vegetation with evolution of the landscape [36]. Therefore, the high or low value of these indices may not be a good indicator of the suitability or necessity of ecological restoration projects aimed at BHR. In an SES, habitat restoration not only has to restore species diversity but must also consider what causes the value loss of biodiversity over time [37,38] and its evolution and relationship with the social and economic aspects of landscape governance. Indeed, financial constraints, which determine what measures are realistic, and social constraints, which determine whether a given habitat restoration project is acceptable to all stakeholders [33], cannot be ignored, especially if the site of interest is in a private area or an economic productive context. Indeed, the restoration project can produce a dichotomy of interest in social components between the community that can percept positive benefices from the restoration project and involved landowners that can feel that they are not economically adequately recompensated for damages for their activities [15,24]. Therefore, habitat restoration projects must be guided, seeking to reconcile the needs of habitat restoration with local activities and encouraging social involvement, and the acceptance of all stakeholders is a key asset for habitat restoration projects [5,39].

In the light of these considerations, habitat restoration projects cannot be based on mere permutations of indices but on the integration of knowledge, institutions, and stakeholders to ensure the right feedback concerning a range of aspects such as microclimate, habitat connectivity, food resources, predation risk, human disturbance, and others [40–42].

In this paper, starting from a case study, we presented an approach applied to habitat restoration in the Social-Ecological System of the municipality of Campi Salentina, in the Puglia region in Southern Italy. In the developed approach, importance is given to the way of observing and analysing biodiversity. Attention is paid to the capacity to discover and valorise the "Biodiversity Heritage Relicts" found in landscape that in other ways can be definitively lost in time.

The project focused on an intermittent watercourse called the "Lacrima", seeking to restore a habitat that had never been surveyed, containing patches of vegetation attributable to 92A0 "*Salix alba and Populus alba galleries*" in consideration of the biotic and abiotic components that characterised the area.

Furthermore, in the context of the public funding received, we reported an analysis of the relevant administrative issues that have affected the development of the project.

2. Materials and Methods

2.1. Study Area

The study focused on the municipality of "Campi Salentina", in the province of Lecce, Italy characterised by a mainly agricultural landscape, with vineyards, olive groves, and arable fields fragmented by rural roads (Figure 1).

The Lacrima River was strongly influenced by human activities as its entire course was waterproofed and as its surroundings are characterised by agricultural production on both the sides. The Lacrima River was never subject to conservation actions or direct management activities regarding the vegetation until the situation developed significantly. The only activities carried out were the hydraulic maintenance for hydrogeological mitigations, and the main institutional rules applied are focused on the restriction of urbanization growth to keep the hydrogeological risk low. Many parts of the area of the Lacrima River focused on in this project were characterised by private lands with limited access for the community to walk.

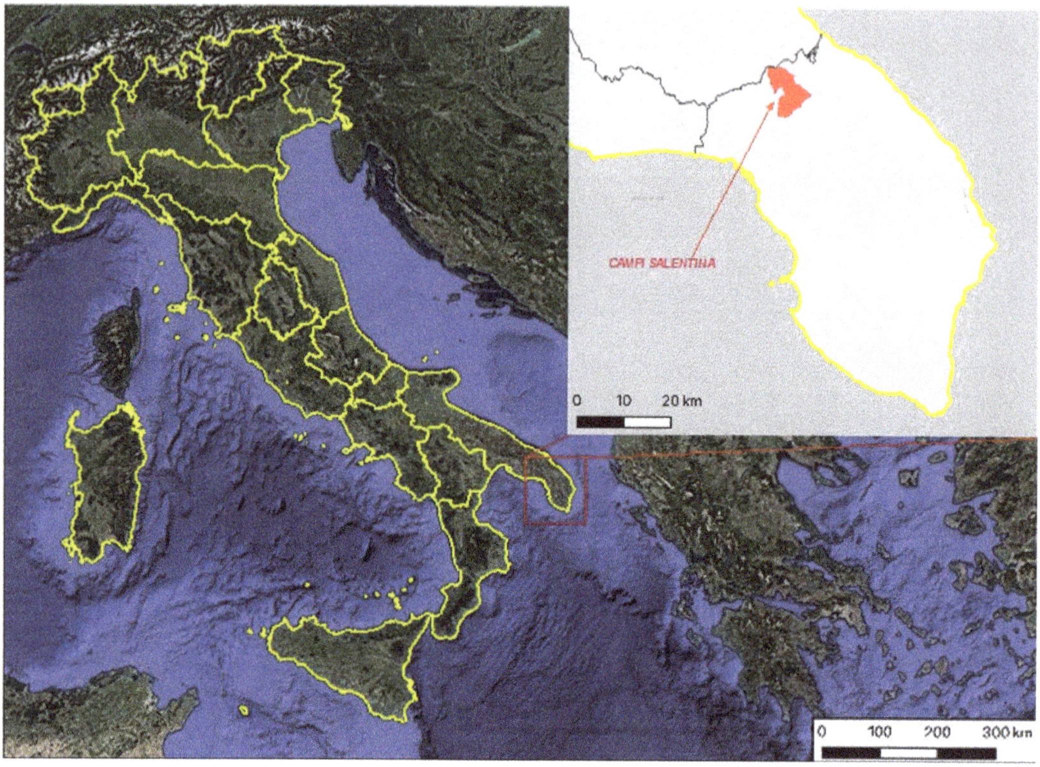

Figure 1. Location of study area.

In this area, the European Biodiversity Directive 92/43/EEC did not point to any presence of habitat and vegetations of interest in terms of biodiversity conservation until the current project development, and it was not highlighted as an important area of glaciation of floristic refuges in southern Italy [43].

The idea for a habitat restoration project was prompted by a call for project proposals to be funded as part of Puglia Regional Operational Program 2014–2020 ("POR PUGLIA 2014–2020, Asse VI–Azione 6.6—Sub-Azione 6.6.a, "Interventi per la tutela e la valorizzazione di aree di attrazione naturale", Realizzazione di progetti per la rete ecologica regionale"). There were to be a total of 10 projects, with a maximum budget of 1.3 million euros for each [44].

In this paper, we present the methodology adopted for the project called "The Unknown Past And Aware Future Of The Landscape Mosaic Of the Lacrima River: A Project For the Ecological Restoration Of The Riparian Forest". The project focused on a stretch of the river located in the municipality of "Campi Salentina" and was financed by Puglia Regional Administration. The main aims of the POR Puglia 2014–2020 axis VI is Environmental protection and promotion of natural and cultural resources and some activities promoted are: redevelopment and reconstruction of degraded landscapes; removal of detractors of landscape quality, relocation of network infrastructures and related spaces, naturalistic restoration; production of specific representations of the landscape values described for each reference area, and creation of innovative systems and services for the use of resources.

2.2. Methods

One of the important phases of this ecological restoration project was to decide the goals, aims, and priority actions [31,45] (Figure 2). Therefore, the starting point of the methodology applied was to site analysis combining GIS analysis with field activities to pinpoint the potential sites of interest for the project, i.e., those that were connected with specific elements of the landscape that were already known. This included the phytosociological analyses of the vegetational characteristics of the areas identified in the first step, and assigning at each representative species a coverage value according to the Braun-Blanquet ordinal scale [46].

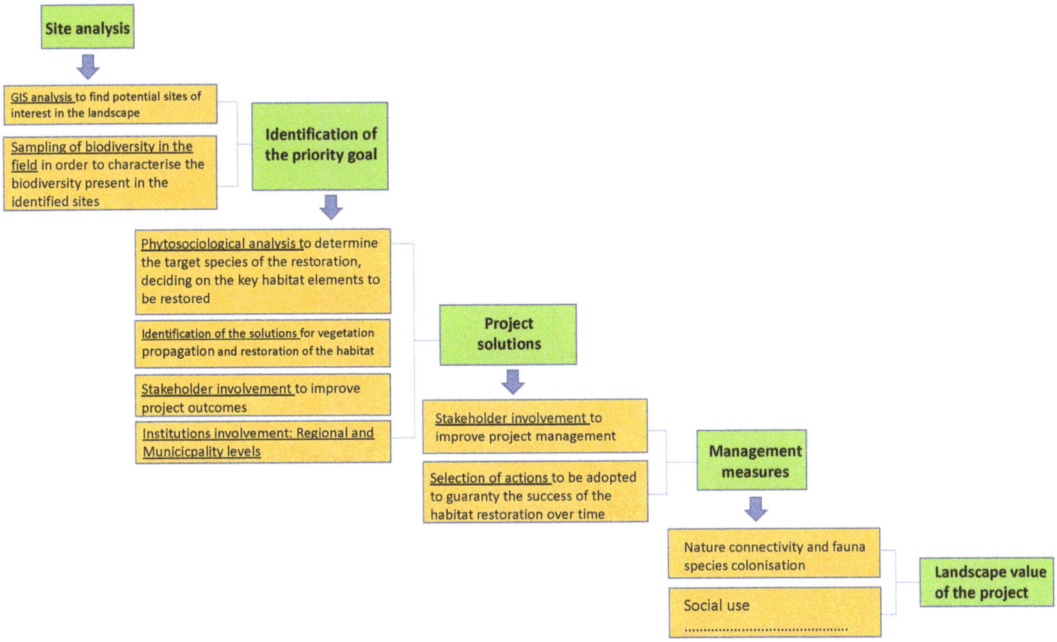

Figure 2. Methodological flowchart of the ecological restoration project realised in the municipality of Campi Salentina.

After that, an important action was to identify a reference habitat for the restoration project starting from the phytosociological analysis results of the residual vegetation discovered in relation to the abiotic components of the area. This represented the model for planning the habitat restoration project [25]. Therefore, the next phase of the ecological restoration entailed the use of phytosociological analysis results to study groups of species of plant that usually compose one specific habitat [47]. This is important in order to understand the existing vegetational composition and to correctly plan all the restoration measures, including the choice of species to be used for habitat restoration and assessment of the potential biodiversity of the area selected in consideration of the environmental conditions and landscape characteristics. As previously mentioned, this analysis helped to understand how the vegetation might evolve under natural conditions, and to select the most suitable vegetation to be used in the restoration [48–51].

Naturally, the introduction of new vegetation is not a trivial matter but must take account of the expected results and the capacity of the vegetation to take root. Therefore, an important aspect was to plan vegetation management measures over time in order to guarantee high resilience during the initial phase of plant development. Moreover, it was necessary to consider the various stakeholders and types of land use that might constrain the project, using this knowledge to develop measures to reduce human disturbance over time, as well as drivers and pressures that might result in habitat degradation.

Lastly, we analysed the potential influence of the habitat restoration project on the social and ecological components of the landscape on a bigger scale than the site area, with a view to future developments. Therefore, a qualitative impact analysis of the land cover change generated by the project was carried out analysing the variation of ecosystem services provision defined like the benefits to humans provided by the natural environment and from healthy ecosystems [52,53]. We gave an indication of the type of ecosystem services improved or reduced from the land cover change and from the functional aspect of vegetation introduced considering the bibliography information and different experts involved in the project [2,24,54]. This phase helps to set secondary project goals in terms of the creation of positive ecological and social externalities [55,56].

In any Social-Ecological System, it is important to consider the social and institutional components that can influence the ecological and economic aspects of the landscape [55]. Therefore, we reported the administrative feedback on the applied restoration project realization. In this regard, we analysed specific measures carried out by the administration in certain project phases ranging from the initial draft of the project to the awarding of the contracts for the necessary work.

3. Results

3.1. Site Analysis

The GIS analysis highlighted four forest areas characterised by the presence of *Quercus ilex* L., which is the potential natural vegetation in the Salento, widespread in other sites. After that, aerial orthophotos were analysed to find areas of potential interest that were not highlighted by the previously acquired information. Along the Lacrima River, which is an element of ecological interest for the Landscape Plan of the Puglia Region, this technique highlighted a small and unrecorded vegetation patch, not cited in the Landscape Plan or any other biodiversity study or plan (Figure 3).

To explore the vegetation along the Lacrima River, a botanical survey was conducted to characterise its flora and vegetation. During the sampling, two patches of hygrophilous vegetation were found (Figure 4).

The patches are characterised by the presence of *Populion albae* Br.-Bl. ex Tchou 1948 consisting of specimens of, *Populus alba* L., *Populus tremula* L., *Ulmus minor* Mill. and occasionally *Quercus virgiliana* Ten. These represent unique testimony and evidence of past mature riparian habitat in the Province of Lecce (Table 1).

Table 1. Phytosociological analysis of the arboreal vegetation.

<table>
<tr><th>Species</th><th>Braun-Blanquet Scale</th><th>Description</th><th>Extension (mq)</th></tr>
<tr><td>Populus alba L.</td><td>4</td><td>species forming colonies extending over more than half of the relief surface</td><td rowspan="2">3500</td></tr>
<tr><td>Populus tremula L.</td><td>1</td><td>isolated individuals</td></tr>
<tr><td>Quercus virgiliana Ten.</td><td>2</td><td>individuals gathered in groups</td><td rowspan="2">2500</td></tr>
<tr><td>Ulmus minor Mill.</td><td>4</td><td>species forming colonies extending over more than half of the relief surface</td></tr>
</table>

Figure 3. Landscape Elements of interest protected under the Puglia Regional Landscape Plan [44].

Figure 4. A "biodiversity heritage relict" in the Municipality of Campi Salentina.

The presence of this habitat, which is typical of rural areas with a good natural water supply, is consistent with the historical map, which shows the presence of a natural vegetations in 1800 in addition to the Lacrima River, how showed by the Plate from the Geographical Atlas of the Kingdom of Naples in the period 1788–1822 by Antonio Rizzi-Zannoni [57].

Therefore, this nucleus of vegetation might correspond to a BHR of the riparian habitat that is unique in the Province of Lecce and has never been detected before. It is located in the agricultural landscape pattern characterised mainly by arable lands, olive groves, and vineyards (Table 2) (Figure 4).

Table 2. Land cover characterisation of the Municipality of Campi [44].

Land Cover	ha	%
Forests	8	0.2
Grasslands	66	1.4
Arable lands	1646	36.3
Urban areas	440	9.7
Olive groves	1221	27.0
Vineyards	975	21.5
Other	174	3.8
Total	4529	100.0

3.2. Identification of the Priority Goals of the Restoration Project

In this area, although only seven adult plants were discovered, numerous young plants and many seedlings were found, demonstrating the suitability of the environmental conditions for the development of this phytocenosis. Indeed, they are stable cenoses until the hydrological conditions of the stations on which they develop change. The aim of the habitat restoration project was to focus on recovering the damaged habitat, starting from the elements already partially present in accordance with the scope of the call for project proposals. This involves: strengthening and functioning of the connection function of ecological corridors in the landscape, contrasting the processes of fragmentation and degradation of the biodiversity, and increasing ecological functionality and regional biodiversity levels.

Therefore, the restored vegetation proposed in this project will help to develop and expand the habitats cited in Directive 92/43/EEC, Annex I, 92A0—"*Salix alba and Populus alba galleries*" which have been reduced to a fraction of their former extent [37].

This habitat is widespread in Italy, but on a regional scale, being mainly concentrated in the north and not reported in the south of Puglia, including the Province of Lecce (Figure 5). Therefore, the project will enhance and enrich biodiversity in southern Puglia by creating natural patches that ensure conservation and ecological enhancement in an otherwise purely agricultural landscape.

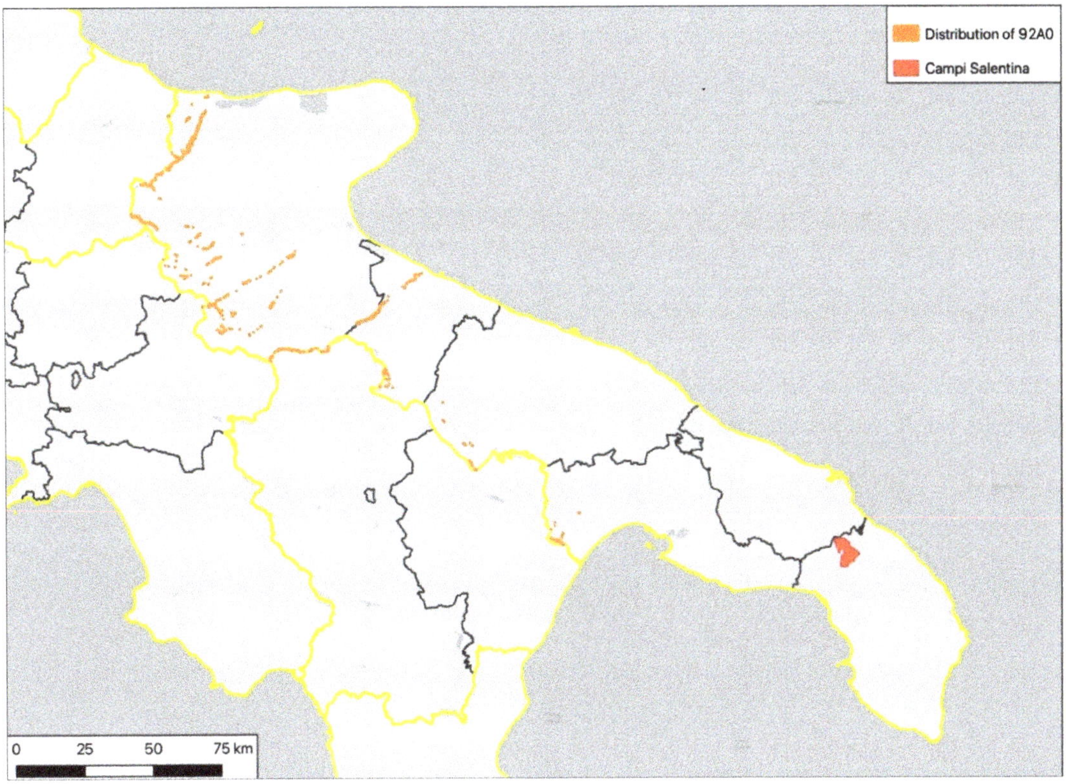

Figure 5. Distribution of *Salix alba and Populus alba galleries* in the Puglia Region [45].

3.3. Project Solutions

In this project, the "biodiversity heritage relict" was amplified and enhanced by including botanically compatible species typical of this area.

By introducing plant elements and further consolidating the embankments, over time, the restoration measures will favour the expansion of the existing spontaneous vegetation, which is typical of natural ecosystems in which periods of considerable water flow alternate with drier periods.

An important aspect of the project was the budget, which was €1.3 million including administrative costs. Therefore, a key aim of the project was to eliminate the sources of negative pressure acting on the vegetation which can lead to its disappearance as a result of the contextual expansion of the agroecosystems. In addition, there was an attempt to develop institutional strategies that would guarantee the management of the habitat over time and encourage social and cultural initiatives.

Considering the amount allocated for the restoration project, the initial phase of the project planned to create two groups of vegetation with arboreal species located in the initial and final part of the project area and to perform recovery measures on the banks of the Lacrima River, removing ruderal vegetation and sowing grassland vegetation such as *Paspalum distichum* L. Along the river, a pathway for river maintenance was planned and the accessibility of the community in the area was increased (Figure 6A).

Thanks to the involvement of Puglia Regional Administration and the local landowner, it was possible to integrate the project with a strip of vegetation along the two banks of the river. In addition, the Municipality of Campi Salentina held meetings with the owner that enabled this vegetation to be introduced without increasing the costs of land acquisition.

Therefore, the final project includes two arboreal-shrubby (the area I is about 4 ha, and the area II is about 1 ha) patches hosting riparian forest habitat in two specific places along the river that will be connected by herbaceous and shrubby vegetation along the banks (Figure 6B). The total surface of new vegetation development is 3.5 ha (excluding the vegetation already present) with 5600 new plants introduced (excluding the replacement of dead or unrooted plants).

Great importance was given to the choice of plant species to be used for the restoration of riparian habitats and the enhancement of the specific vegetation forming these natural habitats. By means of a phytosociological study and a careful analysis of the vegetation present on the site and in the neighbouring areas, a list of species suitable for the intended purpose was drawn up. Phytosociology was used to analyse the relationship in terms of quantity and coverage between the various potential species of the association and to gather information on the plants to be included in the restoration. The following tree-shrub species were used for the riparian forest: *Populus alba*, *Ulmus minor*, that characterized the 20% of the vegetation planted in the proximity of the river side, *Quercus virgiliana*, *Fraxinus angustifolia* subsp. *oxycarpa* (M. Bieb. ex Willd.) Franco and Rocha Afonso, *Crataegus monogyna* Jacq., *Ligustrum vulgare* L., *Sambucus nigra* L. and *Vitex agnus-castus* L., while for the garrigue, maquis and woodland, *Quercus ilex* L., *Pistacia lentiscus* L., *Smilax aspera* L., *Lonicera implexa* Aiton, *Rosa sempervirens* L., *Phlomis fruticosa* L. and *Cistus* sp. pl. were planted. Unfortunately, the current restrictions designed to curb the spread of the *Xylella fastidiosa* bacterium in the Salento prevented the planting of other Mediterranean scrub and garrigue species.

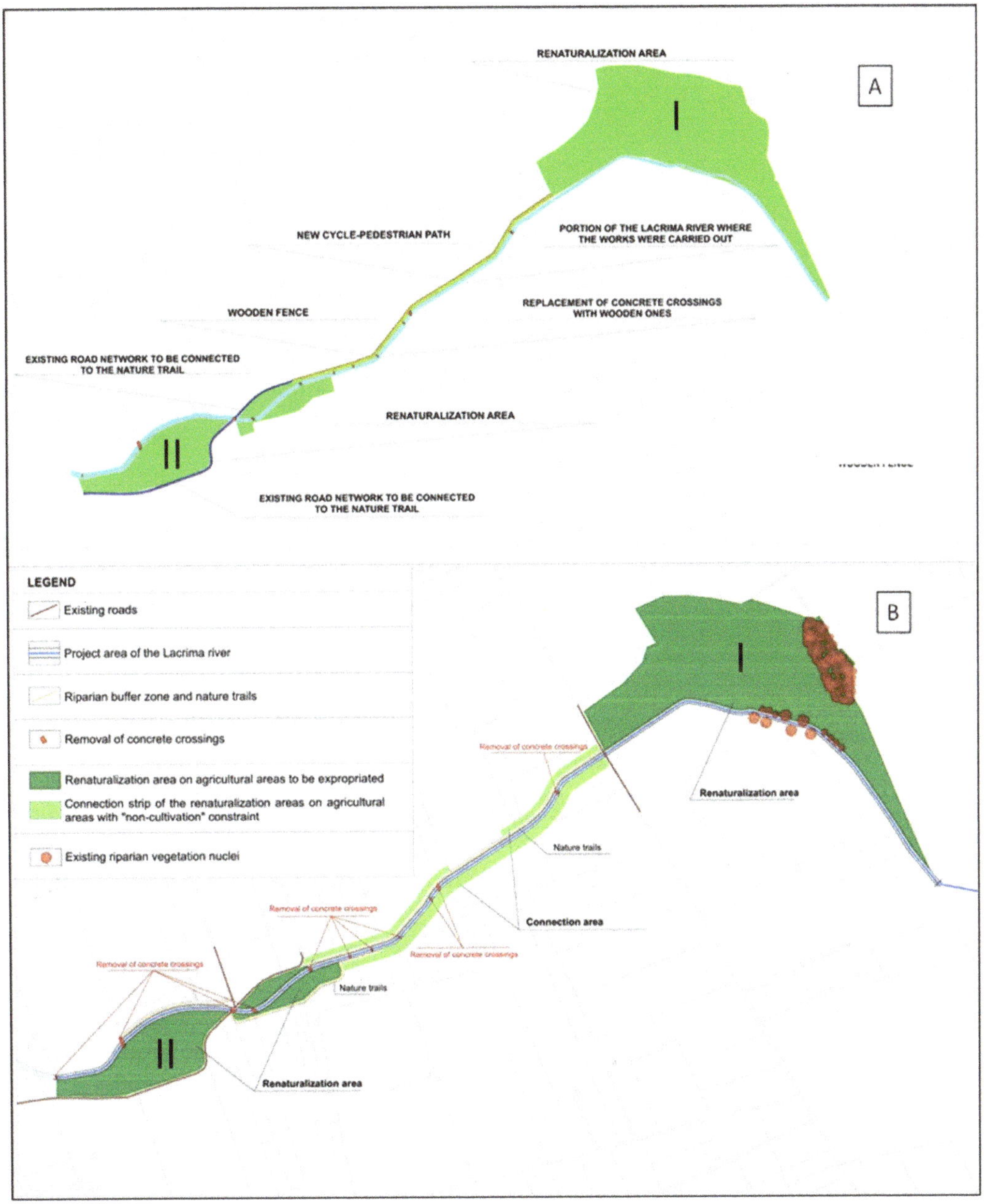

Figure 6. Plan of the ecological restoration project. (**A**) First draft of the project and (**B**) second evolution of the project.

In Table 3, the frequency (quantity) of each species used for riparian forest and Mediterranean scrub/garrigue are reported.

Table 3. Percentage of species used for the restoration of the riparian forest and Mediterranean scrub/garrigue.

SPECIES	*QUANTITY USED (%)*
Ulmus minor	8.3
Quercus Virgiliana	5.8
Populus alba	11.7
Fraxinus angustifolia subsp. *oxycarpa*	2.5
Crataegus monogyna	1.7
Ligustrum vulgare	3.3
Sambucus nigra	1.3
Vitex agnus-castus	1.3
Cornus sanguinea	1.5
Quercus ilex	14.2
Pistacia lentiscus	20.0
Smilax aspera	5.8
Lonicera implexa	5.8
Rosa sempervirens	3.3
Cistus sp. pl.	6.7
Phlomis fruticosa	6.7

The planting of the vegetation sought to achieve seriation similar to the natural habitat, particularly concerning the *Salix alba and Populus alba galleries*. In this regard, the intention was to create a gradual transition from hygrophilous plants (near the channel) to increasingly smaller xerophilous species (scrub and garrigue) towards the surrounding agricultural land (Figure 7). To simulate a natural ecosystem, the planting was performed randomly, with no geometrical shape or pattern.

The perceptual value of the landscape was increased, since all the detractors were eliminated. These included walkways and concrete elements that were replaced with structures made of wood or other natural materials that were conducive to a "beautiful landscape".

Vegetation development in the landscape is the result of the interaction of genetic factors and environmental conditions [58]. Therefore, seeds and cuttings of local ecotypes were used as propagation material to preserve genetic fitness [25]. For *Populus alba, Quercus virgiliana* and *Ulmus minor*, propagation material was collected from the biodiversity heritage relict itself. This served to avoid genetic contamination caused by alien species and ecotypes from different ecological contexts and to ensure that the introduced vegetation was already adapted to the natural environment of the project. This helps avoid introducing ecotypes to the environment that are not from Puglia, which might hybridise with indigenous ecotypes, causing extraneous genes to flow into the local populations. Such genes risk altering the characteristics of the local flora, selected by the area's distinctive environmental conditions, which is in perfect harmony with the climate and able to react in specific ways to regional environmental stressors typical of the Salento [25].

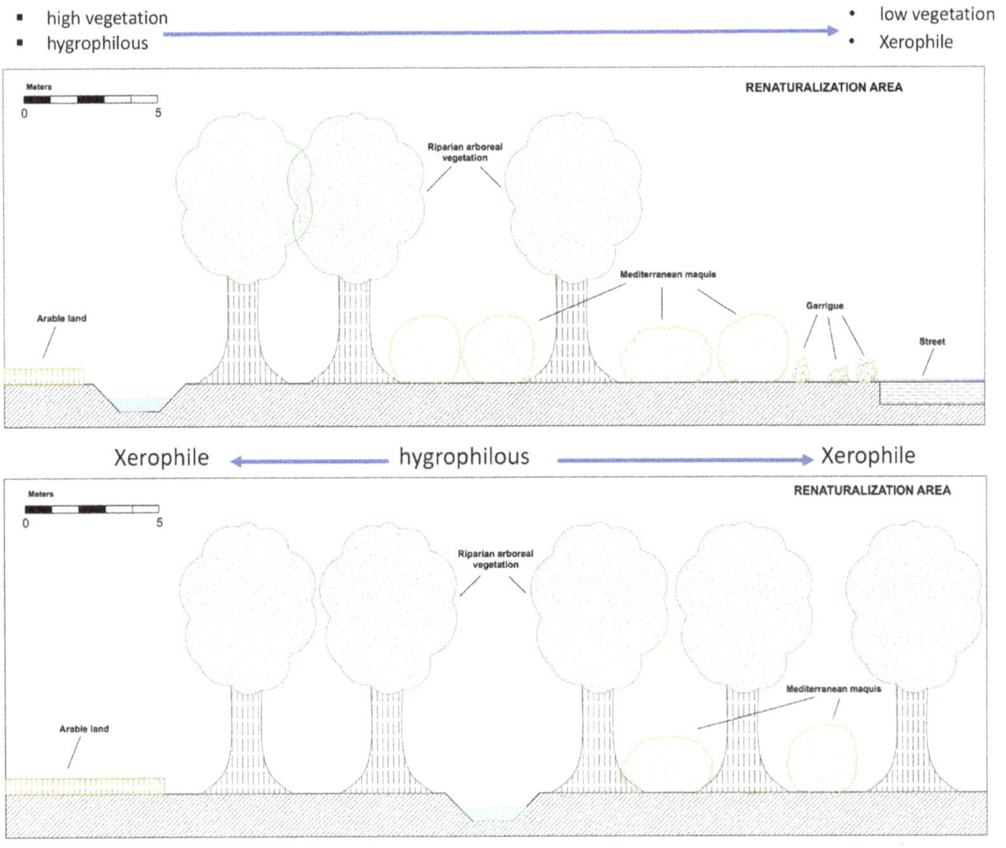

Figure 7. Cross-sections of the plant structure adopted for the ecological restoration along the channel.

It is of primary importance that the plant material is gathered and selected from autochthonous germplasm by skilled personnel who have expert knowledge of when to harvest the seeds and/or cuttings, and the most reliable techniques for the harvest, storage, and multiplication of the propagation material. With this aim, portions of plants were collected on site to be used as cuttings (Figure 8).

3.4. Management of the Habitat Restoration and Stakeholder Involvement

The management of the project over time is the responsibility of the Municipality of Campi Salentina, which will guarantee the success of the project and its ongoing maintenance. The project in question does not require significant economic resources for its management as it uses species that are adapted to the local climate. Indeed, only the first three or four years are critical, as it is necessary in this period to carry out emergency summer irrigation for the planted areas in order to favour the rooting of plants. Therefore, during the initial phase of the project, the municipality of Campi Salentina has planned irrigation activities to guarantee recovery of the vegetation from water stress and if necessary to remove and replace dead and unrooted plants. After the first three to four years, the vegetation should be self-sustaining and fully integrated with the newly formed habitat. The use of local genetic material will help to support the success of the vegetation's development because it will be adapted to the environmental context of the project.

Figure 8. Cuttings prepared from *Populus alba, Quercus virgiliana* and *Ulmus minor* plants.

In the tender for the awarding of the contract for the works, the municipality included the condition that the company planting the vegetation must guarantee the replanting of at least 20% of the dead and unrooted plants without additional costs for the municipality.

In Social Ecological Systems, another important factor in the success of ecological restoration projects is the involvement of the owners of the affected land. This was essential to finding an agreement for acquisition areas affected by the project and gaining the understanding and involvement of the landowners who undertake agricultural activities at the edge of the project area and manage the vegetation that has been introduced. Specifically, the municipality of Campi Salentina is seeking to develop a management model for the areas of interest in partnership with landowners, who will hopefully agree to take care of the maintenance of the planted vegetation on the basis of appropriate incentives. This should result in lower municipal expenditure on management, greater protection of the area from degradation and vandalism and greater acceptance of the measures on the part of stakeholders. Indeed, the intention was not to impose the measures but to develop the project in agreement with the community that it directly concerns, specifically the landowners.

The municipality launched a direct involvement campaign by sending out an information letter and holding a meeting with the landowners. However, this has not yet produced the desired results, partly because of the large number of landowners involved. Indeed, the project site, which covers an area of about 5 hectares, is divided into 40 individual land parcels belonging to about 250 registered owners [59,60]. This strong fragmentation of ownership is aggravated by the fact that many owners do not reside in the municipality of interest but in completely different regions, making their involvement and interest even more problematic. The fact that many plots of land have multiple owners can create internal conflicts among the owners themselves regarding the choice of how to use the land.

The project schedule made it impossible to further develop this strategy and involve more stakeholders. In the end, a partial result was achieved, whereby some secondary areas were planted without having to expropriate them.

Moreover, the format of the project did not include the possibility to developed actions after the realization of the projects like the monitoring restoration result. In any case, the municipality will carry out the monitoring of the area to detect dead and unrooted plants the substitutions of which need to be organised. Perhaps, after 10 years, when the plants have more high structure and the habitat introduced are more structured, it will be possible to carry out the analysis of the social and cultural perception of the project to understand how to continue the landscape restoration project valorisations. However, a website was developed to sponsor the project in the community and where it will be possible to highlight new actions or social events connected with the knowledge of the project.

3.5. Ecosystem Services Impacts

The habitat restoration project can have a positive impact in terms of ecosystem services because it will act on various ecological components passing from arable lands to forest riparian habitat (Table 4).

Table 4. Summary of qualitative ecosystem services impacts passing from arable lands to forest riparian habitat [12,24,54,55,61–63]. "+" indicates positive variation and "−"indicates negative variation.

Ecosystem Services		Impact
"Regulation functions"	*Gas regulation*	+
	Climate regulation	+
	Disturbance prevention	+
	Water regulation	+
	Water supply	NA
	Soil retention	NA
	Soil formation	NA
	Nutrient regulation	+
	Waste treatment	NA
	Pollination	+
	Biological control	+
"Habitat functions"	*Refugium function*	+
	Nursery function	+
"Production functions"	*Food*	−
	Raw materials	−
	Genetic resources	+
	Medicinal resources	NA
"Information functions"	*Aesthetic information*	+
	Recreation	+
	Cultural and artistic information	+
	Spiritual and historical information	+
	Science and education	+
"Carrier functions"	*Habitation*	NA
	Tourism-facilities	+

The relevance of regulation functions is limited in consideration of the extensions of the habitat restoration. Instead, the Habitat functions, Information functions and Carrier functions can have high relevance independently from the dimension of the area restored. These habitats will constitute reserves of biodiversity within an agricultural matrix, where autochthonous and nesting species of birds, reptiles, and amphibians typical of riparian habitats will find refuge. Furthermore, these habitats will also be a source of biodiversity because plants such as *Quercus virgiliana* will be able to continue to propagate, guaranteeing both the sustenance and regeneration of the forest over time and the spread of coverage to neighbouring areas where, in conditions of non-human disturbance, they can develop other small, wooded areas. Furthermore, these wooded areas and new anthropic elements introduced to replace the old types will constitute hotspots of natural vegetation in a strongly anthropized matrix, providing a series of natural "stepping-stones" for the transit and refuge of fauna [64,65] (Figure 9).

Figure 9. The red circle highlights an *Natrix natrix* (Linnaeus, 1758) detected during a survey in the realisation phase of the project in one bridge that was constructed to replace an old cement bridge and sides.

Using birds as ecological indicators, it can be said that a successful intervention will favour the colonisation of the area by at least three species. The first is *Sylvia atricapilla*, a forest species that needs, at least in the reproductive period, areas with arboreal elements and shrub species forming a rich undergrowth. The second species is *Luscinia megarhynchos*, which is closely linked to cool and humid woodlands with abundant undergrowth. The third species is *Remiz pendulinus*, typical of humid freshwater environments, which nests on tree branches suspended over the water [66] (Figure 10).

In these terms, although it is not the main purpose of this project, the measures can be framed in terms of "Green infrastructure", since they favour the creation of a network that has a positive impact not only on biodiversity in the context of the municipality of Campi Salentina, but on the entire landscape context to which it belongs, becoming ecologically part of the Natura 2000 network. Indeed, the EU defines green infrastructure as "a network of natural and semi-natural areas planned at a strategic level with other environmental elements, designed and managed in such a way as to provide a broad spectrum of ecosystem services" [67].

Figure 10. The signage with the masterplan of the project "The Unknown Past And Aware Future Of The Landscape Mosaic Of the Lacrima River: A Project For the Ecological Restoration Of The Riparian Forest" with the indication of Puglia Regional Operational Program 2014–2020 that financed the project.

The existing paths along the channel were improved in order to enable them to be used for the maintenance of the vegetation and the river. However, to support the cultural and social functions of the site, they were designed and structured in such a way as to guarantee safe use by the population, even those with walking difficulties. This should make it possible to develop recreational activities, in harmony with the natural assets, that were not explicitly included in the ecological restoration project. The increase of the community mobility in the project area for recreational activities is an important element to improve the awareness and knowledge of the ecological value of the habitat restored and its value perceptions for human well-being. Indeed, the possibility of an individual to perceive an area of social value and obtain benefits depends on the experience that an individual has in moving within the space, creating occasions for social interactions and enjoyable time [55,68,69]. Given the context of reference, the project could lead to the creation of a "Peri-urban Regional Park" which can be used for educational purposes, but also for scientific research, given the rarity of the restored vegetation in the Salento. The main result of the project lies in its ability to restore community habitats in a purely agricultural context that can be enjoyed by the public by means of a simple walk or bicycle ride. People will be able to immerse themselves in a natural oasis in a purely agricultural context without realising it, during a walk or bike ride that may start several kilometres away.

3.6. *Administrative and Institutional Feedback on the Project Development*

The project was implemented with public funding provided by the regional administration that established the budget, work schedule and reporting deadlines, as well as providing for specific measures such as the expropriation of land on which to implement the works and other initiatives, in accordance with rules indicated by the European Union for the use of international funds (Figure 10).

An important aspect that emerged from the administrative analysis of the project is the inconsistency between the timing of the call for project proposals and the project activities. Therefore, this led to the development of a work schedule designed to comply with the need to report expenses by a set date. This risked being incompatible with the time required for the preparation of the plant material (cuttings and seeds) and the relative planting (preferably to be performed in the periods of vegetative rest from October to March). In this regard, the municipality brought forward the preparation of plant material from autochthonous vegetation with cuttings by starting this activity independently of the financing of the project, before the allocation of financial resources and the start date of the work.

Another important aspect is the management of the project in the first few years, when the vegetation needs more support for development, especially in the summer months due to low rainfall. The funding did not allow for the continuation of the activities beyond the period of time allotted for maintenance. This meant that was impossible to fund maintenance measures that would guarantee the success of the project, such as emergency irrigation, in the first three years after the end of the works. This potentially represents a problem for public administrations that start ecological restoration projects, since they may not have adequate resources to support management and maintenance in the most delicate years for the development of the vegetation. It would have been appropriate to set aside a part of the budget for the first three years after the end of the works, in order to provide emergency irrigation or implement other measures to foster the resilience of the restoration in an integrated way with the available funds.

The last aspect, but not the least important, was to establish tender procedures for awarding contracts for the execution of the ecological restoration measures. The structure and methods adopted for public tenders can have important implications for the introduction of further design ideas, enabling those who carry out the work to improve the basic project with additional measures that were not initially envisaged, provided that that they do not go beyond the maximum budget.

In this case, the municipality chose to issue a tender in which less importance was given to the cost of the works relative to the funding provided by Puglia Regional Administration and more weight was given to the possibility of introducing improvements during the project's implementation.

This prompted companies to introduce new elements consistent with the purpose of the project that were not considered in the initial draft of the project. The tender left open the possibility of integrating new knowledge and experiences in the design phase, which in the first draft of the project may have been limited by the knowledge of the designers, and therefore to maximise the result within the allocated budget.

One measure introduced by the company that won the tender was the development of a website to explain the ecological restoration. This can be seen as an intangible asset serving to increase the visibility of the project and inform the public about the proposed measures and their ecological, social, and cultural value in the landscape.

4. Discussion

Fragmentation is caused overwhelmingly by anthropic pressure, which leads to a progressive loss of surface area, biodiversity, and habitat. Therefore, it is important to combine field analysis and local expert knowledge [5,10,11].

The biodiversity heritage relict considered in this study is not an isolated case limited to the municipality of Campi Salentina. Indeed, other examples are scattered around the province of Lecce and the Puglia region, which is characterised by extensive agricultural landscapes. Forest habitats are mainly confined to patches of less than 1 hectare lying outside protected areas. While there are many policies to preserve natural habitats in protected areas it is necessary to develop strategies to safeguard the unreported habitats lying outside them, and the Lacrima River project can be considered an explorative example. Therefore, the methodology proposed in this study could represent an effective combination of technology, knowledge acquired in the field, and social and institutional involvement that can support habitat restoration policies and projects designed to improve biodiversity in the Social-Ecological System [3,8,70,71].

Certain issues arising from the project, such as the schedule for the activities and budgetary reporting, may yield insights that are useful for improving international European rules governing the funding of restoration projects by enabling them to take account of a broader range of management scenarios. A related concern is that the result of the restoration project cannot be assessed at the time of its completion, because the vegetation needs time to develop and is strongly conditioned by human and environmental conditions. Therefore, ecological restoration projects need to be monitored over time, with the possibility of further intervention at least in the first 3 years. Furthermore, the schedule for project implementation and budgetary reporting must be adapted to the specific needs of the vegetation and not the other way around, because the objective of these projects is not merely compliance with accounting rules but the ecological recovery of the landscape.

The innovative aspect of the project proposal entitled "The Unknown Past And Aware Future Of The Landscape Mosaic Of the Lacrima River: A Project For The Ecological Restoration Of The Riparian Forest" lay in the restoration of a habitat, starting from a small relict of vegetation testifying to its more substantial presence in the past and thus to what "might have been" in the absence of agricultural activities and other anthropogenic pressures. In short, a habitat of community importance, currently rare in the context of the Salento, has been reconstituted.

This project is therefore not limited to enhancing habitats already present and recorded, as often happens in protected areas, but aims to enhance, by means of specific measures, habitats that are currently not recorded but have intrinsically high conservation value and that represent "biodiversity heritage relict" in the landscape.

Therefore, this project adopted an approach based on expertise and the ability to interpret the ecological aspects in the landscape and not on the use of ecological indices. It may be that on the basis of vegetation indices alone, this area would not have been considered for ecological restoration because of its small size and low number of species. However, the particular species found along the river have great significance in terms of biodiversity conservation (Habitats Directive), and the importance of the project thus goes beyond the simple interpretation of numbers. Naturally, this is particularly suitable for SESs where the remaining habitat has been heavily compromised by human actions over time. In contrast, this approach may not be appropriate in natural contexts where the selection of vegetation is mainly determined by disturbance unrelated to anthropic activity, and the reduction of the habitat is the result of the natural evolution of the landscape. In this case, it might not make sense to intervene in a "biodiversity heritage relict" because it would distort the natural evolution of the system and have poor results, meaning that any economic expense would be wasted.

The main aspect that emerges from the implementation of restoration ecology projects in a highly anthropized context concerns the relationship between social and environment factors, since they are of equal importance and are to be understood as integral parts of a single system in dynamic equilibrium. In an ecological restoration project, the tension between social and natural components needs to be replaced by a holistic vision characterised by the co-evolution of society and the environment, understood as strictly interdependent systems, in continuous dynamic interaction and subject to multiple changes, due to the feedback mechanisms existing between them. Landscape transformation processes must therefore promote social inclusion and the development of knowledge. Such processes and knowledge must not be the exclusive preserve of the "experts" but must involve all interested parties who live or operate in some way in the territory. The municipality of Campi Salentina has sought to develop a management strategy for the areas affected by the project in collaboration with certain landowners who are expected take care of the maintenance of the planted vegetation. However, this highlights the need to develop inclusive regional planning strategies that can facilitate and encourage ecological restoration projects. This may be difficult to implement organically in the design phase of the restoration project, although it should be more effective in the development phase.

A bibliographic analysis was carried out to verify the existence of similar interventions, with little evidence. Although it is not possible to exclude the existence of similar projects in other contexts not referenced in the scientific literature, surely this design hypothesis is among the first such projects in Italy to reconcile past landscape vegetation, of which few traces remain today and which have not been well publicised, with the future evolution of the social-ecological landscape. Indeed, this project was included as an example of the good national practices and good projects in the national annual report of land-use loss [72].

5. Conclusions

This study provided a point of reference and reflection for other habitat restoration projects in strong damaged habitat, since creating and sharing experiences increases the ability to develop other projects through a learning-by-doing approach. The approach adopted, the experience gained, and the solutions developed in the course of the project produce knowledge that can form the starting point for new initiatives.

In a habitat restoration project conducted within an SES, an additional goal is to create the ecological, social, and cultural conditions that enable further opportunities to be exploited independently of the primary purpose of the project. This does not require a precise design but rather flexibility, making it possible to adapt to requests that may arise by chance and without planning. This aspect involves the project designers, who are aware that there are aspects that cannot be developed in the project directly for economic reasons or because they do not fall within the scope of the project. Therefore, when designing a habitat restoration project in an SES, the designers can leave certain questions open by creating the conditions for their development in the future. This is what was done in the Lacrima project regarding the potential use of the area for social use and institutional management in the time.

If the proposed measures are successful in recreating the desired community habitat, the site of the intervention could then be proposed as an SCI (Site of Community Importance) or be included in a network of natural areas (NATURA 2000) to improve the local conservation practices. This would be helpful for the further development of the project because it would be managed as a "Protected Area". This could help to draw on specific forms of financing for the implementation of other initiatives that could enhance biodiversity by guaranteeing high resilience over time in terms of the socio-economic evolution of the context of reference. In this regard, an ecological restoration project in an SES must be based on institutions that value creativity and flexibility, seeing design as an experiment from which to learn and acquire new knowledge, in which participation is recognised as the key to success.

Author Contributions: Conceptualization, T.S., A.T., S.A. and P.M.; methodology, T.S., A.T., S.A. and P.M.; validation, R.T.; formal analysis, T.S., A.T., S.A. and P.M.; investigation, T.S., A.T., S.A., P.M., R.D., G.L.G. and R.T.; data curation, T.S., A.T., S.A., G.L.G., R.D. and P.M.; writing—original draft preparation, T.S.; writing—review and editing, A.T. and S.A.; supervision, P.M.; project administration, R.T. and R.D.; funding acquisition, R.T. All authors have read and agreed to the published version of the manuscript.

Funding: This research received no external funding.

Institutional Review Board Statement: Not applicable.

Informed Consent Statement: Not applicable.

Data Availability Statement: The project information is available on https://www.comune.campi-salentina.le.it/amministrazione/attivita/bandi-di-gara-e-contratti/item/por-puglia-f-e-s-r-f-s-e-2014-

2020-realizzazione-di-progetti-per-la-rete-ecologica-regionale-oasi-naturalistica-itinerante-del-bosco-ripariale-della-lacrima (accessed on 23 August 2021).

Acknowledgments: This work describes habitat restoration in Campi Salentina in the Puglia Region. The Project was funded under POR PUGLIA 2014–2020, Asse VI–Azione 6.6—Sub-Azione 6.6.a, "Interventi per la tutela e la valorizzazione di aree di attrazione naturale" Realizzazione di progetti per la rete ecologica regionale". We thank, the Regional Landscape Department and Parks service and biodiversity protection Department contributed to the additional improvement of the final project. We thank Paolo Perrino, and Giuseppe Caputo who were part of the project development team.

Conflicts of Interest: The authors declare no conflict of interest.

References

1. Antrop, M. Landscape Change: Plan or Chaos? *Landsc. Urban Plan.* **1998**, *41*, 155–161. [CrossRef]
2. Semeraro, T.; Gatto, E.; Buccolieri, R.; Vergine, M.; Gao, Z.; De Bellis, L.; Luvisi, A. Changes in Olive Urban Forests Infected by *Xylella fastidiosa*: Impact on Microclimate and Social Health in urban areas. *Int. J. Environ. Res. Public Health* **2019**, *16*, 2642. [CrossRef]
3. Semeraro, T.; Giannuzzi, C.; Beccarisi, L.; Aretano, R.; De Marco, A.; Pasimeni, M.R.; Zurlini, G.; Petrosillo, I. A constructed treatment wetland as an opportunity to enhance biodiversity and ecosystem services. *Ecol. Eng.* **2015**, *82*, 517–526. [CrossRef]
4. Semeraro, T.; Gatto, E.; Buccolieri, R.; Catanzaro, V.; De Bellis, L.; Cotrozzi, L.; Lorenzini, G.; Vergine, M.; Luvisi, A. How Ecosystem Services Can Strengthen the Regeneration Policies for Monumental Olive Groves Destroyed by *Xylella fastidiosa* Bacterium in a Peri-Urban Area. *Sustainability* **2021**, *13*, 8778. [CrossRef]
5. Semeraro, T.; Scarano, A.; Buccolieri, R.; Santino, A.; Aarrevaara, E. Planning of Urban Green Spaces: An Ecological Perspective on Human Benefits. *Land* **2021**, *10*, 105. [CrossRef]
6. Antrop, M. Why landscapes of the past are important for the future. *Landsc. Urban Plan.* **2005**, *70*, 21–34. [CrossRef]
7. dos Santos, J.S.; Dodonov, P.; Oshima, J.E.; Martello, F.; de Jesus, A.S.; Ferreira, M.E.; Silva-Neto, C.M.; Ribeiro, M.C.; Collevatti, R.G. Landscape ecology in the Anthropocene: An overview for integrating agroecosystems and biodiversity conservation. *Perspect. Ecol. Conserv.* **2021**, *19*, 21–32. [CrossRef]
8. Virapongse, A.; Brooks, S.; Metcalf, E.C.; Zedalis, J.G.; Kliskey, A.; Alessa, L.A. Social-ecological systems approach for environmental management. *J. Environ. Manag.* **2016**, *178*, 83–91. [CrossRef] [PubMed]
9. Drudi, E. *"Non ha Dato Prova di Serio Ravvedimento" Gli Ebrei Perseguitati Nella Provincia del Duce*; Giuntina: Firenze, Italy, 2014.
10. Cavallo, F.L. *Terre, Acque, Macchine: Geografia della Bonifica in Italia Tra Ottocento e Novecento*; Diabasis: Parma, Italy, 2011.
11. Snowden, F.M. *La Conquista Della Malaria. Una Modernizzazione Italiana 1900–1962*; Einaudi: Torino, Italy, 2008.
12. Petrosillo, I.; Zaccarelli, N.; Semeraro, T.; Zurlini, G. The effectiveness of different conservation policies on the security of natural capital. *Landsc. Urban Plan.* **2009**, *89*, 49–56. [CrossRef]
13. Petrosillo, I.; Semeraro, T.; Zurlini, G. Detecting the "Conservation Effect" on the maintenance of natural capital flow in different natural parks. *Ecol. Econ.* **2010**, *69*, 1115–1123. [CrossRef]
14. Sardegna Foreste. Available online: https://www.sardegnaforeste.it/notizia/come-nel-medioevo-le-foreste-diventarono-una-risorsa-di-primaria-importanza (accessed on 4 August 2021).
15. Semeraro, T.; Pomes, A.; Del Giudice, C.; Negro, D.; Aretano, R. Planning ground based utility scale solar energy as green infrastructure to enhance ecosystem services. *Energy Policy* **2018**, *117*, 218–227. [CrossRef]
16. Chiarucci, A.; Araujo, M.B.; Decocq, G.; Beierkuhnlein, C.; Fernandez-Palacios, J.M. FORUM The concept of potential natural vegetation: An epitaph? *J. Veg. Sci.* **2010**, *21*, 1172–1178. [CrossRef]
17. Loidi, J.; Fernandez-Gonzalez, F. Potential natural vegetation: Reburying or reboring? *J. Veg. Sci.* **2012**, *23*, 596–604. [CrossRef]
18. Tomaselli, G.; Russo, P.; Riguccio, L.; Quattrone, M.; D'Emilio, A. Assessment of landscape regeneration of a Natura 2000 site hosting greenhouse farming by using a dashboard of indicators. A case in Sicily through the territorial implementation of a "pilot project" at farm level. *Land Use Policy* **2000**, *92*, 104444. [CrossRef]
19. Antrop, M. Background concepts for integrated landscape analysis. *Agric. Ecosyst. Environ.* **2000**, *77*, 17–28. [CrossRef]
20. Naveh, Z. Interactions of Landscapes and Cultures. *Landsc. Urban Plan.* **1995**, *32*, 43–54. [CrossRef]
21. Palang, H.; Fry, G. *Landscape Interfaces. Cultural Heritage in Changing Landscapes*; Kluwer Academic Publishers: Dordrecht, The Netherlands, 2003.
22. UNESCO. Operational Guidelines for the Implementation of the World Heritage Convention. Available online: Whc.unesco.org (accessed on 10 November 2007).
23. Cumming, G.S.; Epstein, G. Landscape sustainability and the landscape ecology of institutions. *Landsc. Ecol.* **2020**, *35*, 2613–2628. [CrossRef]
24. Winter, K.B.; Lincoln, N.K.; Berkes, F.; Alegado, R.A.; Kurashima, N.; Frank, K.L.; Pascua, P.; Rii, Y.M.; Reppun, F.; Knapp, I.S.S.; et al. Ecomimicry in Indigenous resource management: Optimizing ecosystem services to achieve resource abundance, with examples from Hawaii. *Ecol. Soc.* **2020**, *25*, 26. [CrossRef]
25. Society for Ecological Restoration International. Available online: www.ser.org (accessed on 10 April 2021).

26. Batista, T.; Manuel de Mascarenhas, J.; Mendes, P.; Pinto-Gomes, C. Assessing Vegetation Heritage Value: The Alentejo Central (Portugal) as a Case Study. *Land* **2021**, *10*, 307. [CrossRef]
27. UNESCO. World Heritage Centre's Natural Heritage Strategy. Available online: https://whc.unesco.org/en/naturalheritagestrategy/ (accessed on 20 May 2021).
28. Marafa, L. Integrating natural and cultural heritage: The advantage of feng shui landscape resources. *Taylor Fr. Online* **2010**, *9*, 307–323. [CrossRef]
29. Leitão, L. Bridging the Divide between Nature and Culture in the World Heritage Convention: An Idea Long Overdue? *Georg. Wright Forum* **2017**, *34*, 195–210.
30. Aung, M.Z.N.; Shibata, S. Vegetation Conditions in Sacred Compounds at Myanmar's Bagan Cultural Heritage Site. *Heritage* **2019**, *2*, 2745–2762. [CrossRef]
31. Liu, X.; Liu, S. Introduction to the special issue: Biodiversity mechanism in natural succession and ecological restoration. *Ecol. Eng.* **2020**, *143*, 105614. [CrossRef]
32. Gilbert, O.L.; Anderson, P. *Habitat Creation and Repair*; Oxford University Press: Oxford, UK, 1998.
33. Mille, J.R.; Hobbs, R. Habitat Restoration—Do We Know What We're Doing? *Restor. Ecol.* **2007**, *15*, 382–390. [CrossRef]
34. Johnson, D.H. The comparison of usage and availability measurements for evaluating resource preference. *Ecology* **1980**, *61*, 65–71. [CrossRef]
35. Mitchell, M.S.; Powell, R.A. Linking fitness landscapes with the behavior and distribution of animals. In *Landscape Ecology and Resource Management. Linking Theory with Practice*; Bissonette, J.A., Storch, I., Eds.; Island Press: Washington, DC, USA, 2003; pp. 93–124.
36. Hou, X.; Liu, S.; Zhao, S.; Zhang, Y.; Wu, X.; Cheng, F.; Dong, S. Interaction mechanism between floristic quality and environmental factors during ecological restoration in a mine area based on structural equation modelling. *Ecol. Eng.* **2018**, *124*, 23–30. [CrossRef]
37. Natura 2000. Available online: https://ec.europa.eu/environment/nature/natura2000/index_en.htm (accessed on 1 May 2021).
38. The Habitats Directive. Available online: https://ec.europa.eu/environment/nature/legislation/habitatsdirective/index_en.htm (accessed on 1 May 2021).
39. Semeraro, T.; Aretano, R. Landscape regeneration in social ecological system: A process without time. In *Paradise Lost of the Landscape-Cultural Mosaic. Attractiveness, Harmony, Atarassia, Proceedings of the 21st IPSAPA/ISPALEM International Scientific Conference, Venice, Italy, 6–7 July 2017*; Ipsapa/Ispalem: Udine, Italy, 2018; ISBN 978-88-942329-3-6.
40. Zhang, Y.; Liu, X.; Lv, Z.; Zhao, X.; Yang, X.; Jia, X.; Sun, W.; He, X.; He, B.; Cai, Q.; et al. Animal diversity responding to different forest restoration schemes in the Qinling Mountains, China. *Ecol. Eng.* **2019**, *36*, 23–29. [CrossRef]
41. Zhang, Y.; Wu, Y.; Zhang, Q.; Ran, J.; Price, M. Distribution of a Giant Panda Population Influenced by Land Cover. *J. Wildl. Manag.* **2018**, *82*, 1199–1209. [CrossRef]
42. Ferretti, V. Framing territorial regeneration decisions: Purpose, perspective and scope. *Land Use Policy* **2021**, *102*, 105279. [CrossRef]
43. Regione Puglia, BURP. Available online: http://burp.regione.puglia.it/documents/10192/37071690/DET_227_27_11_2018.pdf/3e96fad1-213c-4830-8e09-8135f934a80f;jsessionid=9151347260F5451E7691534A6B366407 (accessed on 15 April 2021).
44. SIT Puglia. Available online: http://www.sit.puglia.it/ (accessed on 5 August 2021).
45. Lackey, R. Societal values and the proper role of restoration ecologists. *Front. Ecol. Environ.* **2004**, *22*, 45–46.
46. Jongman, R.H.G.; Ter Braak, C.J.F.; Van Tongeren, P.F.R. *Data Analysis in Community and Landscape Ecology*; Cambridge University Press: Cambridge, UK, 1995.
47. Dengler, J.; Chytrý, M.; Ewald, J. Phytosociology. In *Encyclopedia of Ecology*, 2nd ed.; Jørgensen, S.E., Fath, B.D., Eds.; Academic Press: Cambridge, MA, USA, 2008; pp. 2767–2779. [CrossRef]
48. Saxena, A.K.; Singh, J.S. A Phytosociological Analysis of Woody Species in Forest Communities of a Part of Kumaun Himalaya. *Vegetation* **1982**, *50*, 3–22. [CrossRef]
49. Rao, D.S.; Prayaga, M.; Kumar, O.A. Plant Biodiversity and Phytosociological Studies on Tree Species diversity of Khammam District, Telangana State, India. *J. Pharm. Sci. Res.* **2015**, *7*, 518–522.
50. Decocq, G. Moving from Patterns to Processes: A Challenge for the Phytosociology of the Twenty-First Century? In *Vegetation Structure and Function at Multiple Spatial, Temporal and Conceptual Scales*; Box, E.O., Ed.; Springer: Berlin/Heidelberg, Germany, 2016; pp. 407–424.
51. Semeraro, T.; Arzeni, S.; Turco, A.; Medagli, P. Ecosystem Services in Strategic Environmental Assessment: A Case Study of an Urban Development Plan in Gallipoli City. *IOP Conf. Ser. Mater. Sci. Eng.* **2020**, *960*, 022018. [CrossRef]
52. De Groot, R.S.; Alkemade, R.; Braat, L.; Hein, L.; Willemen, L. Challenges in integrating the concept of ecosystem services and values in landscape planning, management and decision making. *Ecol. Complex* **2010**, *7*, 260–272. [CrossRef]
53. De Groot, R.S.; Wilson, M.; Boumans, R. A typology for the description, classification and valuation of Ecosystem Functions. *Goods Serv. Econ.* **2002**, *41*, 393–408.
54. Semeraro, T.; Radicchio, B.; Medagli, P.; Arzeni, S.; Turco, A.; Geneletti, D. Integration of Ecosystem Services in Strategic Environmental Assessment of a Peri-Urban Development Plan. *Sustainability* **2021**, *13*, 122. [CrossRef]
55. Semeraro, T.; Zaccarelli, N.; Lara, A.; Sergi-Cucinelli, F.; Aretano, R. A Bottom-Up and Top-Down Participatory Approach to Planning and Designing Local Urban Development: Evidence from an Urban University Center. *Land* **2020**, *9*, 98. [CrossRef]

56. Ullah, A.; Sam, A.S.; Sathyan, A.R.; Mahmmod, N.M.; Zeb, A.; Kachele, H. Role of local communities in forest landscape restoration: Key lessons from the Billion Trees Afforestation Project, Pakistan. *Sci. Total Environ.* **2021**, *772*, 145613. [CrossRef] [PubMed]
57. Atlante Geografico del Regno di Napoli. Available online: https://www.gruppoarcheologicokr.it/biblioteca/g-a-rizzi-zannoni-atlante-geografico-del-regno-di-napoli-tavola-n-31-1808/ (accessed on 20 April 2021).
58. Wang, L.; Zhao, X.; Song, Q.; Wang, J.; Kou, Y.; Jiang, X.; Shao, X. Morphological traits of *Bryum argenteum* and its response to environmental variation in arid and semi-arid areas of Tibet. *Ecol. Eng.* **2019**, *136*, 101–107. [CrossRef]
59. Catasto, Servizi Catastali Online del Network Catasto. Available online: https://www.catasto.it/ (accessed on 8 March 2019).
60. Maggiore, G.; Semeraro, T.; Aretano, R.; De Bellis, L.; Luvisi, A. GIS Analysis of land-use change in threatened landscapes by *Xylella fastidiosa*. *Sustainability* **2019**, *11*, 253. [CrossRef]
61. Burkhard, B.; Kroll, F.; Nedkov, S.; Müller, F. Mapping ecosystem service supply, demand and budgets. *Ecol. Ind.* **2012**, *21*, 17–29. [CrossRef]
62. Strohbacha, M.W.; Haaseb, D. Above-ground carbon storage by urban trees in Leipzig, Germany: Analysis of patterns in a European city. *Land. Urban Plan.* **2012**, *104*, 95–104. [CrossRef]
63. Semeraro, T.; Aretano, R.; Pomes, A. Green infrastructure to improve ecosystem services in the landscape urban regeneration. *IOP Conf. Ser. Mater. Sci. Eng.* **2017**, *245*, 082044. [CrossRef]
64. Schadt, S.K.; Kaiser, T.S.; Frank, K.; Wiegand, T. Analyzing the effect of stepping stones on target patch colonisation in structured landscapes for Eurasian lynx. *Landsc. Ecol.* **2011**, *26*, 501–513. [CrossRef]
65. Semeraro, T.; Aretano, R.; Barca, A.; Pomes, A.; Del Giudice, C.; Gatto, E.; Lenucci, M.; Buccolieri, R.; Emmanuel, R.; Gao, Z.; et al. A Conceptual Framework to Design Green Infrastructure: Ecosystem Services as an Opportunity for Creating Shared Value in Ground Photovoltaic Systems. *Land* **2020**, *9*, 238. [CrossRef]
66. La Gioia, G. *Atlante Degli Uccelli Nidificanti in Provincia di Lecce (2000–2007)*; Edizioni del Grifo: Lecce, Italy, 2009; p. 176.
67. European Commission (EC). Green Infrastructure (GI)—Enhancing Europe's Natural Capital. Communication from the Commission to the European Parliament, the Council, the European Economic and Social Committee and the Committee of the Regions, COM 249 Final Brussels. 2013. Available online: https://eur-lex.europa.eu/resource.html?uri=cellar.d41348f2-01d5-4abe-b817-4c73e6f1b2df.0014.03/DOC_1&format=PDF (accessed on 9 April 2020).
68. Haase, D. Urban ecology of shrinking cities: An unrecognised opportunity? *Nat. Cult.* **2008**, *3*, 1–8. [CrossRef]
69. Masys, A. *Disaster Management: Enabling Resilience*; Springer: Berlin, Germany, 2008; Volume 3, pp. 1–8. [CrossRef]
70. MEA. *Ecosystems and Human Well-Being: Current State and Trends*; Island Press: Washington, DC, USA, 2005.
71. O'Neill, R.; Kahn, J.R. Homo Economicus as a Keystone Species. *Bioscience* **2000**, *50*, 333–337. [CrossRef]
72. Sistema Nazionale per la Protezione dell'Ambiente. Consumo di Suolo, Dinamiche Territoriali e Servizi Ecosistemici. 2021. Available online: https://www.snpambiente.it/2021/07/14/consumo-di-suolo-dinamiche-territoriali-e-servizi-ecosistemici-edizione-2021/ (accessed on 1 August 2021).

land

MDPI

Article

Fostering Carbon Credits to Finance Wildfire Risk Reduction Forest Management in Mediterranean Landscapes

Fermín Alcasena [1,2,*], Marcos Rodrigues [2,3], Pere Gelabert [2,3], Alan Ager [4], Michele Salis [5], Aitor Ameztegui [2,3], Teresa Cervera [6] and Cristina Vega-García [2,3]

1 USDA Forest Service International Visitor Program, College of Forestry, Oregon State University, 321 Richardson Hall, Corvallis, OR 97331, USA
2 Department of Agricultural and Forest Engineering, University of Lleida, Avenida da Rovira Roure 191, 25198 Lleida, Spain; marcos.rodrigues@udl.cat (M.R.); perejoan.gelabert@udl.cat (P.G.); aitor.ameztegui@udl.cat (A.A.); cristina.vega@udl.cat (C.V.-G.)
3 Joint Research Unit CTFC-AGROTECNIO, Forest Science and Technology Centre of Catalonia, Ctra. Sant Llorenç de Morunys km 2, 25280 Solsona, Spain
4 Missoula Fire Sciences Laboratory, Rocky Mountain Research Station, USDA Forest Service, 5775 US Highway 10W, Missoula, MT 59808, USA; alan.ager@usda.gov
5 National Research Council (CNR), Institute of BioEconomy (IBE), Traversa La Crucca 3, 07100 Sassari, Italy; michele.salis@ibe.cnr.it
6 Forest Ownership Centre, Government of Catalonia, Santa Perpètua de Mogoda, 08130 Santa Perpètua de Mogoda, Spain; tcervera@gencat.cat
* Correspondence: ferminalcasena@eagrof.udl.cat

Abstract: Despite the need for preserving the carbon pools in fire-prone southern European landscapes, emission reductions from wildfire risk mitigation are still poorly understood. In this study, we estimated expected carbon emissions and carbon credits from fuel management projects ongoing in Catalonia (Spain). The planning areas encompass about 1000 km^2 and represent diverse fire regimes and Mediterranean forest ecosystems. We first modeled the burn probability assuming extreme weather conditions and historical fire ignition patterns. Stand-level wildfire exposure was then coupled with fuel consumption estimates to assess expected carbon emissions. Finally, we estimated treatment cost-efficiency and carbon credits for each fuel management plan. Landscape-scale average emissions ranged between 0.003 and 0.070 T CO_2 year^{-1} ha^{-1}. Fuel treatments in high emission hotspots attained reductions beyond 0.06 T CO_2 year^{-1} per treated ha. Thus, implementing carbon credits could potentially finance up to 14% of the treatment implementation costs in high emission areas. We discuss how stand conditions, fire regimes, and treatment costs determine the treatment cost-efficiency and long-term carbon-sink capacity. Our work may serve as a preliminary step for developing a carbon-credit market and subsidizing wildfire risk management programs in low-revenue Mediterranean forest systems prone to extreme wildfires.

Keywords: wildfire risk; landscape management; ecosystem services; carbon credits; green deal

1. Introduction

Carbon emissions from fires in European countries, and Mediterranean areas, in particular, represent about 4.03 Tg C per year [1,2]. However, a few extreme fires account for the bulk of burned areas and carbon dioxide emissions to the atmosphere [3,4]. These are extreme escaped fires that easily overwhelm firefighting capacity, and the containment is restricted to strategic locations in backing and flanking fire spread areas [5,6]. This is the reason why recent studies advocate for a comprehensive long-term solution to better coexist with fire, and forest fuel management is emerging as a fundamental strategy complementing fire suppression and ignition prevention in populated areas [7–9].

Fuel treatments mitigate wildfire intensity on treated locations and reduce wildfire spread and overall wildfire likelihood at the landscape scale [10–12]. Commonly implemented prescriptions in Mediterranean forests are low pruning, thinning from below, prescribed fires, mechanical mastication, and targeted grazing [13–15]. On the other hand, the most frequent treatments at the landscape scale consist of fuel break networks dividing the landscape into large planning areas, systematic buffer clearings within the wildland–urban interface, and intensively managed scattered treatment units known as strategic management points (SMPs) [16–18]. Wildfire managers design SMPs based on expert criteria, heading-fire pathway analysis, and fire-weather synoptic conditions [19]. The implementation of SMPs is gaining extensive adoption in many fire-prone Mediterranean areas, such as Catalonia, where fuel treatments have become the primary risk reduction strategy [20].

Primary fuel treatment objectives in Mediterranean areas are protecting human communities, timber harvesting, restoring fire-adapted ecosystems, reducing extreme wildfire potential, and facilitating safe and effective fire suppression [21,22]. Moreover, some studies have shown that forest management increased carbon sequestration [23,24] and developed management prescriptions under different wildfire hazard scenarios based on stand structure conditions [25,26]. However, protecting carbon stocks from extreme fires was often considered a secondary objective in southern European regions, and the reduction in emissions remains unknown in large-scale ongoing fuel reduction programs. Previous works conducted in the western US implemented quantitative assessment methods that coupled fire simulation modeling with stand-level carbon loss functions to estimate the effect of fuel treatments in reducing carbon emissions from wildfires [27–29]. These studies found that thinning weight plus prescribed fire treatment intensity and the frequency and severity of future wildfires determine the benefit of increasing carbon pools in managed forests.

Multifunctional Mediterranean forests currently provide meager economic revenues from timber harvesting. Over the last 50 years, limited management plus the prevalent fire exclusion policy favored fast fuel buildup in open woodlands and dense forest regeneration in clear areas [30,31]. Moreover, large-scale afforestation campaigns planted extensive areas over marginal lands with pines to facilitate forest development on poor soils [32]. As a result, the fires rapidly evolved from fuel-limited short-events to weather-driven catastrophic events [33]. In central Catalonia (northeastern Spain), these new forest structures fostered stand-replacing high severity wildfires associated with increased mortality rates on the non-serotinous sub-Mediterranean conifers forests [34,35]. Similarly, extreme fires that burned broadleaved forests triggered significant changes in stand structures (i.e., conversion to coppice forest or scrublands), hampering woodlands' management and economic viability of many products, such as cork oak production and extraction [36].

The European Union assumed the commitment to reduce emissions by at least 40% by 2030, and forest ecosystems were identified as core natural carbon pools with the potential capacity to store and compensate carbon emissions considerably (COM/2019/640). As a result, the European Union green deal articulated the action plan towards a green and climate-neutral sustainable economy in the EU by 2050. It will mobilize €100 billion over 2021 to 2027 to assist the most affected regions. Among many other actions, promoting a bio-economy in local markets is one of the main strategies for overcoming the fossil fuel-dependent economy [37,38]. Likewise, encouraging the assumption of voluntary carbon credits by large corporations to compensate for their emission may represent an additional funding source to reverse the landscape degradation occurring in many EU rural areas [39,40]. Wildfire managers, in turn, require a quantitative assessment to evaluate carbon sink capabilities and request financial compensation for fuel management programs in low-revenue, poorly managed fire-prone areas.

This study explores the opportunity for carbon-credit-oriented forest management in diverse fire-prone Mediterranean areas across Catalonia (northeastern Spain), where large historical wildfires caused substantial losses in the last decades [36,41]. Providing stand-level and spatially explicit quantitative results is essential to identify high emission hotspots and redirect management efforts at sufficient intensities. Specifically, we addressed the following research questions: (i) What is the expected wildfire carbon emission under the current conditions? (ii) What is the overall reduction in emissions per treated area for planned fuel reductions? and (iii) where could carbon credits help subsidize ongoing risk-reduction fuel management programs? The methods presented in this study may help the small local landowners increase their revenues for the services provided to the community and vindicate large-scale fuel reduction programs in low timber revenue Mediterranean landscapes and elsewhere. Our main goal is to advance the ecosystem service carbon markets that incentivize farmers and rural communities to generate fire-resilient landscapes adapted to extreme wildfires. We ultimately attempt to develop and implement a sound framework for assessing emission reductions and provide the technical blueprint required by large corporations and EU authorities to prove the emission reductions by fuel treatment programs.

2. Materials and Methods

2.1. Study Area

We conducted this study in different planning (n = 6) areas of Catalonia (northeastern Spain) (Figure 1). The extent of the planning areas corresponded to the existing landscape units of Catalonia [42]. The climate is predominantly Mediterranean, with increasing rainfall on pre-littoral mountain ranges, milder winters closer to the coastline to the east, and a transition to sub-Mediterranean and mountain climate in the Pyrenees. Catalonia is one of the most significant fire-prone regions in the Mediterranean Basin and encompasses various physiographic gradients and fire regimes (Figure 1B). On average, 650 fires burn 11.5 thousand ha per year, from which 2% of large fires (>100 ha) account for more than 88% of the burned area (1983 to 2015). A few extreme events (i.e., >1000 ha, catastrophic wildfire episodes of 1986, 1994, 1998, 2003, and 2012) make up the bulk (>65%) of the affected areas [43]. Most fire ignitions (>90%) are caused by humans and show apparent occurrence patterns clustering close to anthropogenic features, such as communication corridors and urban areas [44]. See vegetation description and wildfire history in Appendix A for further details.

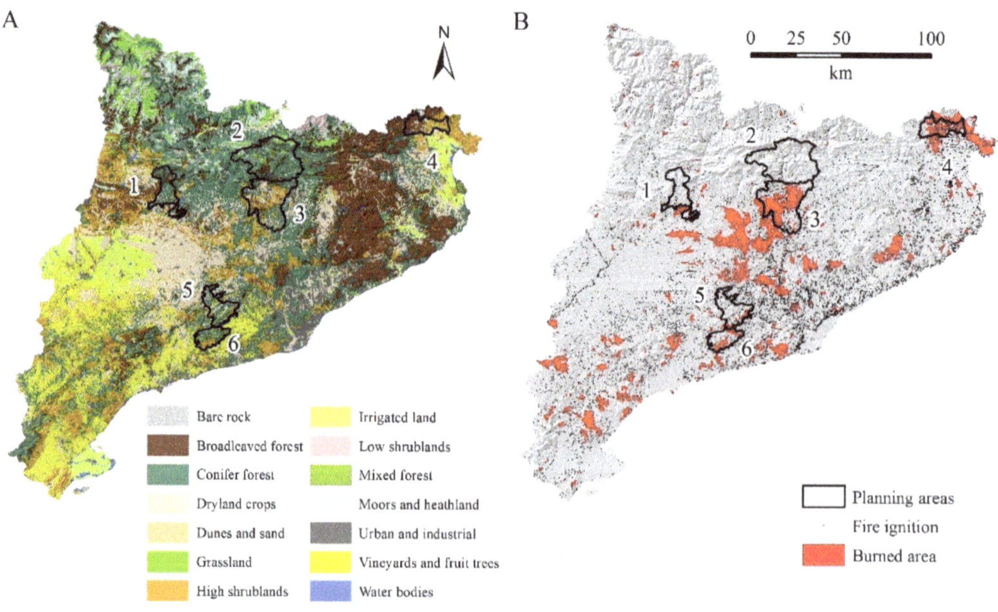

Figure 1. Land cover (**A**) and historic wildfire activity (**B**) in Catalonia (northeastern Spain). In total, the six planning areas encompass about 1000 km^2. The numbers are the planning area codes (see Appendix A).

2.2. Landscape Data

We generated the landscape file assembling topography, surface fuel, and canopy metric grids at a 40 m resolution as required for fire spread simulation modeling using ArcFuels [45]. The modeling domain encompassed the extent of the planning areas plus a 10 km buffer to account for the fires incoming from the neighboring regions and predict realistic wildfire likelihood estimates. Adjacent planning areas (i.e., 2 and 3, and 5 and 6) were merged in the same fire modeling domain. Topographic data, including elevation, aspect, and slope grids, were derived from a 25 m resolution digital terrain model (ign.es). We assigned surface fire behavior models [46] to the 1:5000 scale land-cover vector map of Catalonia [47] to generate the surface fuel model grid. For that purpose, we considered vegetation characteristics, such as species composition, tree cover, thickness, and shrub, plus herbaceous fuel thickness and heights gathered from the 4th National Forest map and the 2012 habitat map of Catalonia [48]. The canopy metrics consisted of canopy height, canopy cover, canopy base height, and canopy bulk density and were obtained from the LiDAR-derived 20 m resolution biophysical variable grids of Catalonia [49].

2.3. Forest Fuel Treatments

Treatment spatial allocation corresponded to the strategic management points (SMP) provided by the Bombers GRAF team (Group of Support to Forest Actions) of the Firefighters of the Catalan Government [22]. These typically locate on watershed divides or ravine bottoms that intersect with primary flow paths to restrict the fire potential and create better opportunities for an effective fire response using a tactical fire [21,50]. Likewise, the design accounts for the existing operational constraints, including the prescribed fire implementation and the accessibility to the site. In addition to treatment units in the planning areas, we also considered the SMPs in neighboring areas within a 10 km buffer (Figure 2).

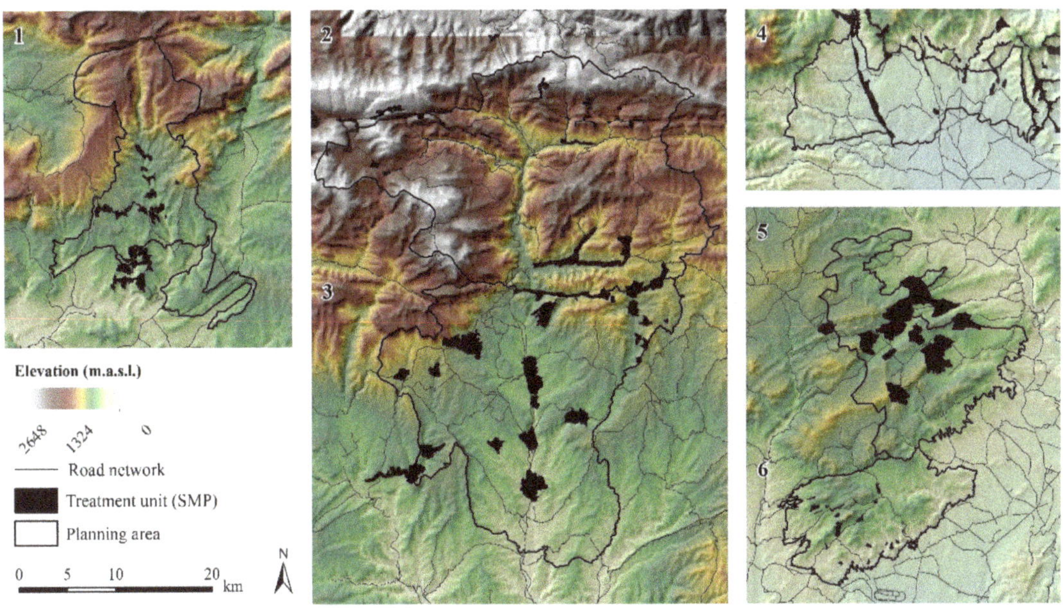

Figure 2. Location of the treatment units or strategic management points (SMPs) in the different planning areas. The numbers refer to the planning area codes (Appendix A; Figure 1). The design was based on expert criteria and provided by the Bombers GRAF team of the Firefighters of the Catalan Government.

We assumed the most widely implemented treatment prescriptions, including a tree thinning followed by a prescribed fire in conifer forests or mechanical mastication in broadleaved forests. Specifically, we considered a heavy-weight thinning from below, where ladder fuels (i.e., suppressed and dominated trees) were cleared to a 60% residual basal area target [26]. The canopy metrics were reduced accordingly to a canopy cover of 60%, a canopy bulk density of 0.09 kg m^{-3}, and a canopy base height of 3.6 m [51]. We assigned timber litter TL_1 and TL_2 surface fuel types for conifer and broadleaved treated forests [46]. The treatment cost was provided by local forest managers [52] (Table 1). We annualized the costs assuming a treatment duration of 8 years for surface fuel reductions and 14 years for the thinning, as described in previous works conducted in the study area [53,54]. The stand-level treatment cost varied between 100 and 228.75 € ha^{-1} per year.

Table 1. Summary table with treatment cost data for fuel reductions in SMPs. The overstory is first reduced to a 60% canopy cover implementing a thinning from below, followed by prescribed fire (in conifer forest) or mechanical mastication (in broadleaved forests). We assumed average cost conditions for ongoing fuel reduction plans.

Type of Treatment	Treatment Location	Treatment Cost (€ ha^{-1})	Effective Duration (yr)	Annualized Cost (€ ha^{-1} yr^{-1})
Prescribed fire	Conifer forests	800	8	100.00
Mechanical mastication	Broadleaf forests	950	8	118.75
Thinning from below	High density forests (cc > 60%)	1000	14	71.43

2.4. Wildfire Occurrence

We trained a machine learning Random Forest model [55] using fire ignition location data (Figure 1B) retrieved from the EGIF Spanish fire database [43] to predict fire ignition probability from a set of geospatial data grids. These layers included proxies for accessibility (distance to roads, distance to forest tracks, and distance to trails), agricultural activities (distance to croplands), human pressure on wildlands (distance to the wildland–urban interface), and potential sparks from power-lines (distance to power lines) [56]. The model was calibrated using a k-fold cross-validation (k = 4) procedure using a sample of 10,835 fires from 1998 to 2015. At each step or fold, we split the sample into a training sample with 75% of fires to fit the model and a 25% validation sample to evaluate the model, calculating the area under the receiver-operating (ROC) curve (AUC) [57]. The sub-models were then combined into the final model used to predict the probability of fire occurrence, presenting an average AUC > 0.73. We generated a 40 m resolution ignition probability grid, where values ranged between 0 and 1 (Figure 3). This raster grid was used as input in fire simulation modeling to locate required fire ignition patterns on the different fire modeling domains.

2.5. Fire-Weather Scenarios

The fire-weather conditions present substantial region-wide differences across Catalonia [58]. Therefore, we considered multiple fire-weather macro-areas or pyromes in the planning areas to accurately replicate the changing gradients on historical burn patterns (Figure 1B). The delimitation of the fire-weather macro-areas was based on climatic and physiographical conditions and historical large fire footprints [59]. The wildfire season was the annual period concentrating 90% of the burned area from large fires (>100 ha; see Appendix A). We considered all the large fire ignitions within a 10 km buffer to compute the wildfire season's duration, and the analysis was conducted together for neighboring planning areas [43].

We then characterized the fire-weather conditions occurring during the wildfire season from hourly temperature records, rainfall, wind speed and direction, relative humidity, and solar radiation from different automatic stations using the Fire Family Plus v 4.2 software [60]. For each macro-area, we identified a representative weather station with a long data record (>15 years). Since containment efforts are very effective under mild weather conditions [61], we assumed 97th percentile extreme weather conditions in wind speed for most frequent wind directions and ERC-G fuel moisture content to generate the fire modeling weather scenarios. Specifically, we developed a set of probabilistic fire-weather scenarios in terms of fire spread duration, wind speed, wind direction, and fuel moisture content combinations (Table 2).

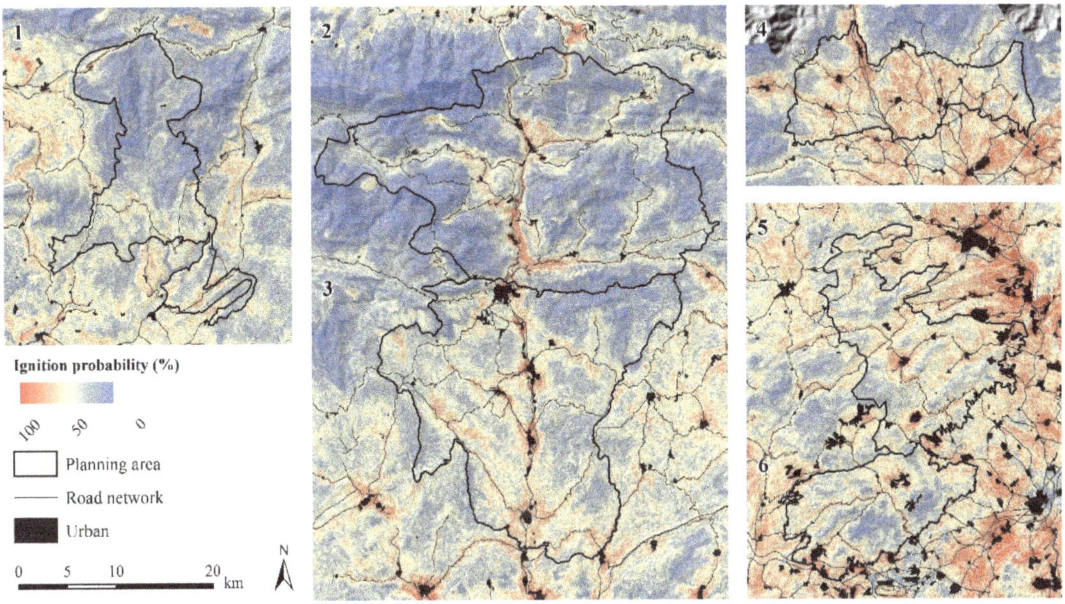

Figure 3. High resolution (40 m) ignition probability grid generated from fire ignitions (1998 to 2015; Figure 1B) and geospatial features using Random Forest. The numbers refer to the planning area codes (Appendix A; Figure 1).

2.6. Fire Spread Simulation Modeling

We used the minimum travel time (MTT) algorithm [62] as implemented in FConstMTT to model fire spread [59]. The algorithm calculates a two-dimensional fire growth at a resolution set by the user by minimizing fire travel time from the cell corners based on Rothermel's fire spread model [63]. The MTT has been widely used in previous studies assessing wildfire transmission and fuel treatment effects in complex terrains worldwide [12,64,65]. The fire ignitions were first distributed within the modeling domain according to the ignition probability grid (Figure 3). Then each fire was independently modeled considering subarea-level 97th percentile extreme fire-weather conditions for the wildfire season (Table 2).

To calibrate the surface fire spread model (see Appendix B), we replicated the actual large fire size (>100 ha) distribution as well as average fire size in every planning area separately, except for adjacent units (Figure 1A). Multiple day events were decomposed into daily blow-up progressions and constant fire-weather conditions during the fire's duration [66]. We set the fire spread durations that better replicated the historical fire size distribution under extreme weather conditions. The fire modeling domains were saturated with thousands of fires that burned each pixel more than 20 times on average and a total area equivalent to 10,000 years of iterations. Fire suppression efforts were omitted due to their limited containment capability during extreme fire events in the study area [67]. As a result, we obtained the pixel-level annual burn probability as [68]:

$$aBP = \frac{n_{xy}}{Y} \tag{1}$$

where aBP is the annual burn probability determined at a 40 m resolution as the n number of times a given xy pixel burned divided by the Y total number of modeled wildfire years or iterations.

Table 2. Fire-weather scenarios for extreme weather conditions (97th percentile) during wildfire season for the different planning areas. While the neighboring planning areas (Figure 1) were merged in a single fire modeling domain or landscape file, we considered multiple fire weather macro-areas. We used wildfire-season automatic weather station data to generate fire-weather scenarios. We set the fire spread duration probabilities that replicated historic fire distributions under constant extreme fire-weather conditions. See modeled and predicted fire size distributions in Appendix B.

Planning Area Code	Macro-Area-Level Weather Station (Municipality)	Fuel Moisture Content, %					Fire Spread Duration, Min (Probability, %)
		1 h	10 h	100 h	LH	LW	
1	Plain of Lleida (Tárrega)	7	8	10	20	65	75 (0.80); 180 (0.20)
	North–south valleys (Torredembarra)	7	9	11	40	70	
	Mountainous areas (Montsec d'Ares)	8	10	12	40	85	
2 and 3	Plain and open valleys (Muntanyola)	7	8	11	20	65	150 (0.25); 300 (0.50); 660 (0.25)
	Mountainous areas (Orís)	9	10	13	25	75	
	Mountain peaks (Cadí-Nord)	11	12	14	30	90	
4	Mountainous area (Cabanes)	8	9	12	40	75	90 (0.70); 540 (0.30)
	Inner Albera (Cabanes)	7	8	11	30	65	
	Coastal Albera (Portbou)	6	7	9	20	60	
5 and 6	Plain of Lleida (Tárrega)	7	8	10	20	65	110 (0.50); 210 (0.50)
	Inner mountains (St. Salvador Guardiola)	9	11	15	40	80	
	Pre-coastal depression (Font-Rubí)	7	8	10	20	65	
	Coastal belt (Torredembarra)	8	9	12	30	75	

Macro-Area-Level Weather Station (Municipality)	Wind Speed Scenario, km h^{-1} (Frequency, %)							
	45°	90°	135°	180°	225°	270°	315°	360°
Plain of Lleida (Tárrega)	-	19 (6)	20 (24)	19 (17)	19 (17)	18 (36)	-	-
North–south valleys (Torredembarra)	14 (12)	8 (8)	12 (43)	12 (20)	11 (8)	12 (9)	-	-
Mountainous areas (Montsec d'Ares)	-	-	-	20 (44)	14 (37)	18 (19)	-	-
Plain and open valleys (Muntanyola)	-	8 (6)	8 (20)	12 (39)	12 (27)	12 (8)	-	-
Mountainous areas (Orís)	-	12 (12)	12 (28)	12 (34)	10 (20)	10 (6)	-	-
Mountain peaks (Cadí-Nord)	-	-	12 (8)	12 (9)	14 (20)	14 (34)	12 (29)	-
Mountainous areas (Cabanes)	-	8 (25)	8 (31)	8 (12)	12 (8)	-	14 (11)	14 (13)
Inner Albera (Cabanes)	-	8 (25)	8 (31)	8 (12)	12 (8)	-	14 (11)	14 (13)
Coastal Albera (Portbou)	32 (14)	-	19 (15)	32 (23)	32 (6)	-	39 (7)	37 (35)
Plain of Lleida (Tárrega)	-	19 (6)	20 (24)	19 (17)	19 (17)	18 (36)	-	-
Inner mountains (St. Salvador Guardiola)	-	10 (5)	15 (32)	10 (41)	10 (13)	10 (9)	-	-
Pre-coastal depression (Font-Rubí)	-	-	14 (7)	19 (58)	19 (23)	18 (7)	19 (5)	-
Coastal belt (Torredembarra)	-	20 (21)	15 (20)	16 (36)	18 (23)	-	-	-

2.7. Expected Carbon Emissions

We estimated pixel-level expected emissions from wildfires combining modeled wildfire likelihood estimates with conditional fire effects from surface fires as [69]:

$$eE_{CO_2} = aBP \times \left(cFC \times EF_{CO_2}\right) \tag{2}$$

where eE_{CO_2} is the expected carbon dioxide gas emission (T ha^{-1} yr^{-1}), aBP is the annual burn probability (%), cFC is the conditional fuel consumption (T ha^{-1}), and EF_{CO_2} is the emission factor (T T^{-1}, dry basis). Emissions are, therefore, expectations estimated as the burn probability times consequences [70]. We used EF for stand-level dominant tree species, ranging between 1415 and 1879 g kg^{-1} [71].

The conditional fuel consumption (*cFC*) for surface fuels was first estimated by fuel category, including litter, duff, 1 h, 10 h, 100 h, and 1000 h using available models from the literature [72], and then summed to a 40-m resolution grid. Next, we assumed extreme fire-weather conditions (Table 1) and stand-level data on the forest floor and above ground dead fuel biomass dry weight per ha for the dominant forest types in the study area to calculate the fuel consumption (see Appendix C). Finally, the reduction in carbon dioxide gas emissions was estimated as the difference between the current conditions and the managed scenario or treated landscape in SMPs (Figure 2). We focused on assessing the reductions due to aBP differences and did not consider emissions from treatments due to the lack of reliable data for ongoing works.

2.8. Cost-Efficiency Analysis

First, we estimated the pixel-level difference, as CO_2 T ha^{-1} per year, between the current conditions and the managed scenario. Then, the CO_2 emission results in managed scenarios were determined after implementing fuel reductions in SMPs (Figure 2), with the treatment cost and prescriptions described in previous sections. Specifically, we modeled the reduction in burn probability for the managed scenarios to assess the treatment effects across the landscape. These results were summarized by the planning area and treatment unit (i.e., SMP). We then estimated the potential revenue from carbon credits considering a €13.18 T^{-1} reference market price for prevented emissions [73]. Finally, we determined the potential contribution of the carbon credits to financing the treatment cost, as a percentage, in each planning area. See the methodological flowchart on Appendix D.

3. Results

3.1. Expected Carbon Emissions

Expected carbon emissions from wildfires varied largely within and among the different planning areas (Figure 4). The average results ranged between 0.003 and 0.060 CO_2 T ha^{-1} per year at the landscape scale. Primarily, burn probability was the main causative factor explaining high expected emissions, despite the existing differences in stand-level conditional fuel consumption. We found the highest contrast between the neighboring forest lands of the planning areas 2 and 3, where similar stand structures (e.g., timber-stage conifer forests) located at less than 15 km of distance showed differences greater than 100 times in expected emissions. The strong north–south climatic gradient explains this difference because southern portions show drier fire-weather conditions (Table 2) and a higher wildfire occurrence (Figure 3). The open plains of planning area 4 and high-fuel load unburned black pine mature forests of central areas in planning area 3 presented the highest emission values from among all forest types. Most black pine mature stands of planning area 4 (i.e., mature timber-stage black pine forests) showing remarkably high conditional and expected emissions (>0.2 CO_2 T ha^{-1} yr^{-1}) were located in unburned patches or islands of the latest extreme fire episodes (Figure 1B). Conversely, recently burned areas in planning area 4, including low pole-stage Aleppo pine and mixed forests, presented a lower potential emission (<0.02 CO_2 T ha^{-1} yr^{-1}). On the other hand, incoming fires arrived from all around in planning areas 5 and 6 (Figure 4), where high emission areas tend to concentrate in the central portions. These areas were primarily dense and mature Aleppo pine stands exposed to long-distance spreading extreme wildfire events.

As expected, we found the lowest values in planning area 2 (Figure 4), where more than 80% of the mountainous areas in the north presented very shallow emission values (<0.001 CO_2 T ha^{-1} yr^{-1}). Except for some rare events occurring in conifer plantations, the wildfires in high-elevation areas (>1500 m.a.s.l) of the Pyrenees hardly ever burn more than 100 ha of forest lands. Winter fires associated with dry and gusty foehn winds can also burn reduced forest patches in mountainous alpine areas, but these are isolated low-frequency episodes with a limited potential compared to Mediterranean fires [74]. Nonetheless, southern wind-driven fires occasionally exposed south-facing lower slopes of the neighboring east–west orientation mountain range in planning area 2. We also found this topographic pattern in planning area 1, but the transition was smoother because the central valley and dominant southern winds occurring during the wildfire season presented the same orientation. In planning area 4 (Figure 4), a large portion to the west showed shallow values (<0.001 CO_2 T ha^{-1} yr^{-1}) mainly because of the lower wildfire occurrence and a northern wind direction channeling the fires through the central valley.

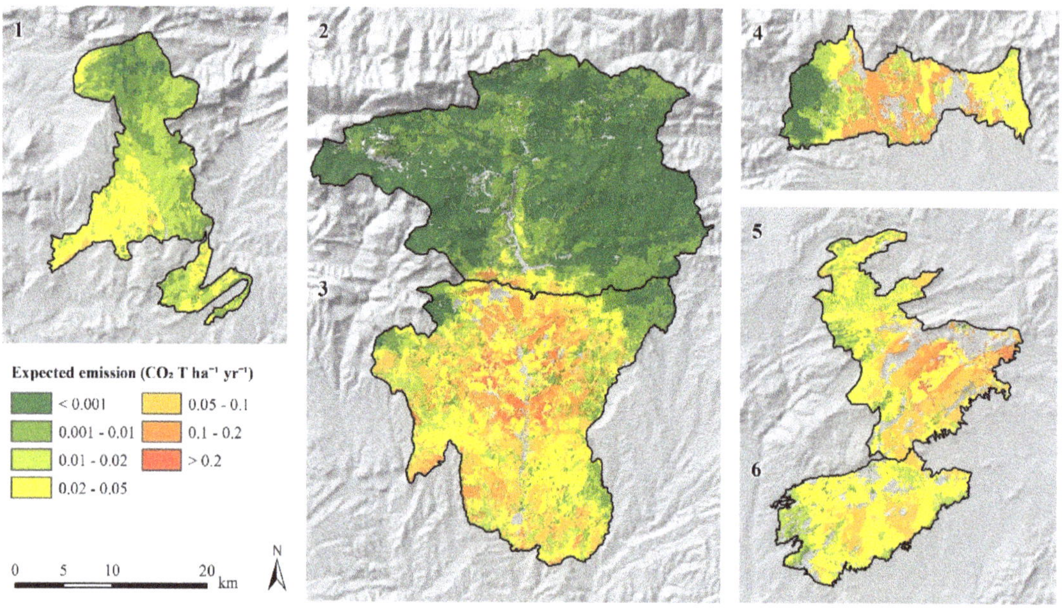

Figure 4. Expected carbon dioxide emissions (CO_2 T ha^{-1} yr^{-1}) from wildfires occurring under extreme fire-weather conditions within the different planning areas. The numbers refer to the planning area codes (Appendix A; Figure 1).

3.2. Carbon Emission Reduction in Managed Scenarios

The landscape-scale emissions in managed scenarios were reduced by between 49 (planning area 6) and 444 CO_2 T per year (planning area 3), which resulted in a reduction between 11 (planning area 6) and 35% (planning area 2) compared to the non-managed conditions (Table 3). Interestingly, a high percentage reduction was not associated with a high decrease in CO_2 T emission per year (e.g., see planning area 2). In planning area 2, most wildfire activity concentrated in the southern portion, where we concentrated many of the treatments, thus explaining the high reduction when presented as a percentage. On the other hand, the highest reduction per planning area extent corresponded to the planning area 4 and the highest reduction per treated area to the planning area 6. The higher overall performance in those units is explained by the high wildfire activity and the higher chance of SMPs encountering a fire in the future. However, we note that the treatment intensity (i.e., % treated area), spatial patterns (i.e., treatment unit size and clustering degree), and topographic position (e.g., water divides vs. valley bottoms) varied among and within the different planning areas. Treatment design is a significant factor affecting emission reductions that should be considered while interpreting the results [10,11].

Emission reduction maps showed very complex patterns and significant differences within the planning areas (Figure 5). Not surprisingly, the treated stands were the highest emission reduction areas, but the performance varied widely within and among the planning areas. As expected, the most notable emission reductions concentrated within a 5 km buffer area of influence around treated SMPs. Overall, the effect reduction decreased with the increase in the distance to the SMPs. Nonetheless, this influence was broader in highly fire-affected regions where the effect resulted very substantially (>0.01 CO_2 T yr^{-1}) far away from treated units (Figure 5). Planning area 3 is a good example where massive catastrophic fires can potentially spread as much as 10 km (Figure 1B).

Table 3. Summary table with landscape-scale expected carbon dioxide emissions from wildfires and emission reductions after treatments. The emission reductions represent landscape-scale results accounting for the fuel treatment effects on treated SMPs (Figure 2) and neighboring lands (Figure 5).

Planning Areas (Code)	Area (ha)	Non-Managed Scenario	Managed Scenario		
		Expected Emission (CO_2 T yr^{-1})	Treated Area (ha)	Emission Reduction	
				(CO_2 T yr^{-1})	(%)
Vall de Rialb (1)	23,467	362	1316	59	16
Capçaleres del Llobregat (2)	62,303	156	3251	55	35
Replans del Bergueda (3)	52,592	2629	5330	444	16
Els Aspres (4)	18,392	1107	2740	246	22
Serres d'Ancosa (5)	22,968	1044	3951	213	20
El Montmell (6)	15,461	454	318	49	11

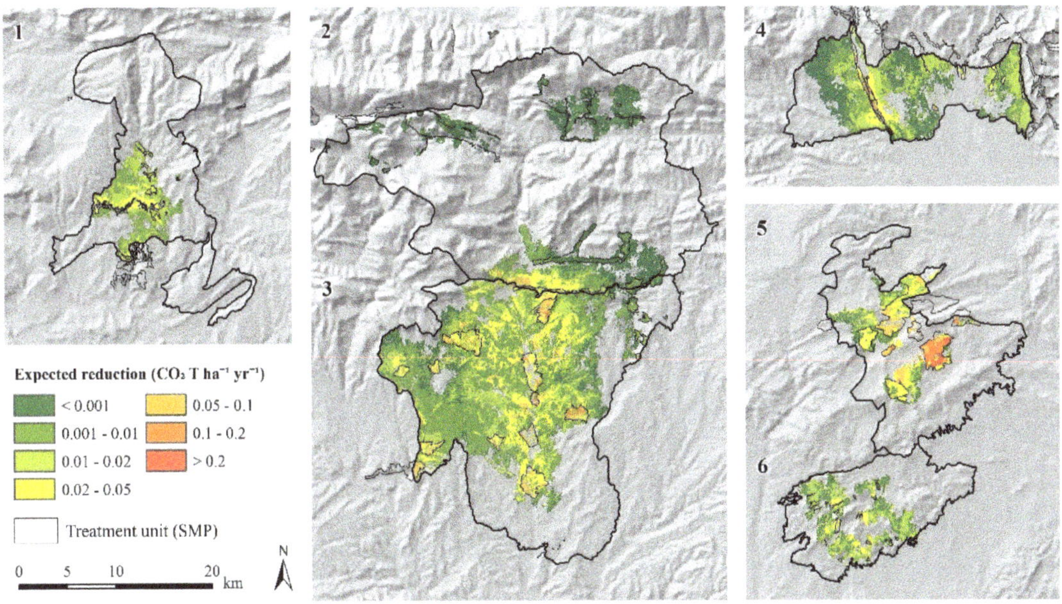

Figure 5. Reduction in carbon emissions (CO_2 T ha^{-1} yr^{-1}) from wildfires occurring under extreme fire-weather conditions in diverse forest systems across the different planning areas. We located the fuel treatments in the strategic management points (Figure 2). The numbers refer to the planning area codes (Appendix A; Figure 1).

The shadow effect in emission reductions obtained in the study areas was directly associated with the lower burn probability of the managed scenario (Figure 6). Specifically, the footprint effect expanded on the opposite side to the large fire arrival border. For instance, the SMPs disrupting southern wind-driven wildfires exhibited a reduction in the treatment's northern side. Previous studies also found this reduced burn probability shades close to treatments due to smaller fire footprints obtained in modeled managed scenarios [69,75]. Nonetheless, very high aBP reductions in large, treated patches did not necessarily produce a significant shade effect in the adjacent lands if the polygons represent a fire sink area. We found this fire-sink effect in the large SMPs of central planning area 5, where the aBP footprint reduction in neighboring regions (Figure 6) was minimal compared to the treated areas and emission reduction (Figure 5). These were the treatment blocks with a low wildfire occurrence encircled by fire source communication infrastructure and urban development areas (Figure 3).

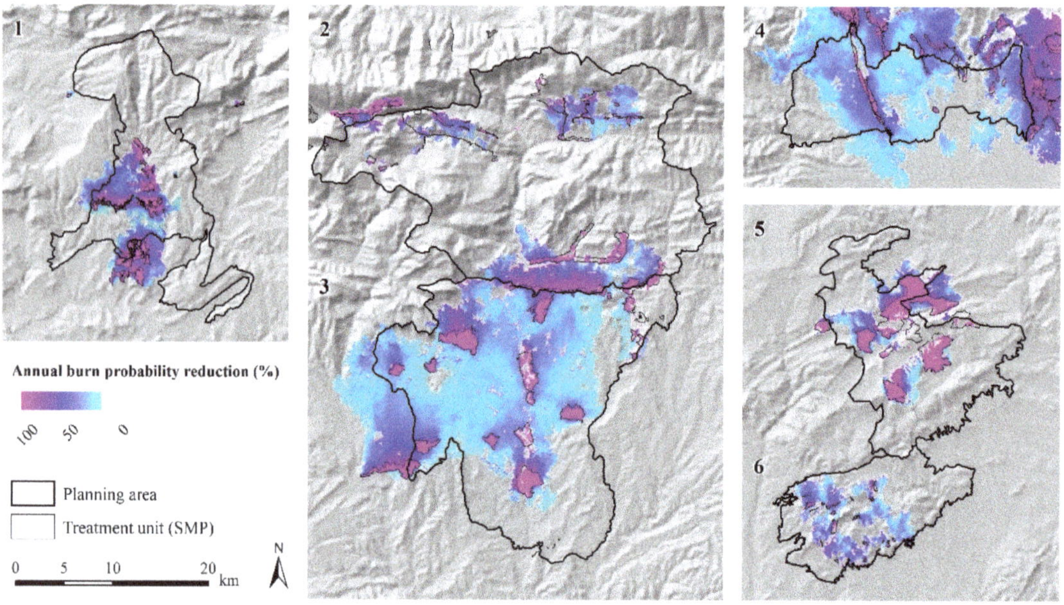

Figure 6. Spatially explicit reduction in the annual burn probability, as a percentage, due to the implementation of treatments in the strategic management points (Figure 2). The numbers refer to the planning area codes (Appendix A; Figure 1).

The treatment unit level emission reductions showed variable results within and among the different planning areas (Figure 7A). They revealed that SMPs were not necessarily allocated in the high-emission hot spots (Figure 4). The bulk of the SMPs attained reduction values between 0.01 and 1 T CO_2 per year on average, and only less than 5% of the treated areas was above 0.08 T CO_2 ha^{-1} per year. The top SMPs in the planning areas 1, 4, and 3 showed reductions greater than 1 T CO_2 per year, but this high performance shifted drastically when we analyzed the reduction per treated area. In planning area 4, for instance, the reduction per treated area varied more than five times between the treatment units with a similar total reduction. Other planning areas, such as the 1 and 2, presented a narrower difference of two times while addressing the performance per treated areas. The treatments implemented in planning area 2 were the worst among the different treatment units and rarely surpassed 0.01 T CO_2 ha^{-1} per year.

The widest variation in performance was obtained in planning area 5 (Figure 7B). Here, the difference between the first and third quartiles was close to 0.05 T CO_2 per year and reflected how the average value alone is not enough to describe the performance results within planning areas. These results presented in box plots could assess the potential performance improvement for the best set of SMP treatment units after excluding those located in rare remote areas with a reduced potential to restrict fire spread and carbon emission (<0.01 CO_2 T ha^{-1} yr^{-1}). Likewise, we found that some planning areas, such as 4 and 5 in particular, contained a significant number of observations above the third quartile (Figure 7B) implemented in very high-emission areas (>0.05 CO_2 T ha^{-1} yr^{-1}) where the project-level average reduction disguised a great allocation. These units were the high-performance strategic management points or fuel break, or treatment barriers implemented in the central part of the planning area 4 (Figure 3).

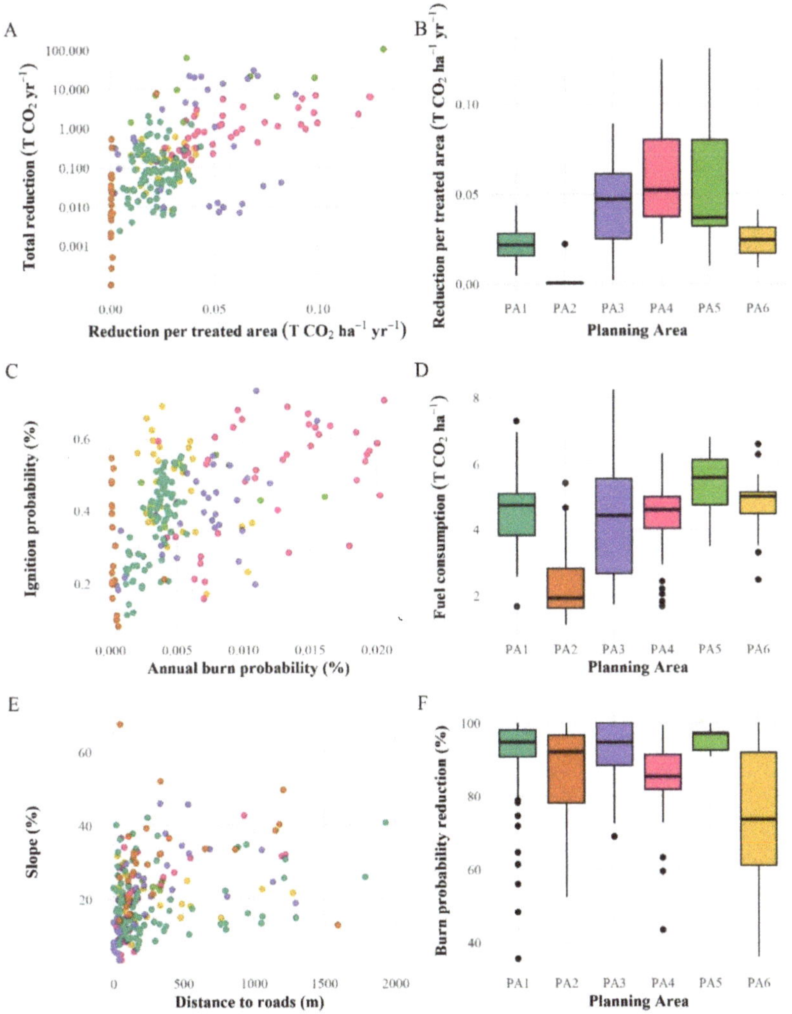

Figure 7. Summary statistics for the strategic management points. The performance is shown by treatment unit (**A**) and planning area (**B**) level. We obtained treatment unit-level wildfire occurrence and wildfire likelihood (**C**). The conditional fuel consumption is presented by planning area (**D**). Treatment implementation (Figure 2) operational constraints in terms of the average slope and distance to rode are presented by treatment unit (**E**). Finally, we show the burn probability reduction in treated SMP (Figure 6) by planning area (**F**). In box plots, the boxes indicate the first/third quartiles, the whiskers indicate 10th and 90th percentiles, the black line is the median, and the dots indicate values below the 10th percentile or above the 90th percentile. The colors refer to planning areas, as shown in boxplots.

Although an increasing wildfire occurrence (Figure 3) correlated with a higher overall wildfire likelihood, high ignition probability values did not necessarily connote a high burn probability (Figure 7C). For instance, many SMPs in planning area 4 with ignition probability values of about 60% sowed a very variable burn probability, even up to three times. These complex patterns were also found in other Mediterranean areas [76,77]. Wildfire burn probability was a significant factor explaining expected emissions because conditional fuel consumption (Figure 7D) alone would have provided a misleading result. While planning areas 1 and 4 presented similar fuel consumption results (Figure 7D), expected emissions were much higher in planning area 4 (Figure 7B). Most treatment units were located close to roads (<250 m) on gentle slopes (<30%) (Figure 7E). Overall, burn probability reductions were higher than 80%, except in planning area 6, where the size of the treatment units was tiny, and large fires easily surpassed the SMPs (Figure 7F).

3.3. Financing Fuel Treatment Cost with Carbon Credits

Forest management provided a significant gross benefit or revenue from preventing wildfire carbon emissions in some planning areas (Table 4). Specifically, the annual income was between 645 and 5853 € for the carbon emission reductions estimated in this study (Figure 5), assuming a €13.18 per CO_2 T market price [73]. While the lowest attained revenue normalized per planning area extent was 0.37 € ha^{-1} in planning area 2, the highest value resulted in 4.70 € per ha in planning area 4. Despite the low total revenue of the planning area 6 (643 € yr^{-1}), the attainment per treated area at the landscape scale was above the average (2.35 € ha^{-1}). Since the treatment allocation showed highly variable stand-level results (Figure 7), we calculated the percentage treated area in high emission hot spots (>0.01 CO_2 T ha^{-1} yr^{-1}). This was an alternative performance metric to determine the SMP area with a significant contribution in reducing emissions. In contrast with the good overall allocation of treatments in planning areas 5 and 6 (>90%), in some planning areas, less than half of the treated area would be considered attractive for carbon credit investment. By contrast, a significant revenue emerged when we computed most carbon credits to a much lower treated area (e.g., 3245 € yr^{-1} to just 29% of the SMP area in planning area 4).

Table 4. Revenue from a potential carbon credit market in the different planning areas. We assumed a fuel treatment effective duration between 8 to 14 years. The wildfire managers implementing fuel reduction programs in the study area provided the treatment cost (Table 2). The SMP area over high emission hotspots (>0.01 CO_2 T ha^{-1} yr^{-1}) was computed as a performance metric for the fuel reduction programs.

Planning Area (Code)	Strategic Management Points		Fuel Treatment Cost			
	Area (ha)	Hotspot Area (%)	Thinning (ha)	Mastication (ha)	Prescribed Fire (ha)	Cost (€ yr^{-1})
1	628	64	164	26	602	74,972
2	1980	0	1099	129	1851	278,880
3	3775	74	1133	711	3064	471,821
4	691	29	24	310	382	76,698
5	3171	96	1657	50	3121	436,342
6	274	93	104	1	273	34,770

Planning Area (Code)	Carbon-Credit Revenue		
	Reduction in Emissions (T CO_2 yr^{-1})	Revenue (€ yr^{-1})	Contribution to Cost (%)
1	58.8	775	1.0
2	55.0	725	0.3
3	444.1	5853	1.2
4	246.2	3245	4.2
5	212.7	2804	0.6
6	48.8	643	1.8

The total treatment cost by planning areas was between 34,770 and 471,821 € per year. The areas requiring a thinning to a greater extent presented the highest cost. These SMPs were predominant in planning areas 2 and 5, where thinning was implemented in more than 50% of the area, and the average cost was above 135 € ha^{-1} per year. On the other hand, the most economical treatments were located in planning area 4 (110 € ha^{-1} yr^{-1}) despite requiring the more expensive mechanical mastication instead of prescribed fire in 45% of the area. The carbon credit revenue's potential contribution in financing the fuel treatment cost was between 0.3 in planning area 2 and 4.2% in planning area 4. Considering only the SMP area allocated in carbon emission hotspots (Table 4), the contribution increased substantially. It would potentially cover up to 14.5% of the cost in the best case (i.e., planning area 4). Conversely, the potential revenue from carbon credits in planning areas 2 and 5 was below 1% in any case. While we expected poor results in planning area 2, the contribution in planning area 5 was exceedingly low for the observed high fire activity (i.e., annually burned area). We presume that this was due to the design of exceeding large treatment blocks in planning area 5 (Figure 2), where much narrower barriers or fuel breaks perpendicular to dominant winds, such as those implemented in planning area 4, would have obtained the same effect for less than half of the cost.

4. Discussion and Conclusions

Despite the European Union's accession to the Paris Agreement [78] and commitment to drastically cut CO_2 emissions, severe wildfires continue to burn vast areas and pose a significant threat to the Mediterranean forests [4,79,80]. As a result, forest management works developing wildfire risk reduction plans have substantially increased in recent years [9,81,82], but the emission reduction effects are still largely unknown. However, sustainable forest management support for preserving carbon pools in fire-prone forest ecosystems lacks specific financing lines. To our knowledge, this work is the first study conducted in Mediterranean areas estimating prevented CO_2 emissions in ongoing risk reduction plans through modeling. Precisely, we assessed the emission reduction effect of strategic management points (SMPs) and revenue from carbon credits in a wide range of fire-prone Mediterranean landscapes. This study provides a valuable baseline for developing a carbon credit market intended to economically compensate small forest landowners for preserving fire resilient cultural landscapes in fire-prone southern European regions.

The fuel consumption quantification from large fires is essential to calculate potential carbon credits in Mediterranean landscapes. Large fires caused substantial losses in the past decades in Catalonia and will likely continue to burn vast portions due to limited management and young forest expansion in marginal agricultural lands [30,41]. Therefore, landscape planning efforts derived from stand-level conditional effects would largely ignore expected CO_2 emissions from large fires ignited elsewhere. In other words, the treatment units or stands present near-impossible odds of getting burned by a fire ignited within the same forest. Indeed, humans ignite most fires close to developed sites and communication infrastructure, and wildfires then hit forest lands after traveling long distances [36,83]. On the other hand, CO_2 emission estimates at regional scales overlook the complex patterns of the burned areas [84,85]. In contrast to most previous works in the Mediterranean region, we implemented stochastic fire simulation to replicate historic fire footprint distributions across the different planning areas and model the annual burn probability at high resolution (40 m). We accounted for the potential sources of uncertainty and variability in model inputs (i.e., ignition probabilities, extreme fire-weather conditions, and fuel loads in forest types) by modeling thousands of iterations (10,000 years) and summarizing the wildfire likelihood pixel-level results to an overall annual burn probability.

We identified the areas where ongoing fuel reduction programs are most likely to generate long-term carbon benefits. The results demonstrated how fuel management programs implemented in fire-prone areas reduced expected carbon emissions on treated stands and the neighboring forest lands. However, many SMPs located in remote regions would hardly encounter a wildfire and presented poor performance results. Thus, we strongly suggest prioritizing SMP treatments in high emission spots and postponing or excluding the fuel reductions in remote areas. Focusing on high emission areas would reduce the fuel treatment program cost drastically. In addition, our high-resolution expected carbon emission maps may help forest managers determine which suitable sites meet these conditions (Figure 4). The SMPs that we tested in this study were designed based on the expert criteria [20,22]. These SMPs were specifically designed to reduce large-fire potential, increase firefighting contention capacity, and did not explicitly address CO_2 emission reductions [21]. Some previous studies estimated encounter rates between fire perimeters and treated areas to evaluate the fuel treatment effectiveness [86], but we decided to implement a high-resolution quantitative assessment based on the burn probability to measure CO_2 emission reductions [69]. Likewise, the burn probability has been widely used in previous works to prioritize and test fuel treatment effects at the landscape scale and select the most convenient design [87–89].

We conducted this analysis as a preliminary step to estimate spatially explicit results for surface fires. Other carbon emissions sources, including emissions during prescribed fires, crown fires, and the use of machinery, were not considered in our analysis. Previous studies assessing the prescribed fire combustion emissions reported a wide range of values between 2 and 10 CO_2 T ha^{-1} depending on the forest systems and treatment intensities [90,91]. Conducting prescribed fires in early spring would represent the year's proper timing for a light burn, preventing duff and heavy log consumption due to a higher fuel moisture content [92]. Considering an eight-year fuel treatment rotation interval, these light burns would represent about 0.375 T CO_2 ha^{-1} per year. Concerning crown fires, these only consume the thin branches and leaves (<6 mm) and modeling these emissions in high resolution at large scales would result in an exceedingly complex calculation. Nonetheless, high-resolution LiDAR and remote sensing-derived estimates would allow for computing the crown fire biomass in the following studies [93,94]. And lastly, we understand that chainsaws and mastication equipment also represent a CO_2 emission. Still, the manually implemented prescribed fire was the dominant treatment type (88% of the area vs. mechanical mastication in 12% of the treated area). We dismissed the use of machinery to extract thinning due to the problematic access to SMPs and the lack of commercial interest. In all, frequent but light surface burns in high emission spots, implemented in spring under high moisture content conditions for the soil, would deliver the most increased net carbon sequestration at the landscape scale [51].

We can conclude from these results that few landowners would be suitable to receive economic compensation for treating SMPs. Furthermore, the potential financial revenue from carbon credits was only substantial in extremely high CO_2 emission areas, such as the central fuel break barrier of the planning area LU4 (Figure 2). At best, the revenue from carbon credits could fund up to 14% of the treatment cost if we focus on SMPs with expected emissions above 0.01 CO_2 T ha^{-1} per year and exclude all the rest. Nonetheless, we need to emphasize that treatment costs will decrease by a factor of three after the first intervention because the following maintenance treatments are much easier to implement [52,54]. Moreover, we noted that results require careful consideration due to the wide range of treated intensities (i.e., % treated areas) and SMP size and shapes, which may affect the modeling outcomes. Likewise, treated SMP in large fire areas interact closely, and individualizing the contribution to the total emission reduction is highly challenging. Indeed, each treatment would require a separate modeling analysis to precisely determine reduced emission at the landscape scale.

We believe that protecting carbon stocks on forest systems may become a significant management objective in future projects as emission neutrality and the promotion of a bio-based economy gain momentum in the European Union. Nonetheless, future efforts should be oriented toward determining the tipping points between carbon benefits and losses from a broader range of forest management actions, which should provide better and more accurate information to compute realistic compensations [95]. However, optimal solutions may compete with other existing objectives (e.g., reducing carbon emissions vs. reducing wildfire transmission to communities), but the trade-off analysis allows assessing treatment co-location opportunities (i.e., forest stands where treatments can meet multiple goals) on vast landscapes [96,97]. Ultimately, treatment prioritization works would provide a valuable set of treatment solutions that would require implementing a risk assessment framework to evaluate cost-efficiency.

Author Contributions: Conceptualization, F.A. and A.A. (Alan Ager); methodology, F.A. and A.A. (Alan Ager); software, F.A.; validation, F.A. and M.R.; formal analysis, F.A.; investigation, F.A.; resources, T.C.; data curation, F.A. and M.R.; writing—original draft preparation, F.A.; writing—review and editing, M.R., M.S., A.A. (Aitor Ameztegui) and C.V.-G.; visualization, P.G.; supervision, C.V.-G.; project administration, T.C.; funding acquisition, T.C. All authors have read and agreed to the published version of the manuscript.

Funding: This study was funded by the LIFE CLIMARK Project (LIFE16 CCM/ES/000065), and the *"Ministerio de Economía y Competitividad. Juan de la Cierva Formación"* research fellows FJCI-2016-31090 and FIJCI-2016-30049).

Institutional Review Board Statement: Not applicable.

Informed Consent Statement: Not applicable.

Data Availability Statement: Not applicable.

Acknowledgments: Special thanks to Joaquim García (CPF), Pere Casals (CTFC), and Jordi García (CTFC) for facilitating litter and duff biomass data for this study. We also thank the Fire Department of Catalonia's GRAF wildfire-fighting specialists for providing the strategic management point cartography. The Meteorological Service of Catalonia (METEOCAT) provided the automatic weather station data. The Centre for National Information on Forest Fires (CCINIF) provided fire data records, the agency responsible for coordinating general forest fires statistics in Spain (EGIF) within the Ministry of Agriculture and Fisheries, Food and Environment (MAPAMA).

Conflicts of Interest: The authors declare no conflict of interest.

Appendix A. Dominant Vegetations Types and Historic Fire Activity

Irrigated agricultural lands, mosaics of low shrublands, and herbaceous xerophytic vegetation cover the central depression below 450 m. The higher altitude and coarse reliefs to the north confine cultivated plots to valley bottoms, with forested areas dominated by Mediterranean oaks and low shrublands on slopes (*Lavandula angustifolia* Mill., *Rosmarinus officinalis* L. and *Quercus coccifera* L.). These shrublands and forests are gradually replaced by tall-shrubland species (*Buxus sempervirens* L. and *Juniperus communis* L.), mid-mountain oak (*Quercus pubescens* Willd.), and conifer species (*Pinus nigra* Arn. and *Pinus sylvestris* L.) first on north-facing slopes, and then on higher elevations (*Pinus uncinata* Ram.). Mosaics of rocky outcrops, low shrublands (*Genista balansae* Boiss.), and pastures cover the high mountain tops above 2400 m. The Mediterranean maquis (*Pistacia lentiscus* L. and *Arbutus unedo* L.) appear in combinations with densely regenerated young Aleppo on the pre-littoral mountain ranges pine cohorts (*Pinus halepensis* Mill.). Silicicolous shrublands (*Cistus* ssp. and *Erica* ssp.) are found in coastal lowlands, sometimes with stone pine (*Pinus pinea* L.). The cork oak (*Quercus suber* L.) forests are limited to the northeastern lowlands.

Table A1. Summary table with main features of the planning areas in the study and observed wildfire activity from 1983 to 2015. See planning area locations in Figure 1. We considered 100 ha as the large fire (LF) threshold. The wildfire activity data was summarized collectively for adjacent planning areas. The wildfire season was determined from the cumulative burned area (Figure A1).

Planning Areas (Code)	Area (ha)	Dominant Vegetation Type and Tree Species (abbreviation)	Burned Area (% yr^{-1})	Fire Occurrence (ip km^{-2} yr^{-1})	LF Burned Area (%)	LF Number (%)	Largest Wildfire
Vall de Rialb (1)	23,467	Evergreen and semi-evergreen oak (Qsp) stands under post-fire regeneration and mature stands of adult black pine (Pn)	0.21	0.009	87	3	4986 ha on 17 July 2009

Table A1. *Cont.*

Planning Areas (Code)	Area (ha)	Dominant Vegetation Type and Tree Species (abbreviation)	Burned Area (% yr^{-1})	Fire Occurrence (ip km^{-2} yr^{-1})	LF Burned Area (%)	LF Number (%)	Largest Wildfire
Capçaleres del Llobregat (2)	62,293	Mature stands of Scots pine (Ps) and black pine (Pn)	0.47	0.013	96	3	25,368 ha on 4 July 1994
Replans del Bergueda (3)	52,591	Post-fire regeneration Aleppo pine (Ph) young stands under and mature stands of Scots pine (Ps)					
Els Aspres (4)	18,392	Post-fire regeneration young stands and adult stands of cork oak (Qs) woodlands managed for cork production	1.49	0.030	95	3	19,612 ha on 19 July 1982
Serres d'Ancosa (5)	22,957	Mature stands of Aleppo pine (Ph)	0.49	0.034	86	2	3852 ha on 6 July 1986
El Montmell (6)	15,461	Stands of Aleppo pine (Ph), evergreen and semi-evergreen oak (Qsp) forests, and post-fire regeneration mixed forests (Mx)					

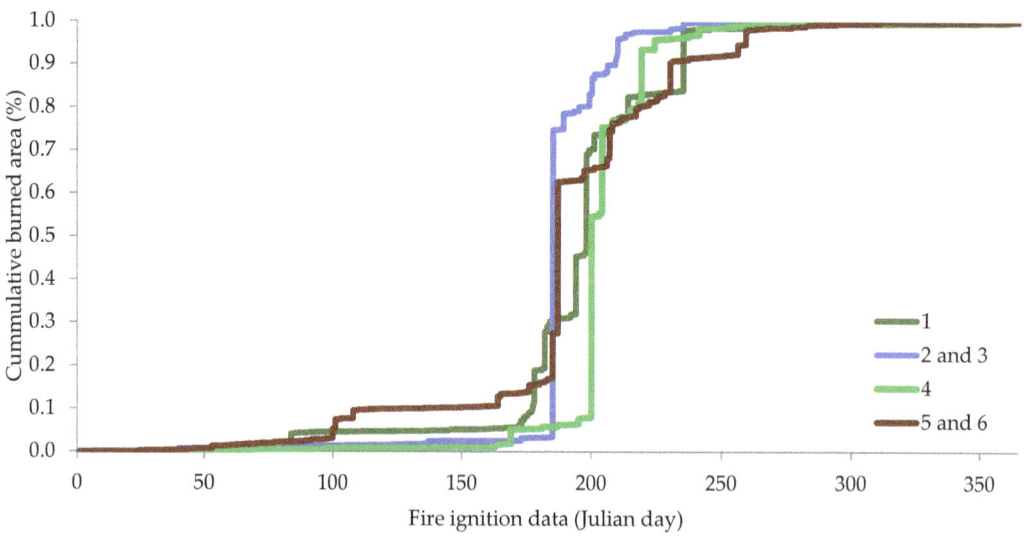

Figure A1. Daily cumulative burned area for the different fire modeling domains. The wildfire season went from 25 June to 23 August in planning area 1, 4 July to 30 July in planning area 2 and planning area 3, 18 June to 13 August in planning area 4, and 10 April to 12 September in planning area 5 and planning area 6.

Appendix B. Fire Spread Model Calibration

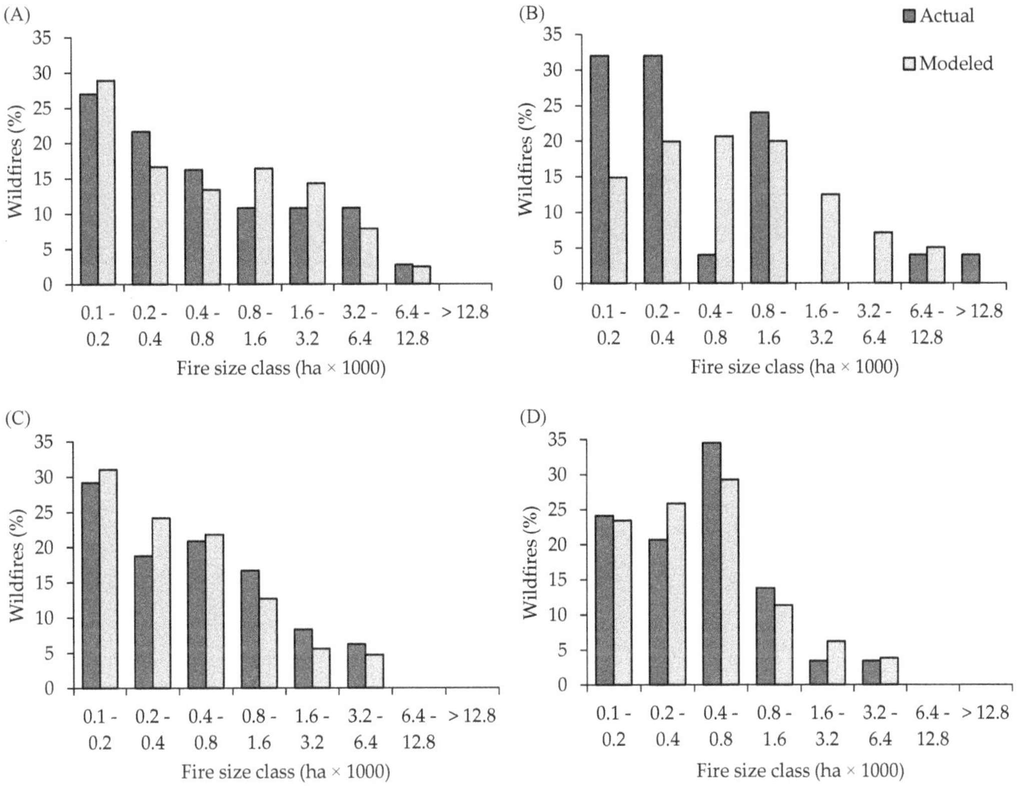

Figure A2. Fire spread model calibration in the different fire modeling domains. Histograms show the fire size distribution for actual (1883 to 2015) and modeled large fires (>100 ha). In planning area 4 (**A**), 29 historical events had a 1244 ha average fire size, and modeled fires had an average fire size of 1216 ha. In planning areas 2 and 3 (**B**), 25 historical events had a 1453 ha average fire size, and the modeled fires 1428 ha. In planning areas 5 and 6 (**C**), 48 historical events had a 780-ha average fire size, and the modeled fires had an average fire size of 779 ha. In planning area 1 (**D**), 29 historical events had a 695-ha average fire size, and the modeled fires had an average fire size of 699 ha.

Appendix C. Dead Fuel Loads in Dominant Forest Types

Required dry biomass data for assessing the FC was obtained from field sampling campaigns conducted on the dominant forest typologies (Table 1; Figure A2) and complemented with fuel loading data from standard fuel models in herbaceous type models and live woody components [46]. We used different methods to estimate dead fuel loadings for the different fuel fractions. The field sampling combined litter and duff extractions on 30 cm radius circular plots (n = 16 plots per forest structure type) and 1, 10, and 100 h fuel extractions on 2 m square plots (n = 3 per forest structure type). All these fuel samples were then oven-dried and added to estimate the dry weight biomass per ha. To calculate the amount of 1000 h dead biomass, we conducted a set of 30 m sampling strips (n = 3 per forest structure type) where we measured the diameter of the fallen logs on the center and the length within a 1 m width. Then, we used species-specific allometric equations [98] to estimate the total amount of biomass from the volume of the logs, i.e., 1000 h fuel category.

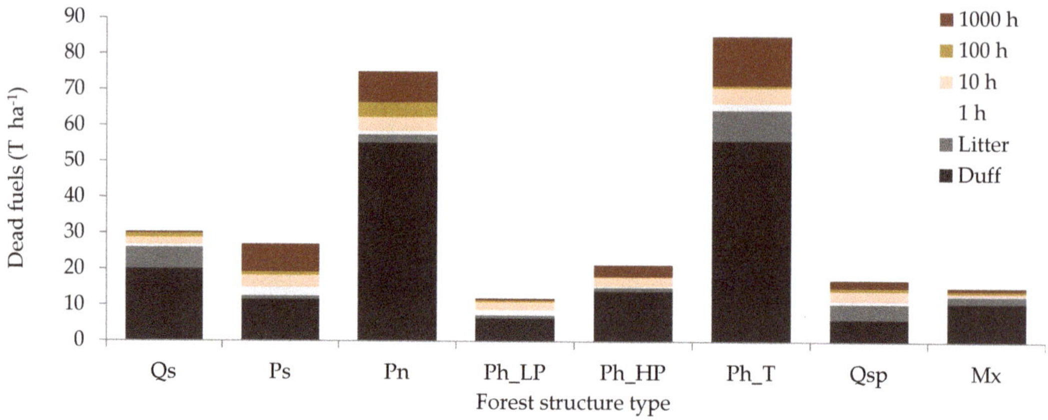

Figure A3. Dry weight (T ha^{-1}) of dead fuel categories, including litter, duff, 1 h, 10 h, 100 h, and 1000 h, fuel load components on the dominant forest types within the LUs. See dominant tree species abbreviations in Table 1. Regular stand ages included low pole-stage (LP), high pole-stage (HP), and timber stage (T).

Appendix D. Methodological Flowchart

The cost-efficiency process was conducted in four main steps (Figure A4). We first assessed expected carbon emissions as the annual burn probability times stand-level conditional carbon emissions for current conditions. Then, we implemented the landscape fuel reduction treatments in SMPs. Next, we predicted the annual burn probability reduction for the treated LCP to assess the expected carbon emission for the managed scenario. Finally, we calculated the carbon emission reduction for every Euro invested in treatments. The carbon credits of reduced carbon emissions were considered as revenue in managed scenarios.

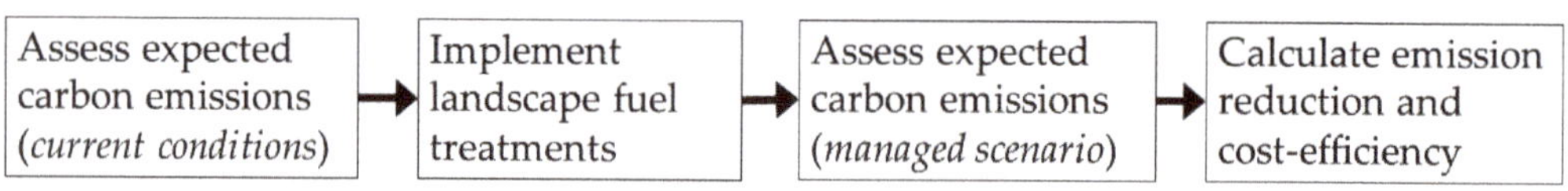

Figure A4. Methodological flowchart of the process conducted to estimate landscape fuel treatment cost-efficiency.

References

1. van der Werf, G.R.; Randerson, J.T.; Giglio, L.; Collatz, G.J.; Mu, M.; Kasibhatla, P.S.; Morton, D.C.; DeFries, R.S.; Jin, Y.; van Leeuwen, T.T. Global fire emissions and the contribution of deforestation, savanna, forest, agricultural, and peat fires (1997–2009). *Atmos. Chem. Phys.* **2010**, *10*, 11707–11735. [CrossRef]
2. Giglio, L.; Boschetti, L.; Roy, D.P.; Humber, M.L.; Justice, C.O. The Collection 6 MODIS burned area mapping algorithm and product. *Remote Sens. Environ.* **2018**, *217*, 72–85. [CrossRef]
3. San-Miguel-Ayanz, J.; Moreno, J.M.; Camia, A. Analysis of large fires in European Mediterranean landscapes: Lessons learned and perspectives. *For. Ecol. Manag.* **2013**, *294*, 11–22. [CrossRef]
4. Balde, B.; Vega-García, C. Estimación de emisiones de GEI y sus trayectorias en grandes incendios forestales en Cataluña, España. *Madera Y Bosques* **2019**, *25*, e2521764. [CrossRef]
5. Rodrigues, M.; Alcasena, F.; Gelabert, P.; Vega-García, C. Geospatial modeling of containment probability for escaped wildfires in a Mediterranean region. *Risk Anal.* **2020**, *40*, 1762–1779. [CrossRef] [PubMed]
6. Tedim, F.; Leone, V.; Amraoui, M.; Bouillon, C.; Coughlan, M.R.; Delogu, G.M.; Fernandes, P.; Ferreira, C.; McCaffrey, S.; McGee, T.K.; et al. Defining extreme wildfire events: Difficulties, challenges, and impacts. *Fire* **2018**, *1*, 9. [CrossRef]
7. Moreira, F.; Ascoli, D.; Safford, H.; Adams, M.A.; Moreno, J.M.; Pereira, J.M.C.; Catry, F.X.; Armesto, J.; Bond, W.; González, M.E.; et al. Wildfire management in Mediterranean-type regions: Paradigm change needed. *Environ. Res. Lett.* **2020**, *15*, 011001. [CrossRef]
8. Curt, T.; Frejaville, T. Wildfire Policy in Mediterranean France: How Far is it Efficient and Sustainable? *Risk Anal.* **2017**, *38*, 472–488. [CrossRef]

9. Palaiologou, P.; Kalabokidis, K.; Ager, A.A.; Day, M.A. Development of comprehensive fuel management strategies for reducing wildfire risk in Greece. *Forests* **2020**, *11*, 789. [CrossRef]
10. Salis, M.; Del Guiudice, L.; Arca, B.; Ager, A.A.; Alcasena, F.; Lozano, O.; Bacciu, V.; Spano, D.; Duce, P. Modeling the effects of different fuel treatment mosaics on wildfire spread and behavior in a Mediterranean agro-pastoral area. *J. Environ. Manag.* **2018**, *212*, 490–505. [CrossRef]
11. Finney, M.A. A computational method for optimizing fuel treatment location. *Int. J. Wildland Fire* **2007**, *16*, 702–711. [CrossRef]
12. Oliveira, T.M.; Barros, A.M.G.; Ager, A.A.; Fernandes, P.M. Assessing the effect of a fuel break network to reduce burnt area and wildfire risk transmission. *Int. J. Wildland Fire* **2016**, *25*, 619–632. [CrossRef]
13. Fernandes, P.M. Scientific support to prescribed underburning in southern Europe: What do we know? *Sci. Total Environ.* **2018**, *630*, 340–348. [CrossRef]
14. Lasanta, T.; Khorchani, M.; Perez-Cabello, F.; Errea, P.; Saenz-Blanco, R.; Nadal-Romero, E. Clearing shrubland and extensive livestock farming: Active prevention to control wildfires in the Mediterranean mountains. *J. Environ. Manag.* **2018**, *227*, 256–266. [CrossRef] [PubMed]
15. Varela, E.; Górriz-Mifsud, E.; Ruiz-Mirazo, J.; López-i-Gelats, F. Payment for Targeted Grazing: Integrating Local Shepherds into Wildfire Prevention. *Forests* **2018**, *9*, 464. [CrossRef]
16. Ascoli, D.; Russo, L.; Giannino, F.; Siettos, C.; Moreira, F. Firebreak and Fuelbreak. In *Encyclopedia of Wildfires and Wildland-Urban Interface (WUI) Fires*; Manzello, S.L., Ed.; Springer: Berlin/Heidelberg, Germany, 2018; pp. 1–9.
17. Elia, M.; Lovreglio, R.; Ranieri, N.; Sanesi, G.; Lafortezza, R. Cost-effectiveness of fuel removals in Mediterranean wildland-urban interfaces threatened by wildfires. *Forests* **2016**, *7*, 149. [CrossRef]
18. Costa, P.; Castellnou, M.; Larrañaga, A.; Miralles, M.; Daniel, K. *Prevention of Large Wildfires Using the Fire Types Concept*; Graf, U.T.d., Ed.; Cerdanyola del Vallès: Barcelona, Spain, 2011.
19. Quilez, R.; Valbuena, L.; Vendrell, J.; Uytewaal, K.; Ramirez, J. Establishing Propagation Nodes as a Basis for Preventing LargeWildfires: The Proposed Methodology. *Front. For. Glob. Chang.* **2020**, *3*, 137. [CrossRef]
20. Romero-Vivó, M.; Soriano, J.L.; Quílez, R.; Gorgonio, E.; Caballero, D.; Larrañaga, A.; Rodríguez y Silva, F.; Blanco, J.; Ortega, G.; López del Río, R.; et al. Definición y Recomendaciones Técnicas en el Diseño de Puntos Estratégicos de Gestión. Available online: https://agroambient.gva.es/documents/162905929/164277177/Dec%C3%A1logo+Valencia+Jornada+PEG/510265c0-fd5e-48b6-bba2-d3038264a134 (accessed on 15 September 2021).
21. Gonzalez-Olabarria, J.R.; Reynolds, K.M.; Larrañaga, A.; Garcia-Gonzalo, J.; Busquets, E.; Pique, M. Strategic and tactical planning to improve suppression efforts against large forest fires in the Catalonia region of Spain. *For. Ecol. Manag.* **2019**, *432*, 612–622. [CrossRef]
22. Castellnou, M.; Prat-Guitart, N.; Arilla, E.; Larrañaga, A.; Nebot, E.; Castellarnau, X.; Vendrell, J.; Pallàs, J.; Herrera, J.; Monturiol, M.; et al. Empowering strategic decision-making for wildfire management: Avoiding the fear trap and creating a resilient landscape. *Fire Ecol.* **2019**, *15*, 31. [CrossRef]
23. Ruiz-Peinado, R.; Bravo-Oviedo, A.; López-Senespleda, E.; Bravo, F.; Del Rio, M. Forest management and carbon sequestration in the Mediterranean region: A review. *For. Syst.* **2017**, *26*, eR04S. [CrossRef]
24. Prada, M.; Bravo, F.; Berdasco, L.; Canga, E.; Martínez-Alonso, C. Carbon sequestration for different management alternatives in sweet chestnut coppice in northern Spain. *J. Clean. Prod.* **2016**, *135*, 1161–1169. [CrossRef]
25. Piqué, M.; Valor, T.; Castellnou, M.; Pagés, J.; Larrañaga, A.; Miralles, M. *Integració del Risc de Grans Incendis Forestals (GIF) en la Gestió Forestal. Incendis Tipus i Vulnerabilitat de les Estructures Forestals al Foc de Capçades*; Generalitat de Catalunya, Departament d'Agricultura, Ramaderia, Pesca, Alimentació i Medi Natural, Centre de la Propietat Forestal, Eds.; Generalitat de Catalunya: Barcelona, Spain, 2011; p. 122.
26. Beltrán, M.; Piqué, M.; Vericat, P. *Models de Gestió per als Boscos de pi Blanc (Pinus halepensis L.): Producció de Fusta i Prevenció D'incendis Forestals*; Centre de la Propietat Forestal, Generalitat de Catalunya: Barcelona, Spain, 2011; 124p.
27. Chiono, L.A.; Fry, D.L.; Collins, B.M.; Chatfield, A.H.; Stephens, S.L. Landscape-scale fuel treatment and wildfire impacts on carbon stocks and fire hazard in California spotted owl habitat. *Ecosphere* **2017**, *8*, e01648. [CrossRef]
28. James, J.N.; Kates, N.; Kuhn, C.D.; Littlefield, C.E.; Miller, C.W.; Bakker, J.D.; Butman, D.E.; Haugo, R.D. The effects of forest restoration on ecosystem carbon in western North America: A systematic review. *For. Ecol. Manag.* **2018**, *429*, 625–641. [CrossRef]
29. Campbell, J.L.; Harmon, M.E.; Mitchell, S.R. Can fuel-reduction treatments really increase forest carbon storage in the western US by reducing future fire emissions? *Front. Ecol. Environ.* **2011**, *10*, 83–90. [CrossRef]
30. Cervera, T.; Pino, J.; Marull, J.; Padró, R.; Tello, E. Understanding the long-term dynamics of forest transition: From deforestation to afforestation in a Mediterranean landscape (Catalonia, 1868–2005). *Land Use Policy* **2019**, *80*, 318–331. [CrossRef]
31. Seijo, F.; Cespedes, B.; Zavala, G. Traditional fire use impact in the aboveground carbon stock of the chestnut forests of Central Spain and its implications for prescribed burning. *Sci. Total Environ.* **2018**, *625*, 1405–1414. [CrossRef]
32. Vadell, E.; De Miguel, S.; Pemán, J. La repoblación forestal en España: Las especies utilizadas desde 1877 a partir de las cartografías forestales. *Hist. Agraria. Rev. Agric. Hist. Rural.* **2019**, 107–136. [CrossRef]
33. Pausas, J.G.; Fernández-Muñoz, S. Fire regime changes in the Western Mediterranean Basin: From fuel-limited to drought-driven fire regime. *Clim. Chang.* **2012**, *110*, 215–226. [CrossRef]
34. Martín-Alcón, S.; Coll, L. Unraveling the relative importance of factors driving post-fire regeneration trajectories in non-serotinous Pinus nigra forests. *For. Ecol. Manag.* **2016**, *361*, 13–22. [CrossRef]

35. Sánchez-Pinillos, M.; Ameztegui, A.; Kitzberger, T.; Coll, L. Relative size to resprouters determines post-fire recruitment of non-serotinous pines. *For. Ecol. Manag.* **2018**, *429*, 300–307. [CrossRef]
36. Salis, M.; Arca, B.; Alcasena-Urdiroz, F.; Massaiu, A.; Bacciu, V.; Bosseur, F.; Caramelle, P.; Dettori, S.; Fernandes de Oliveira, A.S.; Molina-Terren, D.; et al. Analyzing the recent dynamics of wildland fires in Quercus suber L. woodlands in Sardinia (Italy), Corsica (France) and Catalonia (Spain). *Eur. J. For. Res.* **2019**, *138*, 415–431. [CrossRef]
37. Kay, S.; Rega, C.; Moreno, G.; den Herder, M.; Palma, J.H.N.; Borek, R.; Crous-Duran, J.; Freese, D.; Giannitsopoulos, M.; Graves, A.; et al. Agroforestry creates carbon sinks whilst enhancing the environment in agricultural landscapes in Europe. *Land Use Policy* **2019**, *83*, 581–593. [CrossRef]
38. Verkerk, P.J.; Martinez de Arano, I.; Palahí, M. The bio-economy as an opportunity to tackle wildfires in Mediterranean forest ecosystems. *For. Policy Econ.* **2018**, *86*, 1–3. [CrossRef]
39. Errea, M.P.; Arnáez, J.; Ortigosa, L.; Oserin, M.; Ruiz-Flaño, P.; Lasanta, T. Marginación y paisaje en una montaña submediterránea (1956–2001): El ejemplo de Camero Viejo (Sistema Ibérico, La Rioja). *Nimbus* **2007**, *19*, 53–71.
40. Vacchiano, G.; Berretti, R.; Romano, R.; Motta, R. Voluntary carbon credits from improved forest management: Policy guidelines and case study. *Iforest Biogeosci. For.* **2018**, *11*, 1–10. [CrossRef]
41. Retana, J.; Maria Espelta, J.; Habrouk, A.; Ordoñez, J.L.; de Solà-Morales, F. Regeneration patterns of three Mediterranean pines and forest changes after a large wildfire in northeastern Spain. *Écoscience* **2002**, *9*, 89–97. [CrossRef]
42. GENCAT. *Fitxes de les Unitats de Paisatge*; Generalitat de Catalunya; Departament de Territori i Sostenibilitat: Barcelona, Spain, 2016.
43. MAAyMA. *Estadística General de Incendios Forestales. Centro de Coordinación de la Información Nacional Sobre Incendios Forestales*; Ministerio de Agricultura, Alimentación y Medio Ambiente: Madrid, Spain, 2015.
44. Gonzalez-Olabarria, J.R.; Mola-Yudego, B.; Coll, L. Different Factors for Different Causes: Analysis of the Spatial Aggregations of Fire Ignitions in Catalonia (Spain). *Risk Anal.* **2015**, *35*, 1197–1209. [CrossRef] [PubMed]
45. Ager, A.A.; Vaillant, N.M.; Finney, M.A. Integrating fire behavior models and geospatial analysis for wildland fire risk assessment and fuel management planning. *J. Combust.* **2011**, *2011*, 572452. [CrossRef]
46. Scott, J.H.; Burgan, R.E. *Standard Fire Behavior Fuel Models: A Comprehensive Set for Use with Rothermel's Surface Fire Spread Model*; RMRS-GTR-153; USDA Forest Service, Rocky Mountain Research Station: Fort Collins, CO, USA, 2005; p. 72.
47. GENCAT. *Sistema D'informació Geogràfica de Parcel·les Agrícoles (SIGPAC)*; Departament d'Agricultura, Ramaderia, Pesca i Alimentació; Generalitat de Catalunya: Barcelona, Spain, 2016.
48. GENCAT. *Cartografia dels Hàbitats a Catalunya*; Generalitat de Catalunya, Departament de Territori i Sostenibilitat: Barcelona, Spain, 2012; p. 359.
49. ICGC. *Mapes de Variables Biofísiques de L'arbrat de Catalunya*; ICGC: Barcelona, Spain, 2016.
50. Otero, I.; Castellnou, M.; Gonzalez, I.; Arilla, E.; Castell, L.; Castellvi, J.; Sanchez, F.; Nielsen, J.O. Democratizing wildfire strategies. Do you realize what it means? Insights from a participatory process in the Montseny region (Catalonia, Spain). *PLoS ONE* **2018**, *13*, e0204806. [CrossRef]
51. Pique, M.; Domenech, R. Effectiveness of mechanical thinning and prescribed burning on fire behavior in Pinus nigra forests in NE Spain. *Sci. Total Environ.* **2017**, *9*, 316. [CrossRef]
52. Domènech, R.; Piqué, M.; Larrañaga, A.; Beltrán, M.; Castellnou, M. *The Role of Fire in the Conservation of the Black Pine (Pinus nigra Arn.) Habitat. Life+ PINASSA Project (LIFE13 NAT/ES/000724)*; Forest Ownership Center: Catalonia, Spain, 2018; p. 64.
53. Casals, P.; Valor, T.; Besalú, A.; Molina-Terrén, D. Understory fuel load and structure eight to nine years after prescribed burning in Mediterranean pine forests. *For. Ecol. Manag.* **2016**, *362*, 156–168. [CrossRef]
54. Beltrán, M.; Piqué, M.; Cervera, T.; Palero, N.; Camprodon, J. *Best Management Practices for the Conservation of Black Pine (Pinus nigra) Forests. Making Compatible Forest Production and Habitat Conservation. Life+ PINASSA Project (LIFE13 NAT/ES/000724)*; Forest Ownership Center: Catalonia, Spain, 2018.
55. Breiman, L. Random forests. *Mach. Learn.* **2001**, *45*, 5–32. [CrossRef]
56. Rodrigues, M.; De la Riva, J. An insight into machine-learning algorithms to model human-caused wildfire occurrence. *Environ. Model. Softw.* **2014**, *57*, 192–201. [CrossRef]
57. Hanley, J.A.; McNeil, B.J. The meaning and use of the area under a receiver operating characteristic (ROC) curve. *Radiology* **1982**, *143*, 29–36. [CrossRef]
58. Duane, A.; Brotons, L. Synoptic weather conditions and changing fire regimes in a Mediterranean environment. *Agric. For. Meteorol.* **2018**, *253-254*, 190–202. [CrossRef]
59. Alcasena, F.J.; Ager, A.A.; Bailey, J.D.; Pineda, N.; Vega-Garcia, C. Towards a comprehensive wildfire management strategy for Mediterranean areas: Framework development and implementation in Catalonia, Spain. *J. Environ. Manag.* **2019**, *231*, 303–320. [CrossRef]
60. Bradshaw, L.; McCormick, E. *Fire Family Plus User's Guide, Version 2.0*; RMRS-GTR-67WWW; USDA Forest Service: Ogden, UT, USA, 2000.
61. Rodrigues, M.; Alcasena, F.; Vega-García, C. Modeling initial attack success of wildfire suppression in Catalonia, Spain. *Sci. Total Environ.* **2019**, *2*, 323. [CrossRef]
62. Finney, M.A. Fire growth using minimum travel time methods. *Can. J. For. Res.* **2002**, *32*, 1420–1424. [CrossRef]

63. Rothermel, R.C. *A Mathematical Model for Predicting Fire Spread in Wildland Fuels*; INT-115; USDA Forest Service, Intermountain Forest and Range Experiment Station: Ogden, UT, USA, 1972; p. 40.
64. Salis, M.; Arca, B.; Del Giudice, L.; Palaiologou, P.; Alcasena-Urdiroz, F.; Ager, A.; Fiori, M.; Pellizzaro, G.; Scarpa, C.; Schirru, M.; et al. Application of simulation modeling for wildfire exposure and transmission assessment in Sardinia, Italy. *Int. J. Disaster Risk Reduct.* **2021**, *58*, 102189. [CrossRef]
65. Galizia, L.F.; Alcasena, F.; Prata, G.; Rodrigues, M. Assessing expected economic losses from wildfires in eucalypt plantations of western Brazil. *For. Policy Econ.* **2021**, *125*, 102405. [CrossRef]
66. Ager, A.A.; Vaillant, N.M.; Finney, M.A.; Preisler, H.K. Analyzing wildfire exposure and source–sink relationships on a fire prone forest landscape. *For. Ecol. Manag.* **2012**, *267*, 271–283. [CrossRef]
67. Castellnou, M.; Miralles, M. The changing face of wildfires. *Crisis Response* **2009**, *5*, 56–57.
68. Finney, M.A.; McHugh, C.W.; Grenfell, I.C.; Riley, K.L.; Short, K.C. A simulation of probabilistic wildfire risk components for the continental United States. *Stoch. Environ. Res. Risk Assess.* **2011**, *25*, 973–1000. [CrossRef]
69. Ager, A.A.; Finney, M.A.; McMahan, A.J.; Cathcart, J. Measuring the effect of fuel treatments on forest carbon using landscape risk analysis. *Nat. Hazards Earth Syst. Sci.* **2010**, *10*, 2515–2526. [CrossRef]
70. Finney, M.A. The challenge of quantitative risk analysis for wildland fire. *For. Ecol. Manag.* **2005**, *211*, 97–108. [CrossRef]
71. Evtyugina, M.; Alves, C.; Calvo, A.; Nunes, T.; Tarelho, L.; Duarte, M.; Prozil, S.O.; Evtuguin, D.V.; Pio, C. VOC emissions from residential combustion of Southern and mid-European woods. *Atmos. Environ.* **2014**, *83*, 90–98. [CrossRef]
72. Prichard, S.J.; Kennedy, M.C.; Wright, C.S.; Cronan, J.B.; Ottmar, R.D. Predicting forest floor and woody fuel consumption from prescribed burns in southern and western pine ecosystems of the United States. *For. Ecol. Manag.* **2017**, *405*, 328–338. [CrossRef]
73. WorldBank. *State and Trends of Carbon Pricing 2020*; WorldBank: Washington, DC, USA, 2020; p. 347.
74. Wastl, C.; Schunk, C.; Lüpke, M.; Cocca, G.; Conedera, M.; Valese, E.; Menzel, A. Large-scale weather types, forest fire danger, and wildfire occurrence in the Alps. *Agric. For. Meteorol.* **2013**, *168*, 15–25. [CrossRef]
75. Salis, M.; Laconi, M.; Ager, A.A.; Alcasena, F.J.; Arca, B.; Lozano, O.; Fernandes de Oliveira, A.; Spano, D. Evaluating alternative fuel treatment strategies to reduce wildfire losses in a Mediterranean area. *For. Ecol. Manag.* **2016**, *368*, 207–221. [CrossRef]
76. Salis, M.; Ager, A.A.; Alcasena, F.J.; Arca, B.; Finney, M.A.; Pellizzaro, G.; Spano, D. Analyzing seasonal patterns of wildfire exposure factors in Sardinia, Italy. *Environ. Monit. Assess.* **2015**, *187*, 4175. [CrossRef]
77. Alcasena, F.J.; Salis, M.; Ager, A.A.; Castell, R.; Vega-Garcia, C. Assessing wildland fire risk transmission to communities in northern Spain. *Forests* **2017**, *8*, 30. [CrossRef]
78. Nations, U. Paris Agreement. In Proceedings of the Paris Agreement to the United Nations Framework Convention on Climate Change, New York, NY, USA, 12 December 2015.
79. Dupuy, J.; Fargeon, H.; Martin-StPaul, N.; Pimont, F.; Ruffault, J.; Guijarro, M.; Hernando, C.; Madrigal, J.; Fernandes, P. Climate change impact on future wildfire danger and activity in southern Europe: A review. *Ann. For. Sci.* **2020**, *77*, 35. [CrossRef]
80. Lecina-Diaz, J.; Martínez-Vilalta, J.; Alvarez, A.; Vayreda, J.; Retana, J. Assessing the Risk of Losing Forest Ecosystem Services Due to Wildfires. *Ecosystems* **2021**. [CrossRef]
81. Botequim, B.; Fernandes, P.M.; Borges, J.G.; González-Ferreiro, E.; Guerra-Hernández, J. Improving silvicultural practices for Mediterranean forests through fire behaviour modelling using LiDAR-derived canopy fuel characteristics. *Int. J. Wildland Fire* **2019**, *28*, 823–839. [CrossRef]
82. Benali, A.; Sá, A.C.L.; Pinho, J.; Fernandes, P.M.; Pereira, J.M.C. Understanding the Impact of Different Landscape-Level Fuel Management Strategies on Wildfire Hazard in Central Portugal. *Forests* **2021**, *12*, 522. [CrossRef]
83. Gonzalez-Olabarria, J.R.; Mola-Yudego, B.; Pukkala, T.; Palahi, M. Using multiscale spatial analysis to assess fire ignition density in Catalonia, Spain. *Ann. For. Sci.* **2011**, *68*, 861–871. [CrossRef]
84. Migliavacca, M.; Dosio, A.; Camia, A.; Hobourg, R.; Houston-Durrant, T.; Kaiser, J.W.; Khabarov, N.; Krasovskii, A.A.; Marcolla, B.; San Miguel-Ayanz, J.; et al. Modeling biomass burning and related carbon emissions during the 21st century in Europe. *J. Geophys. Res. Biogeosci.* **2013**, *118*, 1732–1747. [CrossRef]
85. Ramo, R.; Roteta, E.; Bistinas, I.; van Wees, D.; Bastarrika, A.; Chuvieco, E.; van der Werf, G.R. African burned area and fire carbon emissions are strongly impacted by small fires undetected by coarse resolution satellite data. *Proc. Natl. Acad. Sci. USA* **2021**, *118*, e2011160118. [CrossRef] [PubMed]
86. Barnett, K.; Parks, S.; Miller, C.; Naughton, H. Beyond fuel treatment effectiveness: Characterizing interactions between fire and treatments in the US. *Forests* **2016**, *7*, 237. [CrossRef]
87. Thompson, M.; Riley, K.; Loeffler, D.; Haas, J. Modeling fuel treatment leverage: Encounter rates, risk reduction, and suppression cost impacts. *Forests* **2017**, *8*, 469. [CrossRef]
88. Parisien, M.-A.; Dawe, D.A.; Miller, C.; Stockdale, C.A.; Armitage, O.B. Applications of simulation-based burn probability modelling: A review. *Int. J. Wildland Fire* **2020**, *28*, 913–926. [CrossRef]
89. Pais, C.; Carrasco, J.; Elimbi Moudio, P.; Shen, Z.-J.M. Downstream protection value: Detecting critical zones for effective fuel-treatment under wildfire risk. *Comput. Oper. Res.* **2021**, *131*, 105252. [CrossRef]
90. Restaino, J.C.; Peterson, D.L. Wildfire and fuel treatment effects on forest carbon dynamics in the western United States. *For. Ecol. Manag.* **2013**, *303*, 46–60. [CrossRef]
91. Sorensen, C.D.; Finkral, A.J.; Kolb, T.E.; Huang, C.H. Short- and long-term effects of thinning and prescribed fire on carbon stocks in ponderosa pine stands in northern Arizona. *For. Ecol. Manag.* **2011**, *261*, 460–472. [CrossRef]

92. Fernandes, P.M.; Loureiro, C. Fine fuels consumption and CO_2 emissions from surface fire experiments in maritime pine stands in northern Portugal. *For. Ecol. Manag.* **2013**, *291*, 344–356. [CrossRef]
93. Arellano-Pérez, S.; Castedo-Dorado, F.; López-Sánchez, C.A.; González-Ferreiro, E.; Yang, Z.; Díaz-Varela, R.A.; Álvarez-González, J.G.; Vega, J.A.; Ruiz-González, A.D. Potential of Sentinel-2A Data to Model Surface and Canopy Fuel Characteristics in Relation to Crown Fire Hazard. *Remote Sens.* **2018**, *10*, 1645. [CrossRef]
94. Fidalgo-González, L.A.; Arellano-Pérez, S.; Álvarez-González, J.G.; Castedo-Dorado, F.; Ruiz-González, A.D.; González-Ferreiro, E. Estimación de la distribución vertical de combustibles finos del dosel de copas en masas de Pinus sylvestris empleando datos LiDAR de baja densidad. *Rev. Teledetec.* **2019**, 1–16. [CrossRef]
95. Campbell, J.L.; Ager, A.A. Forest wildfire, fuel reduction treatments, and landscape carbon stocks: A sensitivity analysis. *J. Environ. Manag.* **2013**, *121*, 124–132. [CrossRef]
96. Alcasena, F.J.; Ager, A.A.; Salis, M.; Day, M.A.; Vega-Garcia, C. Optimizing prescribed fire allocation for managing fire risk in central Catalonia. *Sci. Total Environ.* **2018**, *4*, 872–885. [CrossRef]
97. Ager, A.A.; Houtman, R.; Day, M.A.; Ringo, C.; Palaiologou, P. Tradeoffs between US national forest harvest targets and fuel management to reduce wildfire transmission to the wildland urban interface. *For. Ecol. Manag.* **2019**, *434*, 99–109. [CrossRef]
98. Ruiz-Peinado, R.; Montero, G.; Del Rio, M. Biomass models to estimate carbon stocks for hardwood tree species. *For. Syst.* **2012**, *21*, 42. [CrossRef]

land

Article

Measuring Gross Ecosystem Product (GEP) in Guangxi, China, from 2005 to 2020

Luying Wang [1,†], Kai Su [1,*,†], Xuebing Jiang [2], Xiangbei Zhou [1], Zhu Yu [3], Zhongchao Chen [4], Changwen Wei [1], Yiming Zhang [1] and Zhihong Liao [1]

1 College of Forestry, Guangxi University, Nanning 530004, China; 2109302014@st.gxu.edu.cn (L.W.); xiangbeizhou@st.gxu.edu.cn (X.Z.); 1736110336@st.gxu.edu.cn (C.W.); 2109392044@st.gxu.edu.cn (Y.Z.); 2109302011@st.gxu.edu.cn (Z.L.)
2 School of Mechanical Engineering, Guangxi University, Nanning 530004, China; sherryjiang@gxu.edu.cn
3 Guangxi Forest Inventory & Planning Institute, Nanning 530011, China; 1909302016@st.gxu.edu.cn
4 Guizhou Linfa Survey and Design Co., Ltd., Guiyang 550001, China; 1809303002@st.gxu.edu.cn
* Correspondence: sukai_lxy@gxu.edu.cn; Tel.: +86-130-5117-5199
† These authors contributed equally to this work.

Abstract: The economic and social development evaluation system with the Gross Domestic Product (GDP) as the leading indicator is no longer applicable to the current social progress in China. It is essential to carry out an assessment of the Gross Ecosystem Product (GEP) to integrate ecological benefits into the economic and social evaluation system and promote sustainable socio-economic development. This study took Guangxi, an important province in South China, as the study area. We used four periods of land use and land cover data (LULC), meteorological data, soil data and yearbook statistics to construct a GEP assessment framework based on geographic information system (GIS) and remote sensing (RS) technologies. We accounted for the provisioning services, regulating services, and tourism services provided by Guangxi in 2005, 2010, 2015, and 2020 and analyzed the region's and municipalities' spatial–temporal pattern characteristics and trends of change in GEP. In addition, this study also discusses the relationship between GEP and GDP. The results showed that many important products and services provided by natural ecosystems in Guangxi had enormous economic benefits. GEP had increased from CNY 15,657.37 billion in 2005 to CNY 36,677.04 billion in 2020, and the distribution of GEP showed obvious spatial heterogeneity. The value of ecosystem regulation services was about 65–89% of GEP, which is the main component of GEP. From 2005 to 2020, natural ecosystem protection and socio-economic development have achieved coordinated development in Guangxi. GEP and GDP showed upward trends in general. Although Guangxi is relatively backward in terms of economic development, the scientific quantification of the unrealized value of the services provided by the ecosystem through GEP accounting makes it possible to transform ecological advantages into economic advantages. It could help the local government and people to re-recognize the value of ecological resources and realize the beautiful vision of lucid waters and lush mountains as invaluable assets.

Keywords: ecosystem; ecosystem gross product; accounting; assessment framework; Guangxi

1. Introduction

Through primary and secondary production, ecosystems synthesize organic matter and products that are essential for human survival, provide food and vital energy for humans, create and maintain the Earth's life-support systems, and form the environmental conditions on which human survival depends [1,2]. The connotations of ecosystem services (ESs) can include abundant production of material products, provision of biological habitats, climate regulation, renewal and maintenance of soil fertility, air purification, mitigation of natural disasters, culture and entertainment, and many other aspects [3]. However, the Gross Domestic Product (GDP) growth-oriented development model has led to ecological breakdowns and serious environmental pollution, which is not sufficient as an evaluation indicator for social sustainability and high-quality economic development [4,5]. As the foundation of social development, the value of ecosystems deserves to be an important reference for evaluation. In 2013, the Research Centre for Ecological Environment of the Chinese Academy of Sciences (RCEES) and the International Union for Conservation of Nature (IUCN) proposed the concept of Gross Ecosystem Product (also known as Gross Ecological Product, also known as the value of ESs, GEP). GEP is considered as the products and services value that ecosystems provide for human welfare and economic and social sustainable development in a certain period and region [6,7]. Compared with the traditional GDP accounting system, the GEP accounting system effectively compensates for the failure to measure natural resource consumption, ecological resources, and environmental damage in GDP accounting [8,9]. GEP accounting could provide technical support for the realization of ecosystem value. GEP as a quantitative grasp of ecological assessment together with GDP will become a new yardstick to measure high-quality economic development.

The application and popularization of GEP accounting cannot be separated from its systematic construction, and a standardized and scientific GEP accounting system is the theoretical basis for realizing ecosystem value. GEP as an indicator includes the quantification and valuation of provisioning services, regulation services, and cultural services. Its accounting is carried out from two perspectives: biophysical quantity and value quantity. Finally, the value quantities of individual ESs are added up to represent ecosystem services flow in the form of monetary value [10,11]. In March 2021, the United Nations Statistical Commission officially incorporated GEP into the latest System of Environmental–Economic Accounting, the Framework for Ecosystem Accounting (SEEA-EA) [12]. The biophysical quantity and value of GEP accounting correspond to the stock account and monetary value of the core framework in SEEA-EA. At present, biophysical quantities are mainly obtained by combining geographic information system (GIS), remote sensing (RS) technologies, and InVEST means to calculate the biophysical quantities of agricultural and forestry production and the ecological regulation provided by ecosystems in each period. The monetary values are calculated using both alternative market techniques and simulated market techniques to obtain unit prices corresponding to various ESs.

GEP accounting, as an entry point and breakthrough to improve the evaluation system of economic and social development, has attracted great attention at both the national and regional levels of science and technology development. It has become a research hotspot in ecology in recent years, as well as one of the popular ecological fields supported by national science and technology departments. With the increasing demand for GEP accounting information, Chinese scholars have been successively carrying out pilot work on accounting in many administrative units (global [11], national [13–16], provincial [17,18], municipal [19–21], and county [22,23]). There are also the value calculations of natural geographical units (forests [24], watersheds [25], etc.). At the same time, the accounting of ecosystem value has been adopted and studied extensively in the United States [26,27], the United Kingdom [28,29], Australia [30], India [31], and the Czech Republic [32]. Wherever ecosystem value accounting is being carried out, researchers are actively incorporating their results into local developments in order to provide references for local social and economic development directions. With the continuous exploration of the above research and practice and the continuous improvement of GEP accounting methods, a series of research results have been obtained which play an important role in coordinating economic development and ecological protection. However, there are still some shortcomings, especially in the measurement and selection of the monetary value of ESs. We should adopt a method more in line with the ecological types, environmental conditions, local policies, and economic development level of the study area and master the ontology to make GEP accounting more scientific.

Guangxi is rich in ecosystem types, with valuable forest, ocean, mineral, and biological resources. As an important province in South China, Guangxi has obvious ecological advantages, which not only produce substantial ecological benefits locally but also play a vital role in maintaining the ecological security of neighboring provinces and even the East Asia region. Guangxi is an important ecological barrier in South China. However, Guangxi, with its backward level of economic development, belongs to a backward area of economic growth and still needs to bear colossal opportunity costs to protect the ecosystem, without being able to benefit from it. With the development of society and the intensification of human activities for resource exploitation and destruction, the regional ecosystem cannot maintain various ecological service functions stably. The problems of regional economic development and environment are intertwined, especially in rocky desertification areas, where improvement processes are slow and the unbalanced distributions of ecological resources and ecological benefits are prominent. Since a mature GEP accounting theoretical system and practical mode have not yet been formed, there is limited support for evaluating the benefits of regional ecological protection policies, establishing ecological compensation mechanisms and transforming economic benefits. This is a severe challenge for Guangxi, and it also provides a rare opportunity. However, there has been no overall research on GEP accounting in Guangxi as a region so far and the research on GEP in Guangxi has mainly focused on a certain region [33] or a single ecosystem type, such as forest [34], karst [35], and wetland [36] ecosystem types. A characteristic and systematic theoretical system and practice mode of GEP accounting for Guangxi has not yet been formed. According to the local conditions in Guangxi, it is of great significance to construct a GEP accounting system that fully reflects the regional characteristics and natural features.

The purpose of this study is to solve the research problems mentioned above. Taking Guangxi as the research area and using land use and land cover data (LULC), natural statistics, socio-economic data, and other data, this paper presents a study of the temporal and spatial changes in GEP in Guangxi from 2005 to 2020 in terms of provisioning services, regulation services, and tourism services. According to the regional characteristics of Guangxi, carrying out GEP accounting, optimizing GEP accounting methods, and accurately grasping the value of ecological products and services could not only provide a unified standard for Guangxi's current "two mountains" transformation efficiency evaluation but also provide an index reference for China's future "two mountains" transformation policy path.

2. Study Area

Guangxi is located in the south of China (26° N~21.7° N, 104.5° E~112° E). It belongs to the subtropical monsoon climate zone, with an average annual temperature of 21.50 °C, average rainfall of 1937 mm, and average sunshine duration of 1354 h. The climate is warm, with equal periods of rain and heat. The topography of Guangxi is characterized by extensive mountains and smaller areas of flat land; the surrounding mountains are continuous, and the middle terrain is slightly lower, showing basin-like characteristics. The karst landscape is widely distributed and beautiful in Guangxi and is an important factor in attracting tourists from all over the world. The ecological environment is a golden sign of Guangxi, and the levels of water environment, air environment, and marine environment rank at among the highest in China. The total land area of the whole region is 236,700 km^2, the mainland coastline is about 1500 km long, the maximum span from east to west is about 771 km, and the maximum span from south to north is about 634 km. Guangxi Zhuang Autonomous Region has jurisdiction over 14 prefecture-level cities(Figure 1), including Nanning (NN), Liuzhou (LZ), Guilin (GL), Wuzhou (WZ), Beihai (BH), Fangchenggang (FCG), Qinzhou (QZ), Yulin (YL), Guigang (GG), Baise (BS), Hezhou (HZ), Hechi (HC), Laibin (LB), and Chongzuo (CZ). By the end of 2020, the resident population of Guangxi was 50,126,800, and the GDP was CNY 22,156.69 billion.

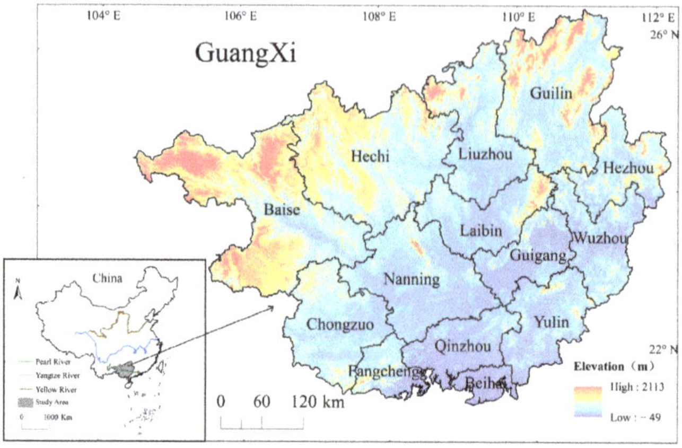

Figure 1. Study area.

3. Materials and Methods

3.1. Index Selection for GEP Accounting

In a natural ecosystem, whether it is a forest, grassland, or agricultural ecosystem, the GEP accounting index should be selected according to the advantages of the ecosystem itself and the characteristics of the basic geomorphological belts and geographical belts in different regions. Guangxi is located in a climate zone that is very favorable to the growth of crops, and there are many kinds of crops grown in the region. In this study, we chose local agricultural products, such as rice, sugar cane, tea, fruit, etc., as statistical objects. Meanwhile, Guangxi has well-developed forestry. The forestry products, such as wood, star anise, and camellia oleifera, were selected for accounting in the study. Husbandry products include meat and milk. Regarding meat, pork and poultry are the main products in Guangxi. In addition, Guangxi is rich in river and sea resources. In this study, we took the quantity of aquatic products as the supply of fishery products. Data on the production and output of agricultural, forestry, husbandry, and fishery products were obtained from the statistical yearbook [37–40]. In the regulation services, in view of Guangxi's karst landscape, which has numerous rivers, well-developed runoff and high forest coverage, we selected four ESs of water conservation, soil conservation, carbon sequestration, oxygen release, and habitat provision as accounting items. Guangxi's tourism industry is relatively developed, with unique landscapes, such as terraced fields, waterfalls, karst landscapes, etc., and strong ethnic artefacts and customs, such as Dong drum towers, Zhuang brocade, and various ethnic costumes, and historically intangible cultural heritages, such as "Liu Sanjie" ballads and "Huashan" murals, etc., which attract large numbers of tourists from home and abroad. Therefore, the numbers of tourists and total tourist consumption are taken as accounting indicators for tourism services, and the total value of the ecosystem due to tourism is counted (Table 1).

3.2. Framework for Accounting GEP

This accounting mainly included three aspects: provisioning services, regulation services, and tourism services (Figure 2). The calculation method used is as follows:

$$\mathrm{GEP} = \mathrm{EPS} + \mathrm{ERS} + \mathrm{ETS} \tag{1}$$

where GEP is gross ecosystem product; EPS is the total value of the four ecosystem provisioning services of agriculture, forestry, husbandry, and fishery; ERS is the total value of the ecosystem regulating services of water conservation, soil conservation, carbon sequestration, oxygen release, and habitat provision services; and ETS is the value of tourism services in the cultural function of ecosystems [11]. Units are calculated in billions of yuan (CNY).

Table 1. Elements of accounting for different ecosystem services. The GEP accounting examined three aspects of ecosystem provisioning services (EPSs), ecosystem regulation services (ERSs) and ecosystem tourism services (ETSs), with agriculture, forestry, animal husbandry, fishery, water conservation service (WCS), soil conservation service (SCS), carbon sequestration and oxygen release service (C/O), habitat provision (HP), and tourism services selected as accounting items.

ES	Accounting Items	Contents	Data Source
EPS	Agricultural	Agricultural products	Annual Statistical Yearbooks
	Forestry	Forestry products	
	Husbandry	Husbandry products	
	Fishery	Fishery products	
ERS	WCS	Annual precipitation, annual evapotranspiration, annual storm water production	China National Environmental Monitoring Centre
	SCS	Soil erosion	National Soil Information Service Platform of China
	C	Aboveground carbon stocks in terrestrial ecosystems, total ecosystem types, biomass	Resource and Environmental Science Data Center of Chinese Academy of Sciences
	O	Carbon sequestration, oxygen emissions	
	HP	Habitat quality, habitat scarcity	NASA
ETS	Tourism	Domestic and inbound tourism arrivals and tourism receipts	Annual Statistical Yearbooks

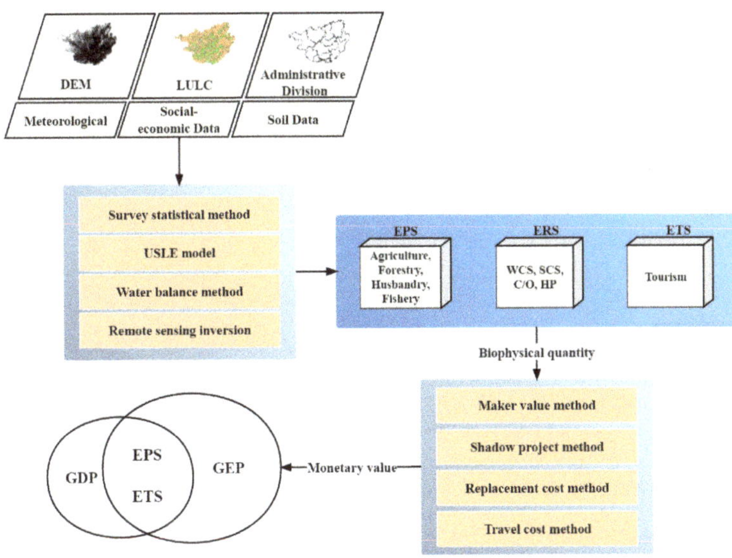

Figure 2. GEP accounting technical route.

3.3. Accounting for Biophysical Quantities

3.3.1. Provision of Ecological Products

According to the characteristics of Guangxi's natural environment and ecosystem, this study counts the total output of material products in the ecosystem as the quantity of provisioning services:

$$E_m = \sum E_i \tag{2}$$

where E_m is the total production of ecosystem products in Guangxi (t); E_i is the production of the ith product (t), i = 1, 2, 3, 4; and i is the product type. The same calculation was made for ecological product provisioning services in other municipalities.

3.3.2. Water Conservation Service

For the calculation of biophysical quantities of water-supporting services, we used the water balance equation. The calculation indexes include annual precipitation, annual evapotranspiration and annual rainstorm yield [41–44]. The calculation formula of the water balance equation is as follows:

$$WC = PRE - ET - QF \tag{3}$$

where WC is water availability, mm; PRE is annual precipitation, mm; QF is storm water runoff, mm; and ET is actual evapotranspiration, mm.

3.3.3. Soil Conservation Service

The biophysical quantity of soil conservation service is characterized by the soil conservation quantity of vegetation [45,46]. The Universal Soil Loss Equation (USLE) is the most widely used and practical remote sensing quantitative model of soil erosion at present, and it has been widely applied in the study of soil conservation in large areas [47,48]. Therefore, USLE was selected in this study to evaluate the soil conservation service of the ecosystem [49]. The calculation formula is as follows:

$$SC = SE_p - SE_a = R \cdot K \cdot LS \cdot (1 - COG) \tag{4}$$

where SC is the soil conservation, $[t/(hm^2 \cdot a)]$; SE_p and SE_a are potential and actual soil erosion, $[t/(hm^2 \cdot a)]$; R is rainfall erosion force factor, $MJ \cdot mm/(hm^2 \cdot h \cdot a)$; K is soil erodibility factor, $t \cdot hm^2 \cdot h/(hm^2 \cdot MJ \cdot mm)$; and LS and COG are the topography factor and vegetation cover factor, respectively, and are dimensionless.

3.3.4. Carbon Sequestration and Oxygen Release Service

The measurements were based on the biomass of each ecosystem and obtained through remote sensing inversions, model simulations, and other technical methods, such as measured data. The main calculation equations are as follows:

$$COS = \sum_{i=1}^{j} AGB_i \times C_i \tag{5}$$

where COS is the aboveground carbon storage of terrestrial ecosystem; i is the type of ecosystem; j is the total number of ecosystem types; AGB_i is the aboveground biomass of the ith ecosystem type; and C_i is the biomass–carbon conversion coefficient of this ecosystem type [50].

The mass of oxygen released by the ecosystem can be measured from the chemical equation for photosynthesis:

$$COP = M_{O_2}/M_{CO_2} \times COS \tag{6}$$

where COP is the oxygen released from the terrestrial ecosystem and $M_{O_2}/M_{CO_2} = 32/44$ is the coefficient of conversion of CO_2 to O_2 [51].

3.3.5. Habitat Provision Service

Habitat provision was reflected through the biological habitat quality index. In this study, the regional habitat was mainly evaluated from two aspects: regional habitat quality and habitat scarcity, which were calculated using the InVEST model [52]. The calculation formulae are as follows:

① Habitat quality

$$D_{xj} = \sum_{r=1}^{R} \sum_{y=1}^{Y_r} \left(\frac{w_r}{\sum_{r=1}^{R} w_r} \right) r_y i_{rxy} \beta_x S_{jr}, \tag{7}$$

$$i_{rxy} = 1 - \left(\frac{d_{xy}}{d_{r\ max}} \right) \tag{8}$$

$$i_{rxy} = 1 - \left(\frac{d_{xy}}{d_{r\ max}} \right) \tag{9}$$

where D_{xj} is the total stress level of raster x in LULC or habitat type j; w_r is the weight of stress factors, indicating the relative destructive power of a stress factor to all habitats; β_x is the accessibility level of grid x; and S_{jr} is the sensitivity of habitat type j to stress factor r. If $S_{jr} = 0$, D_{xj} is not a function of threat r, r_y is the stress factor in raster y, i_{rxy} is the stress effect of stress factor r in raster x on raster y, and the stress effect is divided into a linear attenuation and an exponential attenuation. d_{xy} is the linear distance between raster x and y and $d_{r\ max}$ is the maximum range of threat r [53].

② Habitat scarcity

$$R_x = \sum_{x=1}^{X} \sigma_{xy} R_j, \tag{10}$$

$$R_j = 1 - \frac{N_j}{N_{jbaseline}} \tag{11}$$

where R_x is the scarcity of raster x; R_j is the scarcity index of LULC type j; N_j is the number of rasters of current land use and land cover j; $N_{jbaseline}$ is the number of rasters of LULC type j in the baseline landscape pattern; and σ_{xy} is a binary number, with $\sigma_{xy} = 1$ when raster x is of LULC type j, otherwise $\sigma_{xy} = 0$.

3.3.6. Cultural Services

Concerning the cultural services function, this study only considered the value of leisure tourism. The sum of the total number of international and domestic trips by year was used as an indicator for the evaluation of the cultural services function.

3.4. Accounting for the Value of Monetary

3.4.1. Provision of Ecological Products

The value of ecosystem provisioning services in Guangxi included the total output value of four products: agriculture, forestry, husbandry, and fishery, which was calculated using the alternative market method. According to the statistical yearbook, the calculation range of output value refers to the total amount of products produced by various economic types and modes of operation in the administrative area within the calendar year. The calculated price includes the current price and the constant price. The sum of the output value of each product was used as the total value:

$$V_m = \sum V_i \tag{12}$$

where V_m is the total value of the output of the ecosystem products in Guangxi, in billions of yuan (CNY); V_i is the value of the output of the ith product; i = 1, 2, 3, 4; and i is the product type. The same applies to the calculation of the value for each municipality.

3.4.2. Water Conservation Service

Water conservation is related to the ecological function of water conservation and storage, and its value is calculated by the shadow engineering method [54]. Combined with the local reservoir construction market in Guangxi over the years, by calculating the ratio of the annual fixed investment in water conservancy to the reservoir construction capacity, the average construction project cost of the Guangxi reservoir every five calendar years is taken as the price. The water conservation service value was calculated as follows:

$$C_{WC} = \frac{\sum S_i/O_i}{5}, \tag{13}$$

$$V_{WC} = WC \times C_{WC} \tag{14}$$

where V_{WC} is the value of water connotation; C_{WC} is the average construction project cost of reservoirs in Guangxi; S_i is the total fixed investment in water conservancy for each five-year period from 2000–2020; and O_i is the total construction capacity of reservoirs in Guangxi for each corresponding five-calendar-year period. i = 2000–2020, 2000–2005, 2005–2010, 2010–2015, 2015–2020, corresponding to the calculation of the average construction project cost of reservoirs in 2005, 2010, 2015 and 2020, respectively.

3.4.3. Soil Conservation Service

After the soil conservation was calculated by the USLE, the corresponding value was calculated by the alternative cost method [55]. Based on the statistics for the soil and water loss control project cost in Guangxi small watersheds over the years and the ratio of comprehensive control investment to soil and water loss control area, the soil conservation service value was calculated using the cost of the soil and water loss control project in Guangxi for each calendar year:

$$C_{SC} = E_i/R_i, \tag{15}$$

$$V_{SC} = SC \times C_{SC} \tag{16}$$

where V_{SC} is the soil conservation value; C_{SC} is the cost of erosion control works in Guangxi; E_i is the cost of erosion control works in small watersheds in Guangxi for each year; and R_i is the corresponding erosion control area for each year. i = 2005, 2010, 2015, 2020, corresponding to the calculation of the cost of erosion control works in Guangxi for each calendar year, respectively.

3.4.4. Carbon Sequestration Service and Oxygen Release Service

The market price method was used to calculate the value of the ecosystem carbon sequestration service and the oxygen release service. The corresponding economic prices are carbon transaction price and industrial oxygen price, respectively [56–59].

$$V_{CO} = V_{COS} + V_{COP} \tag{17}$$

where V_{CO} is the total value of carbon sequestration and oxygen release; V_{COS} is the value of carbon sequestration service; and C_{COS} is the value of oxygen release service.

Since the first batch of carbon emissions trading pilot projects was launched in China in 2013, in order to make the calculated value more consistent with the situation of China's carbon trading market, the carbon trading prices of major countries in the world in 2005 and 2010 were taken as the unit price of carbon sequestration value in this study. The average carbon trading prices of seven pilot cities in China were adopted in 2015 and 2020. The calculation formula is:

$$V_{COS} = COS \times C_{COS} \tag{18}$$

where V_{COS} is the value of carbon sequestration service and C_{COS} is the carbon trading price.

The price of industrial oxygen over the years was taken as the unit price of oxygen release service value, and the formula is:

$$V_{COP} = COP \times C_{COP} \tag{19}$$

where V_{COP} is the value of oxygen release service and C_{COP} is the price of industrial oxygen.

3.4.5. Habitat Provision Service

The habitat provision service value was calculated by taking the conservation value of species per unit area as the price. Referring to the research on methods and guidelines [60,61], the Shannon–Wiener index in the guidelines [62] is rated as IV, and the corresponding value is CNY 26,700/hm^2·a to calculate the total habitat provision service value.

3.4.6. Cultural Services

According to the statistical survey, the sum of total international and domestic tourism consumption each year was taken as the total value of cultural services.

4. Results

4.1. GEP Assessment in Guangxi

4.1.1. 2020 GEP

In 2020, the GEP of Guangxi was CNY 36,677.04 billion, and the GEP per unit area was CNY 0.15 billion per square kilometer. The value of the water conservation service was CNY 16,233.79 billion, which was the greatest value, accounting for 44% of the total value of GEP, followed by tourism services, with a value of CNY 7267.45 billion, accounting for 20% of GEP; this was followed by soil conservation service, agriculture, habitat provision service, carbon sequestration and oxygen release service, husbandry, fishery, and forestry. For the specific composition and proportions, see Figure 3.

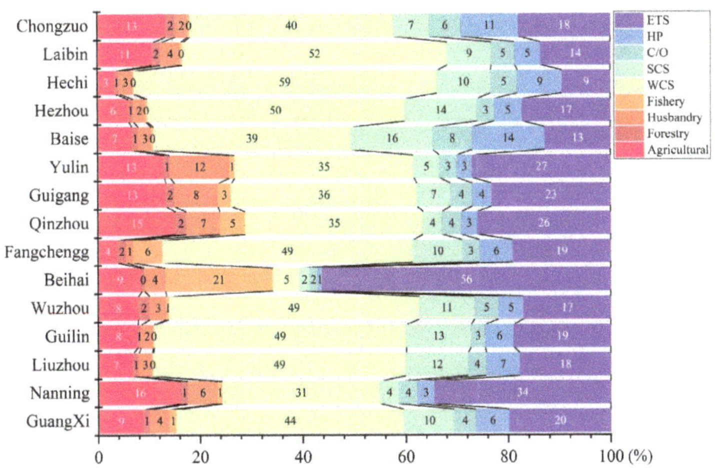

Figure 3. Composition of GEP in 2020.

The value proportions of provisioning services, regulation services, and tourism services in each city were highly unbalanced. LZ, WZ, FCG, BS, HZ, HC, LB, and CZ accounted for a large proportion of provisioning services, reaching more than 65%, of which HC was the most apparent (84%). This showed that ecosystem resources are dominant in these cities. QZ, GG, and YL showed little differences in proportions of GEP structure, the regulation service still reached about 50%, and the proportions of provisioning services and regulation services were basically the same. NN had the most balanced development structure, the value of provisioning services, regulation services, and tourism services accounting for 24%, 41%, and 34%, respectively. Only BH had the smallest share of regulation services, at 10%, while tourism services accounted for the highest proportion (56%).

The value distribution of each accounting sub-index is shown in Figure 4. The distribution of agricultural output value was the same as that for tourism services value, mainly in NN and GL. Forestry was produced primarily in the cities with high forest coverage, such as WZ and BS. NN and YL had the highest value in animal husbandry. Fishery was mainly produced in coastal cities, such as FCG, QZ, and BH. The value of the water conservation service, soil conservation service, and habitat provision service in the north of Guangxi is higher than that in the south, while the distribution pattern of the carbon sequestration and oxygen release service shows an opposite trend, with that in the south higher than that in the north.

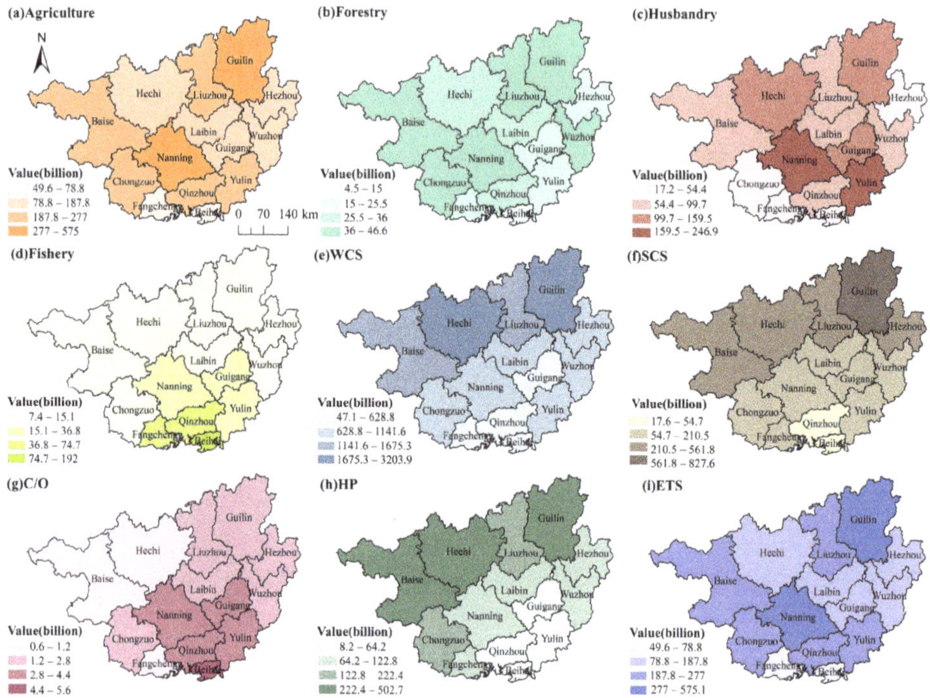

Figure 4. Spatial distribution of sub-indicators in 2020.

4.1.2. Temporal and Spatial Changes of GEP from 2005 to 2020

From 2005 to 2020, the value of provisioning services, regulation services, and tourism services has been increasing, but with differences (Figure 5). The value of regulation services increased the most over the 15 years, by a total of CNY 9846.51 billion. The value of tourism services increased by CNY 6963.72 billion. The value of provisioning services increased by CNY 4209.44 billion.

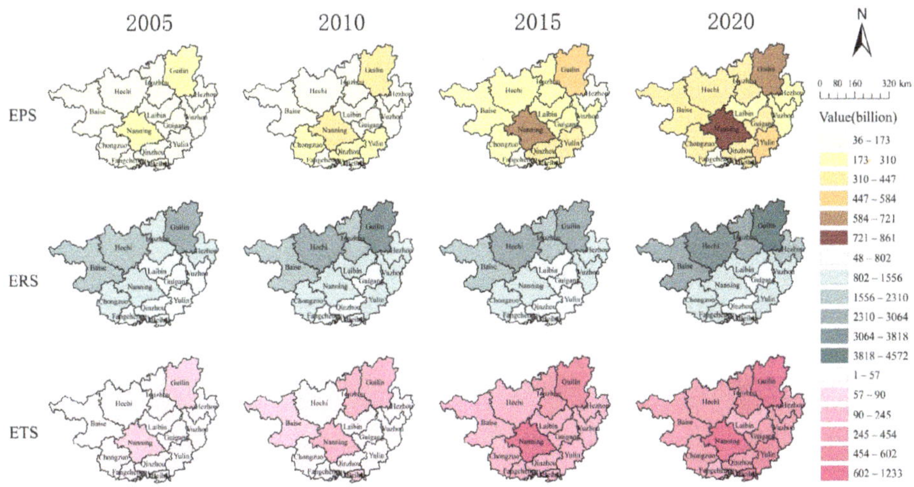

Figure 5. Spatial distribution of the monetary value of terrestrial ecosystem products from 2005 to 2020.

Regarding the value of provisioning services, the largest increase over the 15 years was in NN. From 2010 to 2015, the overall provisioning services value for Guangxi increased the most in these five years, and the provisioning service values of BS, HC, LZ, LB, GG, and WZ exceeded CNY 200 billion in 2015. The distribution of regulation services value showed an increasing trend from southeast to northwest and from the coast to inland. The largest increase in regulating services value during the 15 years was in GL, with a rise of about CNY 2135.36 billion. The value of tourism services increased most obviously between 2015 and 2020, with an increase of CNY 4074.73 billion in five years. Among them, the tourism income of GL and NN in 2020 have both reached more than CNY 1200 billion. YL had substantial growth from 2015 to 2020, with an increase of CNY 551.40 billion, and in 2020 it became the region with the highest tourism services value after GL, NN, and LZ (Figure 5).

In 2005, 2010, and 2015, the value of regulation services accounted for the largest proportion, provisioning services the second, followed, finally, by tourism services. However, in 2020, the value of tourism services exceeded the value of provisioning services and ranked second. In 2020, the total value of the four provisioning services of agriculture, forestry, husbandry, and fishery in Guangxi was CNY 5638.22 billion, of which the output value of agricultural products was CNY 3268.80 billion, accounting for 58% of GEP. The proportion of the value of regulation services gradually decreased, from 89% in 2005 to 65% in 2020 in GEP. The value of tourism services was CNY 7267.45 billion, accounting for 20% of GEP in 2020, while in 2005, the value of tourism services was only CNY 303.73 billion, accounting for only 2%.

In general, the GEP in Guangxi was dominated by water-related services (Figure 6), and the value of water-related ESs, such as water conservation, soil conservation, and agriculture, were high. In the four accounting periods, water conservation service always maintained the most significant contribution. In 2020, the total value of water conservation service in Guangxi was CNY 16,233.79 billion, which increased by CNY 6504.03 billion compared with 2005. The northwest of Guangxi is a karst area with severe soil erosion, so the function of soil conservation is particularly crucial. The total value of soil conservation service in Guangxi increased from CNY 900.58 billion in 2005 to CNY 3612.81 billion in 2020, which indicates that ecological engineering plays a vital role in controlling soil erosion and improving the ecological environment in Guangxi. In 2020, the total value of carbon sequestration and oxygen release service in Guangxi was CNY 1554.24 billion, of which the values for the carbon sequestration service and oxygen release service were

CNY 91.25 billion and CNY 1462.99 billion, respectively. Agricultural products accounted for about 53% of the value of provisioning services, followed by husbandry products at about 30%. Forestry was very well developed, but its unit output value and total output value were not as good as those for agriculture, and the value provided by forestry was less compared to the share provided by agriculture, husbandry, and fishery, which was only 13% of agricultural products. During the 15 years, the value of tourism services has changed the most. Compared with 2005, the value of tourism services in 2020 increased by CNY 6963.72 billion, accounting for about 20% of GEP in 2020, which is an important figure that cannot be ignored.

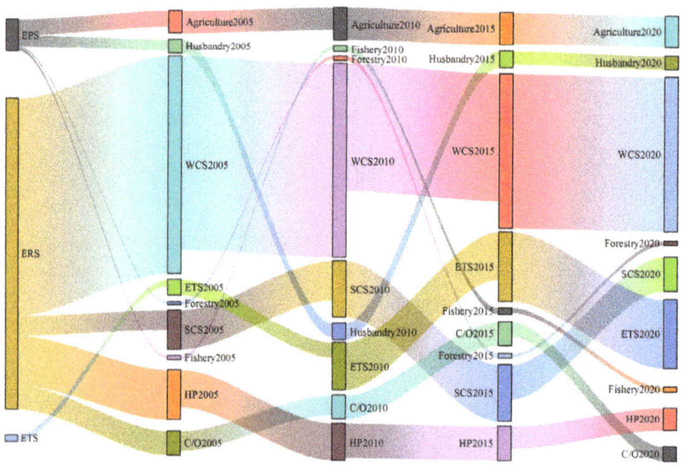

Figure 6. Cumulative change rate (%) of sub-indicators from 2005 to 2020.

4.2. Trends of Ecosystem Regulation Services from 2005 to 2020

The value of ecosystem regulation services was about 65–89% of GEP, which was the main component of GEP. The change in ERSs will have a crucial impact on GEP. From 2005 to 2020, the ecosystem regulation services showed an increasing trend, but there were differences among cities (Table 2). All the ESs in LZ, GG, YL, HC, and LB showed increasing trends, while one or two services decreased in other cities.

Table 2. Changes in regulation services from 2005 to 2020.

ZONE	WCS	SCS	C/O	**Unit: billion t/a** HP
Guangxi	2585.13	5509.52	3950.39	155.09
NN	54.22	198.63	−525.77	7.76
LZ	354.69	164.19	897.07	7.39
GL	1020.05	−461.76	2357.85	6.98
WZ	53.31	−12.80	177.05	1.15
BH	0.96	6.67	−3.05	0.05
FCG	−27.96	6.05	−95.67	1.13
QZ	16.37	38.98	−115.53	1.83
GG	30.23	31.88	53.22	1.38
YL	45.91	40.17	103.64	1.54
BS	246.60	4091.41	−373.67	77.03
HZ	51.43	−60.45	141.80	0.56
HC	690.02	1290.98	1909.42	35.78
LB	109.09	85.03	131.20	3.41
CZ	−59.80	90.55	−707.18	9.10

The extremely significant increase in water conservation service was in the north of GL, the north of LZ, and the middle of LB, which has a typical karst landscape, with GL taking priority, providing 39% of services. The significant reduction in water conservation services occurred in most areas of CZ and FCG, with a decrease of 59.8 billion t/a and 27.96 billion t/a respectively (Figure 7a). The spatial distribution of the carbon sequestration and oxygen release service showed a similar pattern to that of the water conservation service; the law is more prominent, with a trend of decreasing to increasing from southwest to northeast. Among them, CZ in southwest Guangxi had the largest decrease in carbon sequestration and oxygen release service, with a decrease of 707.18 billion t/a, and FCG had a significant decrease, representing arid areas and coastal areas, respectively. GL and HC in northeast Guangxi had the largest increase in the carbon sequestration and oxygen release service, reaching 60% and 48%, respectively, with corresponding values of 2357.85 billion t/a and 1909.42 billion t/a (Figure 7c). Soil conservation service remained basically unchanged in most areas of the whole region but increased significantly only in BS, providing 74% of services, with an increase of 4091.41 billion t/a; the significant decrease was mainly in GL and concentrated in the northern part of GL with a significant change, with a decrease of 461.76 billion t/a, followed by HZ and WZ, with a decrease of 60.45 billion t/a and 12.80 billion t/a, respectively (Figure 7b). Habitat provision service in all cities has increased. The significant increase was still mainly located in BS, and the contribution of BS to the rise of habitat provision service in the whole region has reached 50%. The significant decrease was distributed in the main urban areas of each city, and the degree of decrease was related to the level of urban development. Except for NN, FCG, BH, and YL, the spatial distribution patterns of habitat provision service change and soil conservation service change were similar, and there was a close relationship between them (Figure 7d).

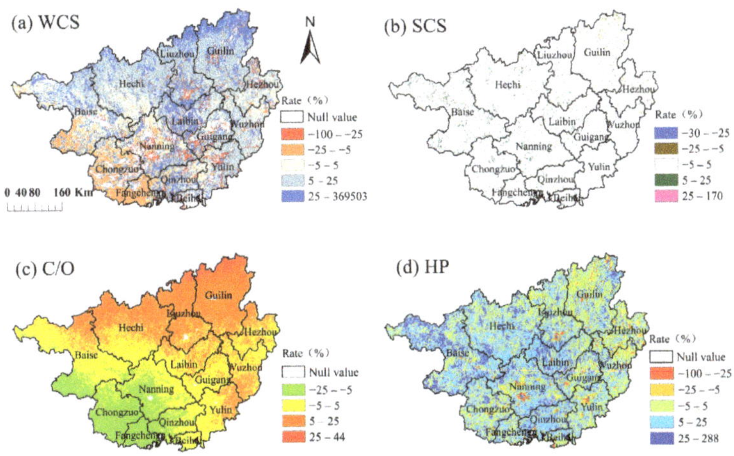

Figure 7. Distribution of ecosystem regulation services changes from 2005 to 2020.

4.3. Comparison between GEP and GDP

From 2005 to 2020, natural ecosystem protection and socio-economic development achieved coordinated development in Guangxi, and GEP and GDP showed upward trends (Figure 8). In 15 years, the GEP and GDP of the whole province increased by 134% and 492%, respectively.

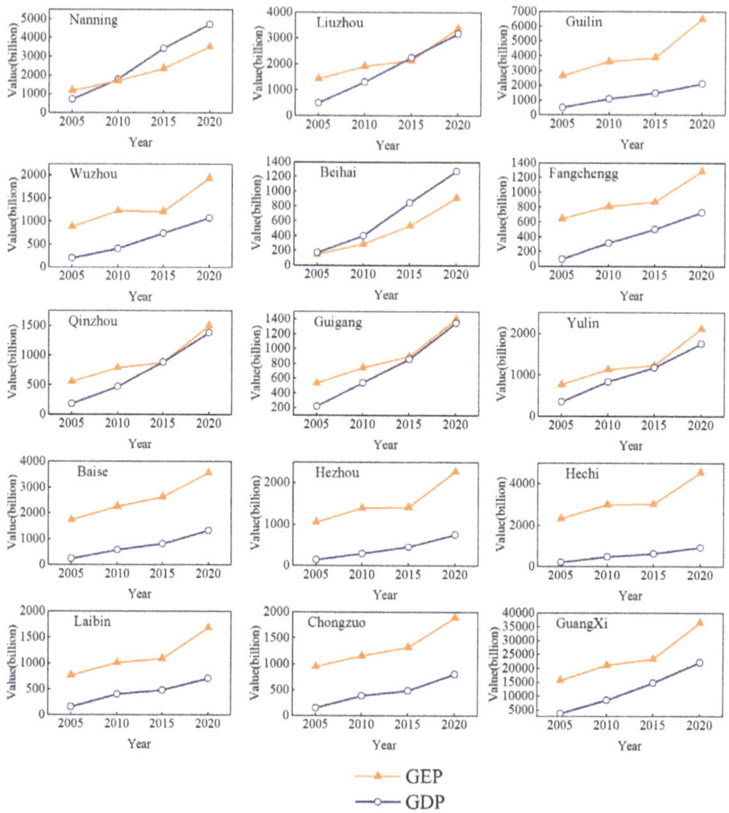

Figure 8. Comparison of GEP and GDP changes in various cities.

Looking at the changes in GEP and GDP in each municipality, although only NN's and BH's GEP were always lower than GDP, GEP and GDP showed an increasing trend during 2005–2020, as did all other municipalities, and the growth rate of GEP was significantly greater after 2015 than from 2005 to 2015, which may be closely related to the work and policy of comprehensively promoting water environment consolidation, strengthening the construction of nature reserves, and carrying out the red line delineation of ecological protection. For NN, the GDP of NN has exceeded GEP since 2010 and continued to grow rapidly. After the growth of GDP, GEP also showed a synchronous growth trend. Over 15 years, the GDP and GEP of NN increased by 373% and 445%, respectively. This shows that socio-economic development has not had a negative impact on NN's natural ecosystem. On the contrary, socio-economic development may promote the protection of the natural ecosystem and make GEP increase. However, BS and HC had extremely high GEP, despite being still relatively economically backward because they are located in a remote inland area, not giving full play to their ecological advantages like NN and GL. Although both GEP and GDP showed a growing trend, the difference between GEP and GDP had gradually decreased since 2015 in LZ, QZ, and GG, which indicates that, compared with other cities, these three cities could more rationally develop and utilize ecosystems, which can promote social and economic development.

5. Discussion

The natural ecosystem is an essential foundation for the sustainable development of the economy and society, and protecting the ecosystem helps to protect the homeland of human beings. Most previous studies have focused on the ecological benefits of ecosystems, but the research on their economic benefits was limited [63]. In this study, the GEP of Guangxi in 2005, 2010, 2015, and 2020 was calculated using LULC, meteorological data, soil data, vegetation coverage, remote sensing data, and socio-economic data. The results show that many important products and services provided by natural ecosystems in Guangxi have tremendous economic benefits. GEP has increased from CNY 15,657.37 billion in 2005 to CNY 36,677.04 billion in 2020. Compared with other provinces, Guangxi is backward in terms of its economy, but GEP accounting makes it possible to transform ecological advantages into economic advantages.

The output values for agriculture, forestry, husbandry, and fishery used in the research and calculations herein refer to the total monetary values of all products of agriculture, forestry, animal husbandry, and fishery, which reflect the total achievements or scales of production in the various industries at that time according to the current price of that year. The calculation of GEP in Guangxi in this study is lower than that in previous studies [64] because of the difference in index selection and evaluation methods. In the selection of indicators, we only calculated the nine indicators listed in this paper. Compared with other GEP studies, we did not evaluate energy and flood storage but calculated the value of habitat provision service. In the evaluation method, especially for the accounting of regulating services value, previous studies mostly adopted the national average price. In contrast, this study determined the unit area cost of reservoir construction and the unit area cost of soil erosion control by collecting and sorting out the local investment amounts and construction areas in statistical yearbooks of various counties and cities in different periods. The difference between the value of the carbon sequestration service and oxygen release service and that of other studies is due to the difference in unit price (such as carbon tax, carbon trading, afforestation cost, industrial emission reduction cost, etc.) in different studies. Since the end of 2019, all industries have been affected by the COVID-19 pandemic. Although the number of tourists and tourism consumption in 2020 has increased significantly compared with 2015, this is due to the rapid development of tourism in Guangxi from 2015 to 2020. In 2015, total tourism consumption in Guangxi amounted to CNY 325.42 billion, CNY 558.04 billion in 2017, CNY 761.99 billion in 2018, and reached three times the 2015 level, with CNY 102.44 billion in 2019.

Guangxi is a famous forestry province, but the results of this study show that the contribution of forestry to GEP is lower than our estimate. Compared with the statistical data published by the Guangxi Zhuang Autonomous Region Forestry Bureau (http://lyj.gxzf.gov.cn/ (accessed on 2 March 2022)), this study's results are low. For example, based on the statistical yearbook, the output value of Guangxi forestry in 2020 was CNY 437.35 billion, while the official statistical result was CNY 7521 billion. The difference in the results mainly lies in the difference of statistical caliber: the officially counted forestry output value is the total output value of the primary industry (forest resources cultivation, forest products output, forest quality, etc.), the secondary industry (wood and economic forest product processing industry, forest product chemical industry, wood pulp and paper industry, etc.), and the tertiary industry (forest health care, forestry tourism and leisure services, eco-cultural industry, forestry exhibition, etc.). The data used in this study come from the statistical yearbook, which mainly counts the forestry output value of timber, medicinal materials, forest products, etc. Moreover, in order to facilitate comparison between different prefecture-level cities, the selected statistical indicators exist in all local cities, so the statistical results of forestry output value in this research are lower than the official statistics.

From the spatial and temporal changes relating to GEP, the values of provisioning services and tourism services are basically the same in terms of spatial distribution, but there are also subtle differences: for example, in 2010, YL had a good income from husbandry, and pig production accounted for a large proportion; QZ's fruit was the greatest contribution to the output value of agriculture, forestry, animal husbandry, and fishery; CZ is known as "China's sugar capital" and sugar cane was its pillar industry. As a result, each city had its characteristic industries to support the development of provisioning services. In the tourism services, BH and FCG reflected the characteristics of coastal areas and had great advantages in developing the tourism industry. GL had the reputation of "The mountains and waters of Guilin are the finest under heaven." Meanwhile, NN is closer to the subtropical region and has a strong Southeast Asian flavor, so the tourism industry became its advantage. At the same time, the tourism development of GL and NN has also played a role in radiating the surrounding areas. With the development over time, the number of tourists and income from other counties and cities are also gradually rising, such as in LZ, BS, and CZ.

By 2020, the GEP sizes of Guangxi cities were ranked as follows: GL > HC > BS > NN > LZ > HZ > YL WZ > CZ > LB > QZ > GG > FCG > BH. GL, NN, LZ, and YL had both advantages in economy and ecology, while HC, BS, HZ, CZ, and LB were underdeveloped in their economies but rich in ecological assets. It can be seen that Guangxi had no consistent development of economy or ecology unilaterally. In addition, the study also found that NN and BH have higher tourism service values, while other cities have higher water conservation service and carbon sequestration and oxygen release service capabilities. Against the background of vigorously promoting ecological construction advocated by the state, especially against the background of global carbon neutrality and ecological compensation, Guangxi and all cities have their advantages; it is necessary to carry out scientific planning of urban development and targeted protection and management of ecosystems and to try to convert the ecological benefits of ecosystems into economic benefits through ecological compensation [65], that is, to convert the "green water and green mountains" into real "golden mountains" to achieve sustainable growth of the regional economies.

It is meaningful to carry out GEP accounting, which could provide a scientific understanding of the unrealized value of the services provided by the ecosystem. According to the GEP accounting results for Guangxi from 2005 to 2020, the following suggestions and countermeasures for ecological protection are put forward: (1) Regular and long-term GEP accounting should be carried out to assess the ecological environment over time, especially in areas with fragile ecological environments and areas with declining ecological service functions found in accounting. GEP accounting and monitoring are helpful to quickly find problems and put forward relevant countermeasures, such as whether it is necessary to focus on regional protection or what kind of ecological engineering construction should be implemented to solve the problems to adjust the ecology. (2) Due to the complexity and differences in geographical regions and ecosystems, the evaluation method should be further improved to form a unified GEP evaluation index system that can be popularized at different scales and in different regions to facilitate a more comprehensive and accurate comparative analysis of GEPs in different places in the later period. What is needed is to be able to simulate and forecast results and changes on the basis of GEP accounting. (3) This study involved comparative accounting in time series and among cities, providing an analysis of GEP results for Guangxi in 2020 and the spatial and temporal changes in GEP over a period of 15 years. In future studies, we will build on this research to conduct more detailed studies using more accurate and indicator-rich ecosystem service datasets, such as GEP accounting for one year for each city, multi-year comparisons, and accounting for smaller administrative areas such as counties and townships.

6. Conclusions

This study used multi-source data and multi-indicators to calculate the changes in GEP in Guangxi from 2005 to 2020. The main conclusions are as follows.

1. Guangxi's natural ecosystem has a considerable value. In 2020, GEP reached CNY 36,677.04 billion, about 1.66 times the GDP. Among the many accounting indicators, the value of water conservation service was the largest, at CNY 16,233.79 billion, accounting for about 44% of GEP. From 2005 to 2020, GEP increased significantly, with the GEP increasing by 134% in 15 years. However, the proportions of the values

of provisioning services, regulation services, and tourism services were obviously different among cities.

2. ERS was the main component of GEP, showing a spatial distribution pattern that was high in the north and low in the south. From 2005 to 2020, ERS generally showed an increasing trend. Among the four ESs, the water conservation service accounted for more than 52% of ERS, followed by the soil conservation service, which accounted for about 15% on average.
3. From 2005 to 2020, GEP and GDP showed a synergistic upward trend, but there were differences among prefecture-level cities. Due to their location in the remote inland area, BS and HC were relatively backward economically but had extremely high GEPs. Through rational development and utilization of ecosystems, as in LZ, QZ and GG, ecological advantages could be transformed into economic advantages to promote economic development.

Author Contributions: Conceptualization, L.W. and K.S.; methodology, L.W. and K.S.; software, L.W.; validation, X.Z., Z.Y. and Z.C.; formal analysis, L.W. and K.S.; resources, K.S.; data curation, L.W., C.W., Z.L. and Y.Z.; writing—original draft preparation, L.W. and K.S.; writing—review and editing, K.S. and X.J.; visualization, L.W., C.W., Z.L. and Y.Z.; supervision, project administration, K.S.; funding acquisition, K.S. All authors have read and agreed to the published version of the manuscript.

Funding: This research was funded by Talent Introduction Program of Guangxi University, grant number A3360051018 and the Youth Science Foundation of National Natural Science Foundation of Guangxi Province (2022GXNSFBA035570).

Institutional Review Board Statement: This study did not require ethical approval.

Informed Consent Statement: Not applicable.

Data Availability Statement: The meteorological data, soil data, land cover data are available in this research, upon any reasonable request, by emailing the authors.

Conflicts of Interest: The authors declare that they have no known competing financial interests or personal relationships that could appear to have influenced the work reported in this paper.

References

1. Sintayehu, D.W. Impact of climate change on biodiversity and associated key ecosystem services in Africa: A systematic review. *Ecosyst. Health Sustain.* **2018**, *4*, 225–239. [CrossRef]
2. Schowalter, T.D.; Noriega, J.A.; Tscharntke, T. Insect effects on ecosystem services—Introduction. *Basic Appl. Ecol.* **2018**, *26*, 1–7. [CrossRef]
3. Machnik, A. Natural Capital and Ecological Ecosystem Services: Methods of Measuring Socio-economic Value of Nature. In *Responsible Consumption and Production*; Leal Filho, W., Azul, A.M., Brandli, L., Özuyar, P.G., Wall, T., Eds.; Springer International Publishing: Cham, Germany, 2020; pp. 511–523.
4. Hickel, J.; Hallegatte, S. Can we live within environmental limits and still reduce poverty? Degrowth or decoupling? *Dev. Policy Rev.* **2022**, *40*, e12584. [CrossRef]
5. Hickel, J.; Kallis, G. Is green growth possible? *New Political Econ.* **2020**, *25*, 469–486. [CrossRef]
6. Edens, B.; Maes, J.; Hein, L.; Obst, C.; Siikamaki, J.; Schenau, S.; Javorsek, M.; Chow, J.; Chan, J.Y.; Steurer, A.; et al. Establishing the SEEA Ecosystem Accounting as a global standard. *Ecosyst. Serv.* **2022**, *54*, 101413. [CrossRef]
7. Ouyang, Z.; Zhu, C.; Yang, G.; Xu, W.; Zheng, H.; Zhang, Y.; Xiao, Y. Gross ecosystem product: Concept, accounting framework and case study. *Acta Ecol. Sin.* **2013**, *33*, 6747–6761. [CrossRef]
8. He, J.; Wan, Y.; Feng, L.; Ai, J.; Wang, Y. An integrated data envelopment analysis and emergy-based ecological footprint methodology in evaluating sustainable development, a case study of Jiangsu Province, China. *Ecol. Indic.* **2016**, *70*, 23–34. [CrossRef]
9. Obst, C.; Hein, L.; Edens, B. National Accounting and the Valuation of Ecosystem Assets and Their Services. *Environ. Resour. Econ.* **2016**, *64*, 1–23. [CrossRef]
10. Wu, C.; Ma, G.; Yang, W.; Zhou, Y.; Peng, F.; Wang, J.; Yu, F. Assessment of Ecosystem Service Value and Its Differences in the Yellow River Basin and Yangtze River Basin. *Sustainability* **2021**, *13*, 3822. [CrossRef]
11. Jiang, H.; Wu, W.; Wang, J.; Yang, W.; Gao, Y.; Duan, Y.; Ma, G.; Wu, C.; Shao, J. Mapping global value of terrestrial ecosystem services by countries. *Ecosyst. Serv.* **2021**, *52*, 101361. [CrossRef]

12. Bagstad, K.J.; Ingram, J.C.; Shapiro, C.D.; La Notte, A.; Maes, J.; Vallecillo, S.; Casey, C.F.; Glynn, P.D.; Heris, M.P.; Johnson, J.A.; et al. Lessons learned from development of natural capital accounts in the United States and European Union. *Ecosyst. Serv.* **2021**, *52*, 101359. [CrossRef]
13. Zang, Z.; Zhang, Y.; Xi, X. Analysis of the Gross Ecosystem Product—Gross Domestic Product Synergistic States, Evolutionary Process, and Their Regional Contribution to the Chinese Mainland. *Land* **2022**, *11*, 732. [CrossRef]
14. Ouyang, Z.; Zheng, H.; Xiao, Y.; Polasky, S.; Liu, J.; Xu, W.; Wang, Q.; Zhang, L.; Xiao, Y.; Rao, E.; et al. Improvements in ecosystem services from investments in natural capital. *Science* **2016**, *352*, 1455–1459. [CrossRef]
15. Jiang, W.; Wu, T.; Fu, B. The value of ecosystem services in China: A systematic review for twenty years. *Ecosyst. Serv.* **2021**, *52*, 101365. [CrossRef]
16. Ma, G.; Yu, F.; Wang, J.; Zhou, X.; Yuan, J.; Mou, X.; Zhou, Y.; Yang, W.; Peng, F. Measuring gross ecosystem product (GEP) of 2015 for terrestrial ecosystems in China. *China Environ. Sci.* **2017**, *37*, 1474–1482.
17. Ouyang, Z.; Song, C.; Zheng, H.; Polasky, S.; Xiao, Y.; Bateman, I.J.; Liu, J.; Ruckelshaus, M.; Shi, F.; Xiao, Y.; et al. Using gross ecosystem product (GEP) to value nature in decision making. *Proc. Natl. Acad. Sci. USA* **2020**, *117*, 14593–14601. [CrossRef]
18. Wu, N.; Chen, H.; Ge, J. The Accounting of Anhui's Gross Ecosystem Product within the "Green GDP2.0" Framework. *J. Anhui Agric. Univ.* **2018**, *27*, 39–49. [CrossRef]
19. Liang, L.-N.; Siu, W.S.; Wang, M.-X.; Zhou, G.-J. Measuring gross ecosystem product of nine cities within the Pearl River Delta of China. *Environ. Chall.* **2021**, *4*, 100105. [CrossRef]
20. Zou, Z.; Wu, T.; Xiao, Y.; Song, C.; Wang, K.; Ouyang, Z. Valuing natural capital amidst rapid urbanization: Assessing the gross ecosystem product (GEP) of China's Chang-Zhu-Tan' megacity. *Environ. Res. Lett.* **2020**, *15*, 124019. [CrossRef]
21. Ouyang, Z.; Lin, Y.; Song, C. Research on Gross Ecosystem Product(GEP): Case study of Lishui City, Zhejiang Province. *Environ. Sustain. Dev.* **2020**, *45*, 80–85. [CrossRef]
22. Dolkar, P.; Yi, X.; Zhiyun, O.; Liyan, W. Assessment of ecological conservation effect in Xishui county based on gross ecosystem product. *Acta Ecol. Sin.* **2020**, *40*, 499–509.
23. Li, F.; Yan, B.; Lu, G.; Li, Z.; Zhu, X. Research on the Premise of Ecological Product Value Realization Mechanism: A Case Study of Gaochun District, Nanjing. *Environ. Prot.* **2021**, *49*, 51–58. [CrossRef]
24. Wu, Z.; Zeng, H. Evaluation of forest ecosystem services value in China based on meta-analysis. *Acta Ecol. Sin.* **2021**, *41*, 5533–5545.
25. Zhang, J.; Zou, Z. Research on calculation and application of GEP in Brahmaputra River Basin. *Ecol. Econ.* **2022**, 1–14. Available online: http://kns.cnki.net/kcms/detail/53.1193.F.20220219.0838.002.html (accessed on 1 June 2022).
26. Jin, D.; Watson, C.; Kite-Powell, H.; Kirshen, P. Evaluating Boston Harbor cleanup: An ecosystem valuation approach. *Front. Mar. Sci.* **2018**, *5*, 478. [CrossRef]
27. Escobedo, F.J.; Giannico, V.; Jim, C.Y.; Sanesi, G.; Lafortezza, R. Urban forests, ecosystem services, green infrastructure and nature-based solutions: Nexus or evolving metaphors? *Urban For. Urban Green.* **2019**, *37*, 3–12. [CrossRef]
28. Raum, S.; Hand, K.L.; Hall, C.; Edwards, D.M.; O'Brien, L.; Doick, K.J. Achieving impact from ecosystem assessment and valuation of urban greenspace: The case of i-Tree Eco in Great Britain. *Landsc. Urban Plan.* **2019**, *190*, 103590. [CrossRef]
29. Bateman, I.J.; Harwood, A.R.; Abson, D.J.; Andrews, B.; Crowe, A.; Dugdale, S.; Fezzi, C.; Foden, J.; Hadley, D.; Haines-Young, R.; et al. Economic Analysis for the UK National Ecosystem Assessment: Synthesis and Scenario Valuation of Changes in Ecosystem Services. *Environ. Resour. Econ.* **2014**, *57*, 273–297. [CrossRef]
30. Raymond, C.M.; Bryan, B.A.; MacDonald, D.H.; Cast, A.; Strathearn, S.; Grandgirard, A.; Kalivas, T. Mapping community values for natural capital and ecosystem services. *Ecol. Econ.* **2009**, *68*, 1301–1315. [CrossRef]
31. Sannigrahi, S.; Pilla, F.; Basu, B.; Basu, A.S.; Zhang, Q.; Wang, Y.; Joshi, P.K.; Chakraborti, S.; Coscieme, L.; Keesstra, S.; et al. Identification of Conservation Priority Zones Using Spatially Explicit Valued Ecosystem Services: A Case from the Indian Sundarbans. *Integr. Environ. Assess. Manag.* **2020**, *16*, 773–787. [CrossRef]
32. Vačkářů, D.; Grammatikopoulou, I. Toward development of ecosystem asset accounts at the national level. *Ecosyst. Health Sustain.* **2019**, *5*, 36–46. [CrossRef]
33. Wei, J.; Yang, Y.; Xie, X.; Liao, L.; Tian, Y.; Zhou, J. Quantifying Ecosystem Service Trade-offs and Synergies in Nanning City Based on Ecosystem Service Bundles. *J. Ecol. Rural. Environ.* **2022**, *38*, 21–31. [CrossRef]
34. Fan, h.; Zhang, Y.; Zou, L.; Pan, L. A study on the baseline value of the Chinese mangrove services and allocation of the value to individual tree. *Acta Ecol. Sin.* **2022**, *42*, 1262–1275.
35. Gao, M.; Hu, Y.; Li, X.; Song, R. Construction of ecological security pattern based on the importance of ecosystem services and environmental sensitivity in karst mountainous areas: A case study in Hechi, Guangxi. *Acta Ecol. Sin.* **2021**, *41*, 2596–2608.
36. Wu, B.; Li, Y.; Zhao, Q.; Zhang, W.; Lu, R. The Service Value and Driving Mechanism of Coastal Wetland Ecosystem in Beibu Gulf of Guangxi. *Ecol. Econ.* **2020**, *36*, 151–157.
37. Guangxi Zhuang Autonomous Region Bureau of Statistics. *Guangxi Statistical Yearbook*; China Statistics Press: Beijing, China, 2006.
38. Guangxi Zhuang Autonomous Region Bureau of Statistics. *Guangxi Statistical Yearbook*; China Statistics Press: Beijing, China, 2011.
39. Guangxi Zhuang Autonomous Region Bureau of Statistics. *Guangxi Statistical Yearbook*; China Statistics Press: Beijing, China, 2016.

40. Guangxi Zhuang Autonomous Region Bureau of Statistics. *Guangxi Statistical Yearbook*; China Statistics Press: Beijing, China, 2021.
41. Wang, Z.; Zhang, Z.; Suo, Y. A new water balance equation introducing dew amount in arid area. *J. Hydraul. Eng.* **2019**, *50*, 710–720. [CrossRef]
42. Wu, X.; Hu, T.; Wang, X.; Jiang, Y.; Li, X. Review of estimating and measuring regional evapotran spiration. *Trans. CSAE* **2006**, *22*, 257–262.
43. Zhang, M.; Mao, R. The Simplified model of salt-water regimes in farmland soil under the processing of evaportranspiration. *Chin. Acad. Sci.* **2003**, *03*, 108–111.
44. Zhao, L.; Zhao, W. Water balance and migration for summer maize in an oasis farmland of northwest China. *Chin. Sci. Bull.* **2014**, *59*, 3430.
45. Montanarella, L.; Panagos, P. The relevance of sustainable soil management within the European Green Deal. *Land Use Policy* **2021**, *100*, 104950. [CrossRef]
46. Renison, D.; Hensen, I.; Suarez, R.; Cingolani, A.; Marcora, P.; Giorgis, M. Soil conservation in Polylepis mountain forests ofCentral Argentina: Is livestock reducing our natural capital? *Austral Ecol.* **2009**, *35*, 435–443. [CrossRef]
47. Verma, S.; Singh, P.; Mishra, S.; Singh, V.; Singh, V.; Singh, A. Activation soil moisture accounting (ASMA) for runoff estimation using soil conservation service curve number (SCS-CN) method. *J. Hydrol.* **2020**, *589*, 125114. [CrossRef]
48. Eckert, S.; Ghebremicael, S.; Hurni, H.; Kohler, T. Identification and classification of structural soil conservation measures basedon very high resolution stereo satellite data. *Environ. Manag.* **2017**, *193*, 592–606.
49. Kong, L.; Zheng, H.; Rao, E.; Xiao, Y.; Ouyang, Z.; Li, C. Evaluating indirect and direct effects of eco-restoration policy on soil conservation service in Yangtze River Basin. *Total Environ.* **2018**, *631*, 887–894. [CrossRef]
50. Deng, Y.; Yao, S.; Hou, M.; Zhang, T.; Lu, Y.; Gong, Z.; Wang, Y. Assessing the effects of the Green for Grain Program on ecosystem carbon storage service by linking the InVEST and FLUS models: A case study of Zichang county in hilly and gully region of Loess Plateau. *J. Nat. Resour.* **2020**, *35*, 826–844. [CrossRef]
51. Ouyang, Z.; Xiao, S.; Zhu, C.; Zheng, H.; Zou, Z.; Song, C.; Bo, W.; Huang, B. *Theory and Methodology of Gross Ecosystem Product (GEP) Accounting*; Beijing Publishing House: Beijing, China, 2021.
52. Yan, L.; Xi, W.; Zhang, Z. Analysis of the Habitat Quality Changes and Influencing Factors in Chuxiong Prefecture under the Background of Landscape Pattern Changes. *Int. J. Ecol.* **2021**, *10*, 655. [CrossRef]
53. Baixue, W.; Weiming, C.; Shengxin, L. Impact of land use changes on habitat quality in Altay region. *J. Resour. Ecol.* **2021**, *12*, 715–728. [CrossRef]
54. Hu, X.; Song, C.; Fan, X.; Xiao, y.; Xu, W.; Ouyang, Z. Main Regulation Services and Value Assessment of Shrub Ecosystem in China. *J. Beijing For. Univ. (Soc. Sci.)* **2021**, *20*, 58–64.
55. Duan, X.; Bai, Z.; Rong, L.; Li, Y.; Ding, J.; Tao, Y.; Li, J.; Li, J.; Wang, W. Investigation method for regional soil erosion based on the Chinese Soil Loss Equation and high-resolution spatial data: Case study on the mountainous Yunnan Province, China. *Catena* **2020**, *184*, 104237. [CrossRef]
56. Economics, V. *State and Trends of Carbon Pricing 2016*; World Bank: Washington, DC, USA, 2016.
57. Santikarn, M.; Churie Kallhauge, A.N.; Bozcaga, M.O.; Sattler, L.; Mccormick, M.S.; Ferran Torres, A.; Conway, D.; Mongendre, L.; Inclan, C.; Mikolajczyk, S. *State and Trends of Carbon Pricing 2021*; World Bank: Washington, DC, USA, 2021.
58. Capoor, K.; Ambrosi, P. *State and Trends of the Carbon Market 2006*; World Bank: Washington, DC, USA, 2006.
59. Linacre, N.; Kossoy, A.; Ambrosi, P. *State and Trends of the Carbon Market 2011*; World Bank: Washington, DC, USA, 2011.
60. Wang, B.; Song, Q. Value assessing methods of species diversity conservation in forest ecosystem. *J. Beijing For. Univ.* **2012**, *34*, 155–160. [CrossRef]
61. Wang, B.; Zhen, G.; Guo, H. Economic Value Assessment of Forest Species Diversity Conservation in China Based on the Shannon-Wiener Index. *For. Res.* **2008**, *21*, 268–274.
62. Chinese Academy of Environmental Planning; Research Center for Eco-Environmental Sciences. *The Technical Guideline on Gross Ecosystem Product (GEP)*; Ministry of Ecology Environment of the People's Republic of China: Beijing, China, 2020.
63. Himes-Cornell, A.; Pendleton, L.; Atiyah, P. Valuing ecosystem services from blue forests: A systematic review of the valuation of salt marshes, sea grass beds and mangrove forests. *Ecosyst. Serv.* **2018**, *30*, 36–48. [CrossRef]
64. Qiu, H.; Hu, B.; Zhang, Z. Study on ecosystem service value of Guangxi in the past 20 years based on land use change. *J. Environ. Eng. Technol.* **2021**, 1–15. Available online: http://kns.cnki.net/kcms/detail/11.5972.X.20211028.1931.009.html (accessed on 1 June 2022).
65. Auffhammer, M. Quantifying economic damages from climate change. *J. Econ. Perspect.* **2018**, *32*, 33–52. [CrossRef]

land

MDPI

Article

Spatiotemporal Distribution and Driving Force Analysis of the Ecosystem Service Value in the Fujiang River Basin, China

Xuemao Zhang [1], Binggeng Xie [1,*], Junhan Li [1] and Chuan Yuan [2]

[1] School of Geographic Sciences, Hunan Normal University, Changsha 410081, China
[2] School of Life Sciences, Southwest University, Chongqing 400715, China
* Correspondence: xbgyb1961@163.com

Abstract: Identification of spatiotemporal changes in ecosystem service value and their drivers is the basis for ecosystem services management and decision making. This research selects Fujiang River Basin (FJRB) as the area of study, using the equivalent factor method to estimate the ecosystem service value (ESV) variation and characteristics of its spatial distribution. The contributions of the drivers of ecosystem service value and their interactions were also explored using the optimal parameters-based geographical detectors (OPGD) model. The results showed the following: (1) the total ESV increased from 104,891.22 $\times$ 10^6 yuan to 105,032.08 $\times$ 10^6 yuan from 2000 to 2020, and displayed an upward trend from the southeast to northwest; (2) The distribution of ESV showed a strong positive spatial autocorrelation. High ESVs were concentrated upstream of the study region with a higher elevation and vegetation coverage, whereas low values were mainly found in the midstream and downstream regions, where frequent human activity occurs; (3) The elevation of natural factors, HAI and LA of human-social factors, and PEL of landscape pattern factors were the main forces leading to ESV differentiation, and the spatial heterogeneity of ESV in the study area resulted from the synergistic effect of natural factors, human socioeconomic activities, and landscape pattern factors. This research reveals the spatial and temporal patterns and drivers of ecosystem service values in the FJRB, and provides a scientific reference for the establishment of land-use planning and ecological environmental protection mechanisms in this region.

Keywords: ecosystem service value; equivalent coefficient; driving factors; heterogeneity; spatial autocorrelation analysis

1. Introduction

Ecosystems are the foundation for the survival and development of human society, and the essence of many environmental problems facing humanity is the destruction and degradation of ecosystem functions and services [1]. Ecosystem services (ESs) are the ability of ecosystems to provide tangible or intangible natural products, environmental resources, and ecological public interests to sustain human activities, such as production, consumption, and circulation [2]. The values of ESs are incredibly high and sometimes immeasurable; thus, maintaining ESs is vital for human welfare [3,4]. However, because the socioeconomic system does not adequately elevate ecosystem assets and their value, ESs are considered a free and inexhaustible public service, resulting their overconsumption and ultimately scarcity [5,6]. Therefore, accurately assessing and quantifying ecosystem service value (ESV) and exploring the drivers that influence the spatial heterogeneity of ESV are vital for maintaining human well-being, promoting a healthy development ecosystem, and providing an essential reference for making reasonable ecological protection policies [7].

Currently, the assessment of ESs primarily includes monetary measurements [8], physical measurements [9], and energy-analysis models [10]. Using monetary units to quantify the ESV allows it to be aggregated and compared across ecosystems. The evaluation findings can also be quickly incorporated into the national accounting system, which is essential for environmental accounting and producing a "green" gross domestic product (GDP) [11]. The main methods for assessing ESV include the market price, travel cost, productivity, and benefit transfer methods [12]. Among them, the benefit transfer method is popular, first proposed by Costanza et al. in 1997. It has the advantages of fast evaluation and low cost of data collection [13]. Based on the global equivalent factor table proposed by Costanza et al. [14], a large number of scholars have evaluated the ESV in different countries and regions [15–17]. Moreover, the study by Xie et al. [18] supplemented and revised the equivalent factor method to provide a reference for ESV studies in China based on the view of Costanza et al. [14]; this method combined the current situation of the ecosystem with the economic and social development of China. However, because of the scale effect in the assessment of ESV, existing value equivalent factors can be used at the macro scale, and it is challenging to address the requirements of ESV in a specific region [19]. In particular, if the value coefficient corresponds to the actual condition in the research region, the scale effect would directly influence the accuracy of the final ESV assessment [20]. Consequently, an equivalence factor per unit area depending on the current local condition for various sizes and areas must be determined [21].

If the description of the spatial pattern of regional ESs is "knowing what it is", exploring the driving force on regional ESs is "knowing why", which is more beneficial in understanding the underlying causes of ecosystem issues and directing local ecological construction [22]. The complex relationship between human and natural ecosystems has attracted the interest of many researchers [23–25], but most studies have focused on the relationship between a single factor and ecosystem services [26], or the confirmation of the main influencing factors and their individual effects on ESV [27]. Such studies disregard the combined effects of factors in ESV changes, and lack a systematic perspective to fully comprehend the driving forces on ESV. Sannigrahi et al. noted that the combined effects of the influencing factors were much higher than their individual effects [28]. The diversity and breadth of influencing factors of ESs show that the impact mechanism of ecosystem services is complex. Related studies in China, especially small-scale studies, are not sufficiently detailed. Changes in ESV typically result from the interaction of several causes [16]. To explain the complex ESV variation accurately and comprehensively, the synergistic effect of these drivers should be considered. The OPGD model can quantitatively detect the single-factor driving factor and the multi-factor interaction driving force, which can compensate for the lack of studies on the driving force of ESV. The influencing factors of ESs can be categorized as natural (terrain, climate, soil, biological), socioeconomic (social economy, land-use), and landscape (landscape index, landscape structure) aspects. Regarding natural factors, topography indirectly regulates soil retention, water supply capacity, and crop production capacity by influencing ecological conditions such as the surface's temperature, the intensity of light, and precipitation [29]. Climate directly regulates hydrothermal conditions and influences ESs [30]. Soil is the background ecological element for organisms to grow and inhabit, and its physical and chemical properties significantly impact ESs [31]. Human factors such as land-use type, changes in land-use intensity, and human activities have an impact on the level of ESs [17,32]. Thus, the synergistic effects of drivers from three aspects (nature, human-social, and landscape) should be considered to provide an accurate and comprehensive explanation of complex ESV changes.

The Fujiang River Basin (FJRB) is a significant headwater and ecological shelter upstream of the Yangtze River Basin and a key ecologically functional region in southwestern China, which has important regional security functions such as water connotation, soil conservation, and biodiversity protection [33]. Jeremy Rayner et al. [34] highlighted that as society develops, the link between the ecological environment and socioeconomic development increases, and there is a growing concern for coordinated or "integrated" policy strategies. Therefore, it is crucial to strengthen the study of ESV in the FJRB and identify its drivers to optimize the ecological structure and coordinate ecological protection and socioeconomic development in the Yangtze River basin. At present, there is a lack of research on ecosystem service valuation and driving force in FJRB. The main research objectives of this study are (1) to assess ESV in different periods (2000, 2010, and 2020), (2) to discuss the spatiotemporal differentiation characteristics of the ESV from 2000 to 2020, and (3) to quantify the drivers of spatial differentiation of ESV. The results of the study will enrich the theoretical knowledge of the complex relationship between drivers and ecosystem services, and provide a theoretical basis for decision making on the regulation of ecosystem functions, the construction of ecological security patterns and the construction of ecological civilization in the FJRB.

2. Materials and Methods

2.1. Study Area

The Fujiang River is the secondary tributary of the Yangtze River and the largest right-bank tributary of the Jialing River. The river is located between 29.10–33.04° N and 103.30–106.30° E, with a total area of 3.92×10^4 km^2. The multi-year average annual precipitation is >800 mm, with significant spatiotemporal differentiation of precipitation, and the multi-year average annual temperature is >15°. The mean elevation of the FJRB is 986.7 m (above sea level) (a.s.l), and the average slope is approximately 14° [33]. Mountains and hills dominate the terrain, and the elevation difference in the basin is 5393 m. The ecological protection of the FJRB is critical to the long-term development of the Yangtze River Basin (Figure 1).

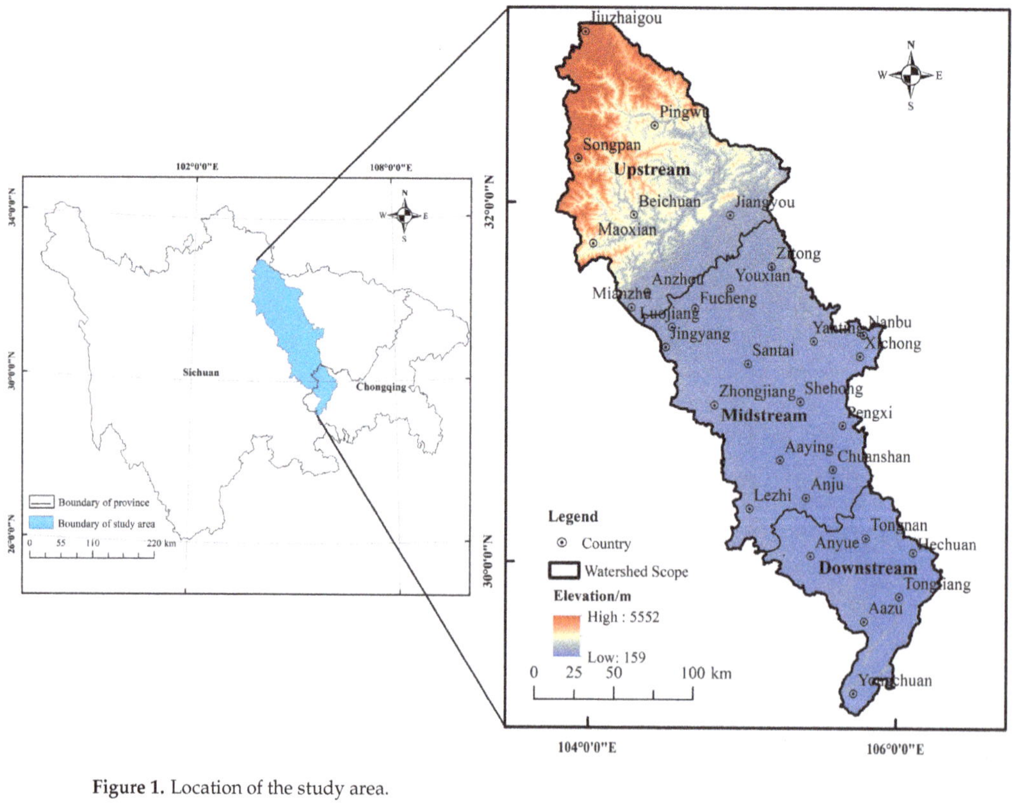

Figure 1. Location of the study area.

2.2. Data Source

The LULC data for the three periods in 2000, 2010, and 2020 were all downloaded from the GlobelLand30 (http://www.globallandcover.com, accessed on 23 May 2022) [35].

Six land-use types were obtained after the "China Land Use/Land Cover Remote Sensing Monitoring data classification system" was used to reclassify each year's land-use types according to the needs of the research: cultivated land, forest land, grassland, waterbody, construction land, and unutilized land (Figure S1). A total of 4658 grids were created by dividing the study area into 3 km by 3 km finishing nets using ArcGIS 10.5. ESV was calculated in these grids. The other geospatial, meteorological, and socioeconomic data are shown in Table 1.

Table 1. Data sources and processing.

	Driving Factor	Sources and Time	Processing
Nature factor	Temperature (Tem) Precipitation (Pre)	https://www.ncdc.noaa.gov/, accessed on 15 March 2022 (2000, 2010, 2020)	Anusplin interpolation model
	Elevation Slop Soil erosion Difference Vegetation Index (NDVI)	https://www.gscloud.cn/, accessed on 18 June 2022 https://www.resdc.cn/, accessed on 18 June 2022 (2000, 2010, 2020)	ArcGIS Spatial analysis tool
Human-social factors	GDP per land (GDP) Population density (POP)	(2000, 2010, 2020) (2000, 2010, 2020)	Kriging method
	Distance from the road (DFR)	https://www.webmap.cn/, accessed on 18 June 2022	Euclidean Distance
	Land use intensity (LA)	[36] (2000, 2010, 2020)	$LA = 100 \times \sum_{i=1}^{n} (A_i \times C_i)$
	Human activity intensity (HAI)	[37] (2000, 2010, 2020)	$HAI = \sum_{i=1}^{n} \frac{A_i P_i}{TA}$
Landscape pattern factors	Landscape Division Index (DIVISION) Contagion Index (CONTAG) Shannon's Diversity Index (SHDI) Landscape Shape Index (LSI) The Proportion of ecological land (PEL)	LULC data (2000, 2010, 2020)	Fragstats software 4.0
Statistic data	Total population Gross Domestic Product (GDP) Total area	Local Bureau of Statistics (2000, 2010, 2020)	

2.3. Calculation of ESV

Costanza et al. [38] originally suggested measuring global ESV using the equivalent coefficient method as demonstrated in the current study. An equivalent factor value is 1/7 of the market price of food produced per hectare of arable land per year [18,39]. Based on the average grain yield per unit area in the FJRB, as well as the lowest price of the indica rice in Sichuan Provence from 2000 to 2020, the equivalent factor value of ESs in the FJRB was 2122.77 yuan/hm^2. The calculating formula of ESV is as follows:

$$\text{ESV} = \sum_{i=1}^{m} A_i \times VC_i \quad (1)$$

$$\text{ESV}_j = \sum_{i=1}^{m} (A_i \times VC_{jm}) \quad (2)$$

where A_i and VC_i denotes the area and ESV coefficient of landscape type i, respectively, ESV_j represents the jth ESV, and VC_{jm} represents the ESV coefficient of jth service of landscape type i. The ESV coefficient of each land-use type in the FJRB was obtained (Table 2).

Table 2. ESV coefficients for each land-use type in the FJRB (yuan/hm^2).

	Type	Cultivated Land	Forest Land	Grassland	Waterbody	Unutilized Land
Regulating service	Gas regulation	1061.39	7429.70	1698.22	3820.99	0.00
	Climate regulation	1889.26	5731.48	1910.49	36,299.37	0.00
	Hydrological regulation	1273.66	6792.86	1698.22	32,902.94	63.69
Supporting service	Soil conservation	3099.24	8278.80	4139.40	3629.94	42.46
	Waste disposal	3481.35	2780.83	2780.83	38,591.96	21.23
	Maintaining biodiversity	1507.17	6920.23	2313.82	5306.93	721.74
Provisioning service	Food production	2122.77	212.28	636.83	636.83	21.23
	Raw material production	212.28	5519.20	106.14	148.59	0.00
Cultural service	Aesthetic landscape	21.23	2717.14	84.91	11,781.38	21.23
	Total	14,668.34	46,382.52	15,368.86	133,118.91	891.57

2.4. ESV Sensitivity Index Analysis

The sensitivity model used in this study calculates the response of ESV to the variety in the value coefficient (VC). The VC for each land-use type was adjusted by 50%, and then the change in ESV as time passed and the level of dependence on the VC were determined [40]. The formula is as follows:

$$CS = \left| \frac{(ESV_j - ESV_i)/ESV_i}{\left(VC_{jf} - VC_{if}\right)/VC_{if}} \right| \quad (3)$$

where CS is the sensitivity index, ESV is the total ESV, and VC refers to the value coefficient; i and j represent the initial and change values, respectively (adjusted up or down by 50%), and f represents the land-use type. If CS > 1, the ESV is elastic to the VC. If CS < 1, elasticity is lacking. Therefore, the larger the CS, the more critical the accuracy of the ESV index.

2.5. Spatial Heterogeneity Analysis of ESV

Spatial autocorrelation analysis is a method used to determine whether there is a correlation in the spatial distribution of an attribute and to calculate the degree of correlation. It can intuitively express the correlation and spatial heterogeneity of a certain spatial phenomenon [41,42]. The formula is as follows:

$$Moran'sI = n\sum_{i=1}^{n}\sum_{j\neq 1}^{n} W_{ij}(x_i - \overline{x})(x_j - \overline{x}) / (\sum_{i=1}^{n}\sum_{j=1}^{n} W_{ij}) \sum_{i=1}^{n}(x_i - x)^2 \quad (4)$$

where n is the number of spatial cells, x_i and x_j are the ESV of the i and j spatial units, respectively, and W_{ij} is the spatial weight. The interpretation of the local indicators of spatial association (LISA) is detailed in Table S1.

2.6. Driving Factor Index System Construction

In the geographical exploration of spatial differentiation mechanisms, the choice of the driving factor indicators is crucial [43]. Based on the research area's actual circumstances and the availability of data principle, three categories and 16 factors were screened out, namely, natural factors (Tem, Pre, elevation, slope, soil erosion, and NDVI), human-social factors (DFR, POP, GDP, HAI, and LA), and landscape pattern factors (DIVISION, SHDI, LSI, CONTAG, and PEL). See Table 1 for details.

2.7. The Optimal Parameter-Based Geographical Detectors Model (OPGD)

The Geographical Detector (Geodetector) is a statistical tool for measuring regionally stratified heterogeneity and performing attribution analysis [44,45]. The OPGD model utilized in this research was created with the GD1.10 package of R, and the spatial discretization and the spatial-scale optimization are both types of parameter optimization [46]. In this study, the q-value was used to measure the independent variable X (16 driving factors) affecting the spatial variation of dependent variable Y (ESVs) at zonal levels in the FJRB. The formula is as follows:

$$q = 1 - \frac{1}{N\delta^2} \sum_{n=1}^{L} N_h {\delta_h}^2 \quad (5)$$

where q is the independent variable (X)'s explanatory power on the dependent variable (Y), ranging from 0 to 1, L is the layer of X, and N_h and δ_h represent the sample size and variable of the ESV in layer h, respectively. The types of interactions are detailed in Table S2.

3. Results

3.1. Analysis of the Spatiotemporal Distribution Characteristics of ESVs in the FJRB

Based on the ESV calculation, the sensitivity index was analyzed according to Formula (3). Considering 2020 as an example, the CS values of the different land-use types were all less than one. Among these, the sensitivity index of forestland was the largest (0.604), while that of farmland, grassland, and waterbody was 0.319, 0.030, and 0.048, respectively. The unused land's sensitivity index was less than 0.001. Therefore, the ESV was not sensitive to the equivalent coefficient of each land-use type.

The total ESVs of the FJRB in 2000, 2010, and 2020 were 104,891.22 × 10^6, 105,032.38 × 10^6, and 105,032.08 × 10^6 yuan, respectively. From 2000 to 2020, the ESV of the FJRB showed a positive growth state (0.13% growth rate), and the total amount of ESV increased by 140.86 × 10^6 yuan. Using the natural break classification, the total ESV in the FJRB was divided into five categories (extremely low, low, medium, high, and extremely high). Table 3 and Figure 2 present the spatial patterns of the ESV from 2000 to 2020. Spatially, the ESV in the FJRB displayed a tendency that increased from southeast to northwest. Among them, the high-value zones are primarily found upstream of the FJRB in the northwest, such as Songpan, Pingwu, and Beichuan, and the highest values of ESV in 2000, 2010, and 2020 were 76.52 × 10^6, 68.80 × 10^6, and 75.02 × 10^6 yuan, respectively. The low-value areas are mainly located in the urban regions of the Fujiang River, such as Youxian, Jingyang, Luojiang, Anyue, and Tongnan.

Table 3. Spatial change tendency of ESV in the FJRB.

Different ESV Grade Zones		Extremely Low	Low	Medium	High	Extremely High
Area (km^2)	2000	664.87	17,308.69	7609.63	4533.36	9082.00
	2010	675.50	16,180.56	8434.22	4622.26	9280.00
	2020	2427.85	15,481.19	7656.17	5356.99	8267.35
Change (%)	2000–2010	1.60	−6.52	10.84	1.96	2.18
	2010–2020	259.41	−4.32	−9.22	15.90	−10.91
	2000–2020	265.16	−10.56	0.61	18.17	−8.97

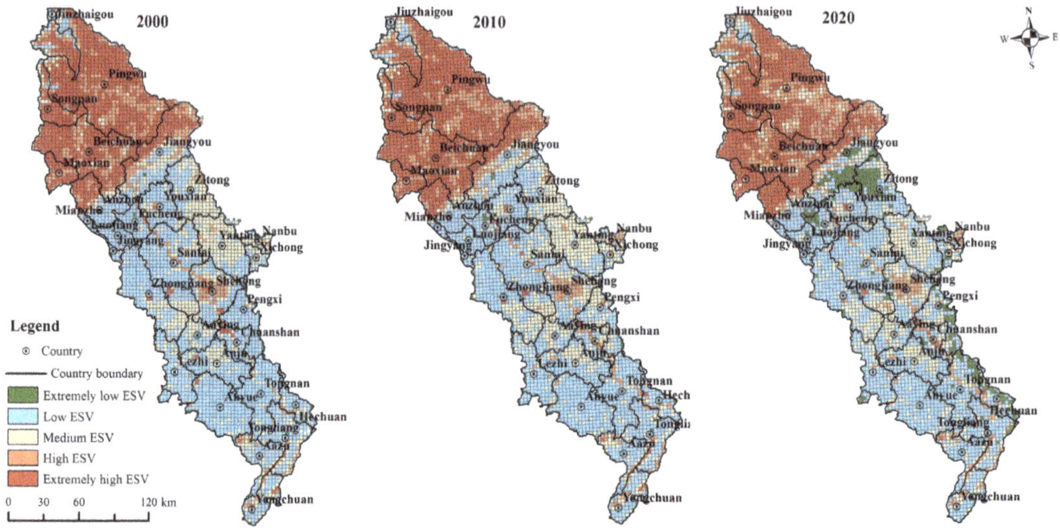

Figure 2. Total ESV of FJRB from 2000 to 2020.

3.2. *Spatial Autocorrelation Analysis of ESVs*

Based on ArcGIS 10.5 and the GeoDa 1.18.0 software, and considering the ESV of the FJRB in 2000, 2010, and 2010 as variables, the spatial autocorrelation analysis of ESVs was performed using the queen's spatial weight method. The Moran's *I* values were 0.755, 0.756, and 0.771 in 2000, 2010, and 2020, respectively, and the Z value was greater than 1.96. The research finding shows that the ESV of the FJRB had a strong positive spatial autocorrelation and spatial agglomeration effect from 2000 to 2020; that is, the areas with high ESV tend to be adjacent in space, and the areas with low ESV also have the tendency of adjacent connections in space.

The LISA clustering chart from 2000 to 2020 (Figure 3) shows the low-low agglomeration of ESV in the midstream and downstream of the study region. These regions have a high level of urbanization and relatively concentrated construction land. The government has strengthened the construction and protection of Jiuhaigo, Xuebaodin, Wanglan, and other national nature reserves in the upstream region. The high-high ESV areas are concentrated in these regions. With the intensive production of non-agricultural industries, the lower-low areas of ESV in the northwest of Fucheng, Youxian, south of Anzhou, northwest of Luojiang, Zhongjiang, southwest of Lezhi, east of Pengxi, Anyue, and Tongnan gradually expanded along the urban development axis. The low-high and high-low ESV areas did not change substantially.

3.3. *Analysis of the Driving Force of the Spatiotemporal Distribution of ESVs*

3.3.1. Factor Detector Analysis

The OPGD model was utilized to calculate the driving forces underlying the spatial differentiation of the ESVs in the FJRB from 2000 to 2020 (Figure 4). The spatial differences of the distribution of ESVs in the FJRB were influenced by natural, human-social, and landscape factors. Among them, the spatial variation of ESV in the FJRB in 2000 was most significantly influenced by PEL, HAI, LA, and elevation, all with q-values > 0.74. In 2010, the contributions of factors such as LA, HAL, elevation, PEL, and Tem were less than those in 2000, all with q-values > 0.62. In 2020, the elevation had the biggest contribution to the spatial heterogeneity of ESV in the FJRB, with a q-value of 0.6833. This was followed by a reduced contribution of Tem, PEL, and LA, all with q-values > 0.50. The influencing factors with the highest contribution during 2000–2020 were elevation, PEL, LA, and HAI, with a high explanatory power (q-values > 65%) and significant influence. Second, the influences of POP, Pre, Slope, and NDVI were approximately 40–60%, which are key factors influencing the spatial differentiation of ESVs within the research area. In addition, both the DFR and the effect of soil erosion were greater than 15%, which are minor factors of importance affecting the spatial variation of ESV in the watershed. Finally, the explanatory power of factors such as GDP, LSI, SHDI, DIVISION, and CONTAG was below 10%.

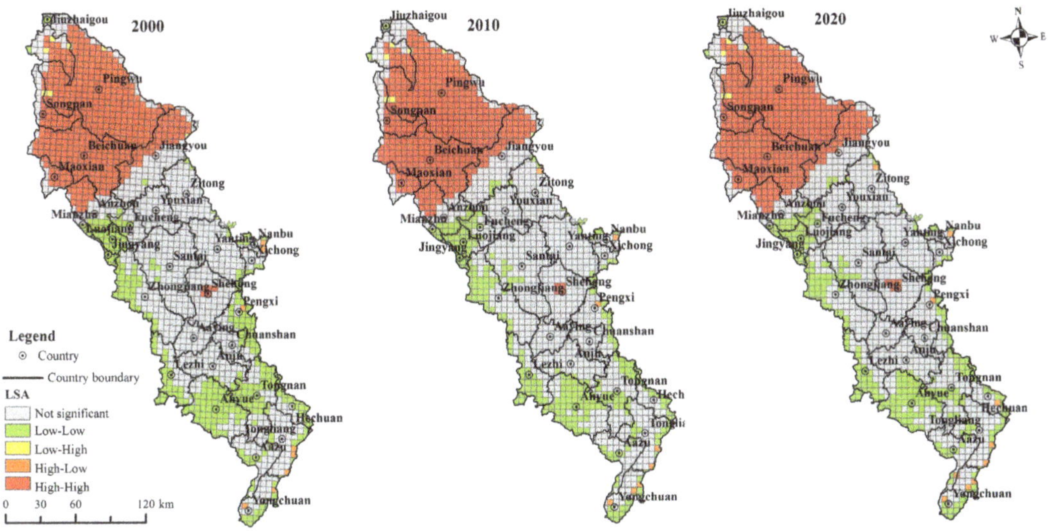

Figure 3. The LISA cluster graph of ESV of FJRB during 2000–2020.

3.3.2. Interaction Detector Analysis

The interaction detector identifies the interaction between different factors on the spatial differentiation of ESV and analyzes whether the dependent variable's explanatory power increases or decreases (Figure 5). The findings of the interaction for the drivers of the spatial heterogeneity of ESVs in the FJRB show that the interaction of any two of the factors was stronger than the impact of a single factor. The types of the two interactions are primarily nonlinear enhancement interaction-enhancement and two-factor enhancement, indicating that the interaction of multiple factors influences the spatial heterogeneity results of ESVs.

In 2000, the interaction of the PEL and NDVI had the most substantial influence on ESV spatial differentiation, with the highest q-value of 0.8829 for factor interaction detection and an explanatory power close to 90%. The LA, HAI, PEL, and elevation interacting with other arbitrary factors on ESV spatial differentiation in the FJRB were above 80%. In 2010, the spatial variation of ESV in the FJRB was most strongly influenced by the interaction between the LA and elevation, with a q value of factor interaction detection reaching 0.8745. Second, the interaction effects of elevation, LA, HAI, and PEL with other factors were all above 70%, and the interaction effects of Tem, Pre, and slope with other factors were all above 50%. In 2020, the most decisive influence on the spatial variation of ESVs in the FJRB was the interaction between HAI and PEL, with q values as high as 0.8520. The interaction reached more than 75%, including LA ∩ HAI (0.8210) and HAI ∩ elevation (0.7555). Although the q-value of the interactions between the LSI, SHDI, CONTAG, Division, and other factors were under 20%, the results revealed that double factors had a greater effect on ESV spatial difference than single factors (Figure 4).

The results of the analysis indicate that even though the q-values of the interactions between the drivers of spatial differentiation ESV in the study region were decreasing from year to year, the q-values of the interactions were still higher than the q-value of the single factor. The complex coupling between different drivers formed a synergistic enhancement effect that jointly influenced the spatial differentiation effect of ESV in the study region. Furthermore, the interaction results between the PEL and human social factors (HAI, LA) and topographic factors (elevation, slope) showed higher q-values, demonstrating the synergistic enhancement effect of multi-factor interaction on the spatial differentiation of ESV in the FJRB. Therefore, the spatial differentiation of ESV in the FJRB results from the interaction between natural, human-social, and landscape factors, and the interactions between natural and human-social factors are more significant.

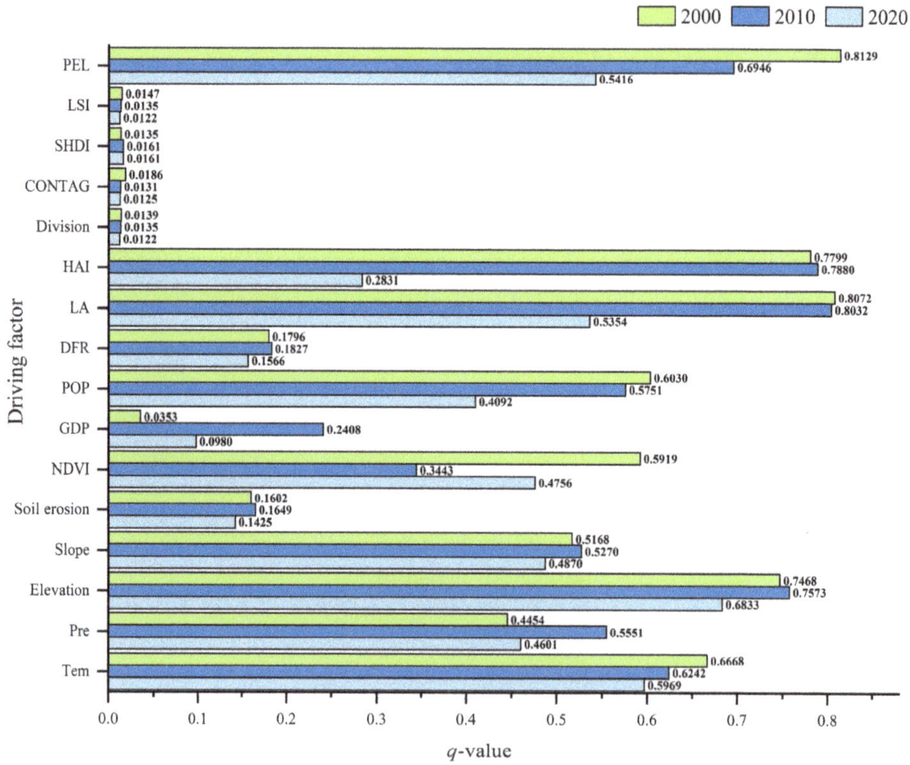

Figure 4. Contributions (q statistics) of different factors to ESV variation from 2000 to 2020.

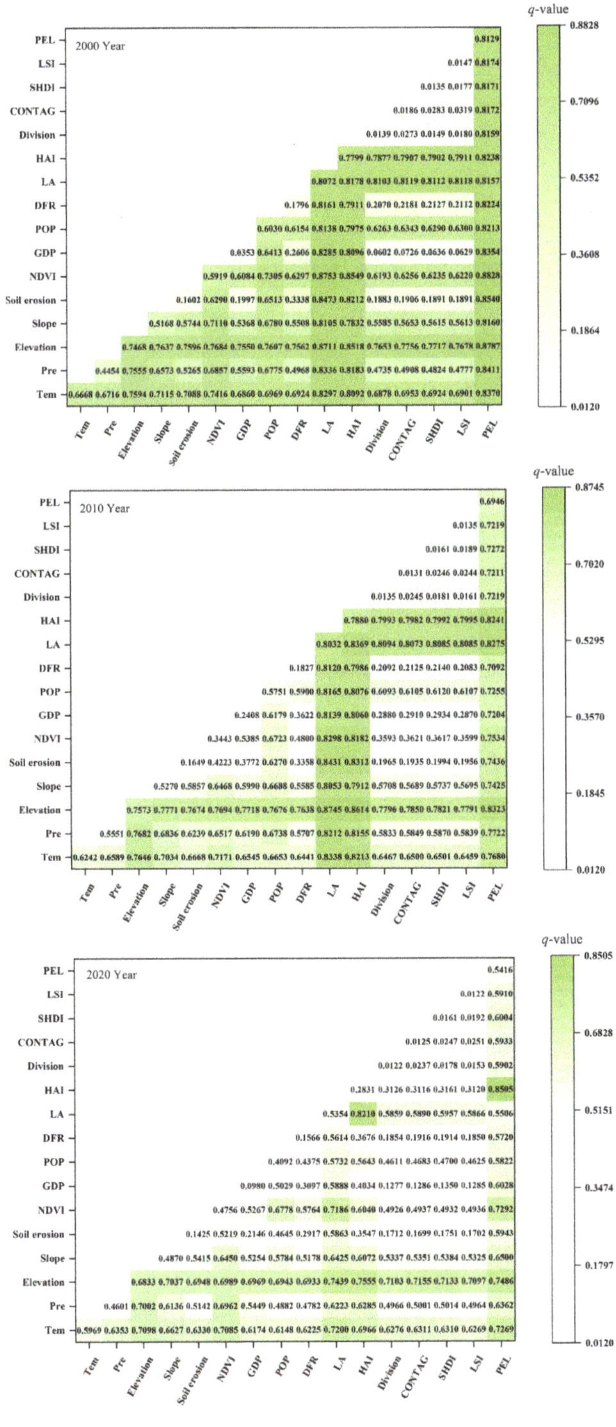

Figure 5. Results of interaction detection from 2000 to 2020.

4. Discussion

4.1. Analysis of the Driving Force of the Spatiotemporal Distribution of ESVs

ESs have been a focal point in geography, ecology, and other fields, serving as a vital link between human well-being and environmental structure [47]. The value assessment of ESs can provide a reliable basis for assessing ecosystem quality changes, formulating payments for ESs policies, and promoting ecosystem protection and ecological civilization construction [48]. As shown by the elevation map analysis, the northwestern part of the research region is situated in the hilly plateau mountainous area of northwest Sichuan Province. The terrain in this region is primarily mountainous, with considerable differences in vegetation changes, apparent terrain fluctuation, less human activities, and forest land with a high vegetation cover as the primary landscape type in these regions. Therefore, the ESV in the northwest is high. Although the southeast part of the research region is situated in the hilly region of the Sichuan Basin, the terrain is relatively flat, the main land-use types are cultivation and construction land, and human activities are relatively intensive, resulting in a low ESV. Therefore, the ESV in the northwest is higher than in the southeast. This study is consistent with previous research findings on ESV distribution in the southwest and the whole of China [16,18].

In this research, the OPGD model was used to explore the driving mechanism of the spatial heterogeneity of the ESV in the FJRB. It can provide a quantitative analysis of the relative importance of the driving force for the ESV in the study region and identify interactions between factors. In 2000–2020, the factors that explained the strongest q-values of spatial variation in ESV in the FJRB were PEL (0.8130), LA (0.8032), and elevation (0.6832). Among them, the most stable influencing factor was elevation, and the influence was stable at approximately 70% for the period 2000–2020. This is due to the significant elevation difference (the difference between the highest and the lowest elevation is 5393 m), and the apparent topographic relief in the study region. This research also investigated the driving mechanism of ESV spatial differentiation in different regions of the upstream, midstream, and downstream of the FJRB in 2000, 2010, and 2020. The findings showed that the dominant drivers of ESV spatial differentiation characteristics in the area differed more prominently in different regions and periods (Figure 6). During 2000–2020, in the upstream region, the differences in the spatial distribution of ESV were primarily impacted by LA, PEL, HAI, and elevation, all of which had an influence higher than 50%, followed by Tem, Pre, and slope, which had influences of 20–30%; the influence of the remaining factors was less than 10%. In the midstream region, the differences in the spatial distribution of ESV during 2000–2020 were primarily influenced by the factors of PEL, LA, and HAI. The explanatory power of the rest of the factors was less than 10%, indicating that the landscape pattern factors and human-social factors influenced the ESVs distribution in the midstream region. In the downstream region, the spatial variation in ESV during the study period was primarily influenced by PEL, LA, HAI, and elevation. These findings are consistent with the study results. Thus, the spatially heterogeneous distribution of ESV in the study area was influenced by a combination of multiple factors, with natural and human-social factors having the most significant influence, followed by landscape pattern factors.

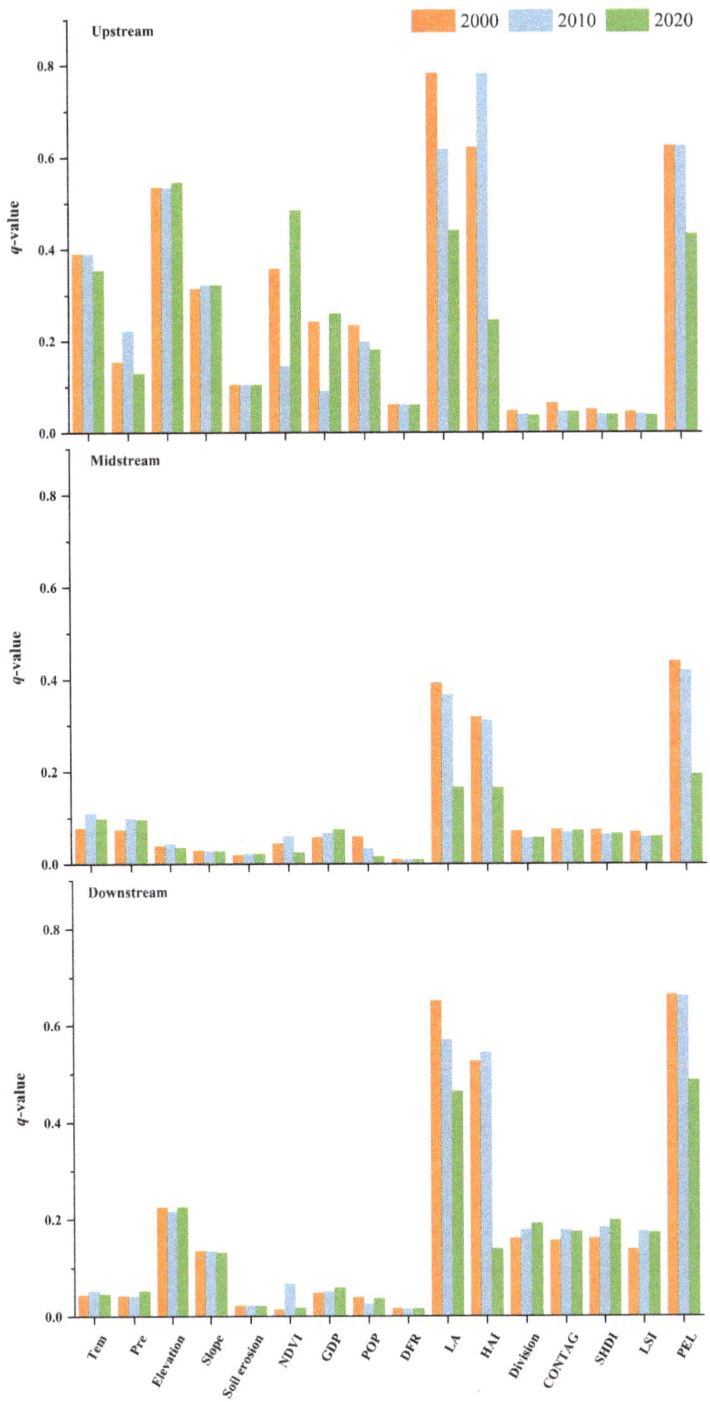

Figure 6. Geographical detection q-values of ESVs in the upstream, midstream, and downstream of the FJRB in 2000, 2010, and 2020.

Overall, based on the OPGD model, 16 drivers were selected from natural, human-social and landscape patterns to investigate the driving force of ESV spatial differentiation, complementing the qualitative and regression analyses that have lacked spatiality in recent years. The result showed that ESV increased with increasing elevation, which is consistent with the studies by Teng et al. in the Qilian Mountains [49]. This is because elevation affects the distribution of temperature, precipitation, and vegetation, which will substantially impact the distribution of ESV. Simultaneously, the interaction of nature and human-social factors might influence the landscape, and the intensity of human activities indirectly changes the distribution and changes of ESVs in the study region. The change in land-use patterns can also modify the spatiotemporal distribution of resources, which further impacts the ecological environment's structure and function [17]. Among the landscape factors, the contribution rate of the PEL was always the highest, up to 0.8120, and the contribution rate of other landscape factors was low, indicating that the change in landscape pattern would also have an impact on the ecosystem function. Furthermore, changes in landscape patterns affect the processes of material cycling and energy flow in ecosystems and eventually lead to changes in regional ESs via interactions with biotic and abiotic processes [33]. The findings demonstrate that the synergistic interactions of natural, socioeconomic, and landscape pattern factors caused spatial differentiation of ESV in the study region. As a result, decision-making institutions should maximize the allocation of natural resources based on local conditions, control the influence of human social and economic activities on the ecological environment, and adapt the allocation of landscape types when developing future ecosystem management policies.

4.2. Policy Suggestions

Land policy determines the strategic direction of the future development of the region [50], and influences the development of regional urban economic construction [51]. Regional planning should not be guided by a prescriptive view of design, but rather by the rationalization of space and the corresponding approach [51]. This study evaluated the ESVs, analyzed the spatial and temporal variability of ESVs, and quantitatively analyzed the drivers of spatial variability of ESVs in the FJRB, thus providing a scientific reference for urban regional planning and environmental protection policies in the study area and over the whole Yangtze River Economic Zone. Considering the above analysis, the government departments should enhance the protection of nature reserves upstream of the study area and reasonably control the encroachment of human activities on forests and water bodies. We advise that ecological restoration projects should be carried out upstream of the study area, such as comprehensive water environment management, mine restoration, ecological replenishment of forest land, and geological disaster prevention and control; midstream and downstream of the study area, projects such as territorial spatial planning, basic farmland protection planning, and habitat improvements should also be implemented to coordinate the development of an ecological-production-living space. Based on the above analysis, an ecological economy should become mainstream. Abundant water, tourism resources, and high vegetation cover in the study area offer the possibility of this eco-economic model. The ecotourism industry, plantations, three-dimensional agriculture and other organic agricultural practices are all examples of ecological economies that could give the study area its own competitive advantage in the construction of the entire Yangtze River Economic Belt. A change in the economic development model can also reduce the contradiction between ecological protection and economic development, and to a certain extent promote regional sustainable development. Further, to enhance policy feasibility, policy makers should pay closer attention to the effects of drivers on the ESV and tailor them to regional conditions. For example, when formulating ecological restoration policies, a diversified and differentiated coordination strategy should be adopted to weigh and coordinate multiple ecosystem service functions; when formulating territorial spatial planning policies, priority should be given to the differences between local geographical conditions and economic development patterns.

4.3. Limitations and Future Work

In this study, the VC was adjusted to reflect the real condition in the research region accurately, and the ESV in the FJRB was calculated using the ESV equivalent coefficient per unit area. Additionally, using the OPGD model, the explanatory power of each driving force on the spatial differential features of ESV was determined from three aspects (natural, human-social, and landscape pattern). In this study, the OPGD method was used to discretize the dependent variable which improves the reliability and accuracy of applying the geographic detector model. However, the distribution of ESV is influenced by the interaction of several factors and is not simply a positive or negative relationship. Meanwhile, because of the difficulty in obtaining and quantifying the data of some factors, policy factors were not significantly considered. Therefore, more factors should be considered from the viewpoint of ecosystem service stakeholders and human welfare, and quantitative and qualitative analysis methods should be combined to explore the spatiotemporal variation features and driving force of ESV. Furthermore, the ecological compensation mechanism in the FJRB should be further studied to provide a scientific basis for formulating reasonable ecological environmental protection policies in the basin.

5. Conclusions

This research assessed the ESV in the FJRB from 2000 to 2020 using the equivalent coefficient method, and systematically analyzed the spatial and temporal evolution patterns of ESVs. Based on the OPGD model, the driving force of spatial heterogeneity of ESVs in the FJRB are revealed from three aspects (nature, human-social, and landscape pattern). The conclusions are as follows:

(1) From 2000 to 2020, the total amount of ESV increased by 140.86×10^6 yuan, and the ESV in the FJRB displayed a tendency that increased from southeast to northwest;
(2) The positive spatial autocorrelation of ESV distribution was significant. The high-value and high-high agglomerated areas were primarily distributed upstream of the FJRB, and the low-value and low-low agglomerated regions were primarily distributed in the midstream and downstream of the FRJB;
(3) The most significant driving factors were elevation, HAI, LA, and PEL, and the synergistic interactions of natural factors, human activities, and landscape pattern factors contributed to the regional heterogeneity of the ESV in the research region.

In future, to optimize the ecosystem function of the FJRB, it is necessary to adopt differentiated regulation models and strategies based on the impacts of various factors and the interaction characteristics and effects of different factors. These findings serve as a scientific guide for formulating management models with accurate, diversified, and differentiated ecosystem functions.

Supplementary Materials: The following supporting information can be downloaded at: https://www.mdpi.com/article/10.3390/land12020449/s1, Figure S1: Map of the land cover pattern of FJRB in 2000, 2010, and 2020; Table S1: Association pattern of Moran's I scatter plot and LISA clustering map; Table S2: Types of interaction between two independent variables and dependent variables.

Author Contributions: X.Z.: Conceived the ideas, designed methodology, writing original draft. B.X.: Review and editing, funding acquisition, project administration, supervision. J.L.: Methodology. C.Y.: Methodology. All authors have read and agreed to the published version of the manuscript.

Funding: This research was funded by National Key R&D Program of China (Project Numbers: 2022YFF1300705) and Joint Funds of the National Natural Science Foundation of China (Project Numbers: U19A2051).

Data Availability Statement: Data is contained within the article.

Conflicts of Interest: The authors declare no conflict of interest.

References

1. Schägner, J.P.; Brander, L.; Maes, J.; Hartje, V. Mapping ecosystem services' values: Current practice and future prospects. *Ecosyst. Serv.* **2013**, *4*, 33–46. [CrossRef]
2. Xiao, R.; Lin, M.; Fei, X.; Li, Y.; Zhang, Z.; Meng, Q. Exploring the interactive coercing relationship between urbanization and ecosystem service value in the Shanghai–Hangzhou Bay Metropolitan Region. *J. Clean. Prod.* **2020**, *253*, 119803. [CrossRef]
3. Li, Y.; Feng, Y.; Guo, X.; Peng, F. Changes in coastal city ecosystem service values based on land use—A case study of Yingkou, China. *Land Use Policy* **2017**, *65*, 287–293. [CrossRef]
4. Xie, G.; Lu, C.; Leng, Y.; Zheng, D.; Li, S. Ecological assets valuation of the Tibetan Plateau. *J. Nat. Resour.* **2003**, *18*, 189–196.
5. Daily, G.C.; Söderqvist, T.; Aniyar, S.; Arrow, K.; Dasgupta, P.; Ehrlich, P.R.; Folke, C.; Jansson, A.; Jansson, B.-O.; Kautsky, N. The value of nature and the nature of value. *Science* **2000**, *289*, 395–396. [CrossRef]
6. Wainger, L.A.; King, D.M.; Mack, R.N.; Price, E.W.; Maslin, T. Can the concept of ecosystem services be practically applied to improve natural resource management decisions? *Ecol. Econ.* **2010**, *69*, 978–987. [CrossRef]
7. Crossman, N.D.; Burkhard, B.; Nedkov, S.; Willemen, L.; Petz, K.; Palomo, I.; Drakou, E.G.; Martín-Lopez, B.; McPhearson, T.; Boyanova, K. A blueprint for mapping and modelling ecosystem services. *Ecosyst. Serv.* **2013**, *4*, 4–14. [CrossRef]
8. Costanza, R.; De Groot, R.; Sutton, P.; Van der Ploeg, S.; Anderson, S.J.; Kubiszewski, I.; Farber, S.; Turner, R.K. Changes in the global value of ecosystem services. *Glob. Environ. Chang.* **2014**, *26*, 152–158. [CrossRef]
9. Gomes, E.; Inácio, M.; Bogdzevič, K.; Kalinauskas, M.; Karnauskaitė, D.; Pereira, P. Future land-use changes and its impacts on terrestrial ecosystem services: A review. *Sci. Total Environ.* **2021**, *781*, 146716. [CrossRef]
10. Zhan, J.; Zhang, F.; Chu, X.; Liu, W.; Zhang, Y. Ecosystem services assessment based on emergy accounting in Chongming Island, Eastern China. *Ecol. Indic.* **2019**, *105*, 464–473. [CrossRef]
11. Fu, B.; Lv, Y.; Gao, G. Major research progresses on the ecosystem service and ecological safety of main terrestrial ecosystems in China. *Chin. J. Nat.* **2012**, *34*, 261–272.
12. King, D.M.; Mazzotta, M.J.; Markowitz, K.J. Ecosystem Valuation. Available online: http://www.ecosystemvaluation.org/default.htm (accessed on 3 February 2023).
13. Liu, Y.; Hou, X.; Li, X.; Song, B.; Wang, C. Assessing and predicting changes in ecosystem service values based on land use/cover change in the Bohai Rim coastal zone. *Ecol. Indic.* **2020**, *111*, 106004. [CrossRef]
14. Costanza, R.; d'Arge, R.; De Groot, R.; Farber, S.; Grasso, M.; Hannon, B.; Limburg, K.; Naeem, S.; O'neill, R.V.; Paruelo, J. The value of the world's ecosystem services and natural capital. *Nature* **1997**, *387*, 253–260. [CrossRef]
15. Chen, T.; Feng, Z.; Zhao, H.; Wu, K. Identification of ecosystem service bundles and driving factors in Beijing and its surrounding areas. *Sci. Total Environ.* **2020**, *711*, 134687. [CrossRef]
16. Luo, Q.; Zhou, J.; Li, Z.; Yu, B. Spatial differences of ecosystem services and their driving factors: A comparation analysis among three urban agglomerations in China's Yangtze River Economic Belt. *Sci. Total Environ.* **2020**, *725*, 138452. [CrossRef]
17. Pan, N.; Guan, Q.; Wang, Q.; Sun, Y.; Li, H.; Ma, Y. Spatial differentiation and driving mechanisms in ecosystem service value of Arid Region: A case study in the middle and lower reaches of Shule River Basin, NW China. *J. Clean. Prod.* **2021**, *319*, 128718. [CrossRef]
18. Xie, G.; Zhang, C.; Zhang, C.; Xiao, Y.; Lu, C. The value of ecosystem services in China. *Resour. Sci.* **2015**, *37*, 1740–1746.
19. Hong, Y.; Ding, Q.; Zhou, T.; Kong, L.; Wang, M.; Zhang, J.; Yang, W. Ecosystem service bundle index construction, spatiotemporal dynamic display, and driving force analysis. *Ecosyst. Health Sustain.* **2020**, *6*, 1843972. [CrossRef]
20. Wu, C.; Chen, B.; Huang, X.; Wei, Y.D. Effect of land-use change and optimization on the ecosystem service values of Jiangsu province, China. *Ecol. Indic.* **2020**, *117*, 106507. [CrossRef]
21. Rao, Y.; Zhou, M.; Ou, G.; Dai, D.; Zhang, L.; Zhang, Z.; Nie, X.; Yang, C. Integrating ecosystem services value for sustainable land-use management in semi-arid region. *J. Clean. Prod.* **2018**, *186*, 662–672. [CrossRef]
22. Xue, C.; Zhang, H.; Wu, S.; Chen, J.; Chen, X. Spatial-temporal evolution of ecosystem services and its potential drivers: A geospatial perspective from Bairin Left Banner, China. *Ecol. Indic.* **2022**, *137*, 108760. [CrossRef]
23. Su, S.; Li, D.; Xiao, R.; Zhang, Y. Spatially non-stationary response of ecosystem service value changes to urbanization in Shanghai, China. *Ecol. Indic.* **2014**, *45*, 332–339. [CrossRef]
24. Gao, F.; Cui, J.; Zhang, S.; Xin, X.; Zhang, S.; Zhou, J.; Zhang, Y. Spatio-Temporal distribution and driving factors of ecosystem service value in a fragile hilly area of North China. *Land* **2022**, *11*, 2242. [CrossRef]
25. Su, K.; Wei, D.-Z.; Lin, W.-X. Evaluation of ecosystem services value and its implications for policy making in China–a case study of Fujian province. *Ecol. Indic.* **2020**, *108*, 105752. [CrossRef]
26. Zhou, D.; Tian, Y.; Jiang, G. Spatio-temporal investigation of the interactive relationship between urbanization and ecosystem services: Case study of the Jingjinji urban agglomeration, China. *Ecol. Indic.* **2018**, *95*, 152–164. [CrossRef]
27. Song, F.; Su, F.; Mi, C.; Sun, D. Analysis of driving forces on wetland ecosystem services value change: A case in Northeast China. *Sci. Total Environ.* **2021**, *751*, 141778. [CrossRef]
28. Sannigrahi, S.; Zhang, Q.; Pilla, F.; Joshi, P.K.; Basu, B.; Keesstra, S.; Roy, P.; Wang, Y.; Sutton, P.C.; Chakraborti, S. Responses of ecosystem services to natural and anthropogenic forcings: A spatial regression based assessment in the world's largest mangrove ecosystem. *Sci. Total Environ.* **2020**, *715*, 137004. [CrossRef]
29. Zhu, M.; He, W.; Zhang, Q.; Xiong, Y.; Tan, S.; He, H. Spatial and temporal characteristics of soil conservation service in the area of the upper and middle of the Yellow River, China. *Heliyon* **2019**, *5*, e02985. [CrossRef]

30. Braun, D.; de Jong, R.; Schaepman, M.E.; Furrer, R.; Hein, L.; Kienast, F.; Damm, A. Ecosystem service change caused by climatological and non-climatological drivers: A Swiss case study. *Ecol. Appl.* **2019**, *29*, e01901. [CrossRef]
31. Wang, Y.; Liu, Z. Vertical distribution and influencing factors of soil water content within 21-m profile on the Chinese Loess Plateau. *Geoderma* **2013**, *193*, 300–310. [CrossRef]
32. Fu, B.; Zhang, L. Land-use change and ecosystem services: Concepts, methods and progress. *Prog. Geogr.* **2014**, *33*, 441–446.
33. Zhang, X.; Du, H.; Wang, Y.; Chen, Y.; Ma, L.; Dong, T. Watershed landscape ecological risk assessment and landscape pattern optimization: Take Fujiang River Basin as an example. *Hum. Ecol. Risk Assess. Int. J.* **2021**, *27*, 2254–2276. [CrossRef]
34. Rayner, J.; Howlett, M. Introduction: Understanding integrated policy strategies and their evolution. *Policy Soc.* **2009**, *28*, 99–109. [CrossRef]
35. Jun, C.; Ban, Y.; Li, S. Open access to Earth land-cover map. *Nature* **2014**, *514*, 434. [CrossRef]
36. Zhuang, D.; Liu, J. Study on the model of regional differentiation of land use degree in China. *J. Nat. Resour.* **1997**, *12*, 10–16.
37. Chen, F.; Ge, X.; Chen, G.; Peng, B. Spatial different analysis of landscape change and human impact in urban fringe. *Sci. Geogr. Sin.* **2001**, *21*, 210–216.
38. Costanza, R.; Patten, B.C. Defining and predicting sustainability. *Ecol. Econ.* **1995**, *15*, 193–196. [CrossRef]
39. Fei, L.; Shuwen, Z.; Jiuchun, Y.; Liping, C.; Haijuan, Y.; Kun, B. Effects of land use change on ecosystem services value in West Jilin since the reform and opening of China. *Ecosyst. Serv.* **2018**, *31*, 12–20. [CrossRef]
40. Kindu, M.; Schneider, T.; Döllerer, M.; Teketay, D.; Knoke, T. Scenario modelling of land use/land cover changes in Munessa-Shashemene landscape of the Ethiopian Highlands. *Sci. Total Environ.* **2018**, *622*, 534–546. [CrossRef]
41. Xie, H.; Kung, C.-C.; Zhao, Y. Spatial disparities of regional forest land change based on ESDA and GIS at the county level in Beijing-Tianjin-Hebei area. *Front. Earth Sci.* **2012**, *6*, 445–452. [CrossRef]
42. Huang, C.; Liu, K.; Zhou, L. Spatio-temporal trends and influencing factors of PM2. 5 concentrations in urban agglomerations in China between 2000 and 2016. *Environ. Sci. Pollut. Res.* **2021**, *28*, 10988–11000. [CrossRef] [PubMed]
43. Liao, J.; Yu, C.; Feng, Z.; Zhao, H.; Wu, K.; Ma, X. Spatial differentiation characteristics and driving factors of agricultural eco-efficiency in Chinese provinces from the perspective of ecosystem services. *J. Clean. Prod.* **2021**, *288*, 125466. [CrossRef]
44. Wang, J.-F.; Zhang, T.-L.; Fu, B.-J. A measure of spatial stratified heterogeneity. *Ecol. Indic.* **2016**, *67*, 250–256. [CrossRef]
45. Wang, J.F.; Li, X.H.; Christakos, G.; Liao, Y.L.; Zhang, T.; Gu, X.; Zheng, X.Y. Geographical detectors-based health risk assessment and its application in the neural tube defects study of the Heshun Region, China. *Int. J. Geogr. Inf. Sci.* **2010**, *24*, 107–127. [CrossRef]
46. Song, Y.; Wang, J.; Ge, Y.; Xu, C. An optimal parameters-based geographical detector model enhances geographic characteristics of explanatory variables for spatial heterogeneity analysis: Cases with different types of spatial data. *Giscience Remote Sens.* **2020**, *57*, 593–610. [CrossRef]
47. Peng, J.; Hu, X.; Zhao, M.; Liu, Y.; Tian, L. Research progress on ecosystem service trade-offs: From cognition to decision-making. *Acta Geogr. Sin.* **2017**, *72*, 960–973.
48. Torres, A.V.; Tiwari, C.; Atkinson, S.F. Progress in ecosystem services research: A guide for scholars and practitioners. *Ecosyst. Serv.* **2021**, *49*, 101267. [CrossRef]
49. Teng, Y.; Zhan, J.; Liu, W.; Chu, X.; Zhang, F.; Wang, C.; Wang, L. Spatial heterogeneity of ecosystem services trade-offs among ecosystem service bundles in an alpine mountainous region: A case-study in the Qilian Mountains, Northwest China. *Land Degrad. Dev.* **2022**, *33*, 1846–1861. [CrossRef]
50. Shahab, S.; Hartmann, T.; Jonkman, A. Strategies of municipal land policies: Housing development in Germany, Belgium, and Netherlands. *Eur. Plan. Stud.* **2021**, *29*, 1132–1150. [CrossRef]
51. Hartmann, T.; Jehling, M. From diversity to justice–Unraveling pluralistic rationalities in urban design. *Cities* **2019**, *91*, 58–63. [CrossRef]

land

MDPI

Article

An Interpretation of Landscape Preferences Based on Geographic and Social Media Data to Understand Different Cultural Ecosystem Services

Yuanting Yang and Wei Duan *

School of Landscape Architecture, Beijing Forestry University, Beijing 100083, China; yuanting@bjfu.edu.cn
* Correspondence: duanwei@bjfu.edu.cn

Abstract: A crucial component of ecosystem services (ES) that represents social and humanities values is the cultural ecosystem service (CES), which refers to the non-material advantages that the environment provides for humans. CES are challenging to deeply understand, and little is known about the interactions between CES and landscape variables, particularly in some remote Chinese cities. In order to assess the dominant landscape variables of different CESs from physical, experiential, intellectual and inspirational aspects, this article investigates the landscape variables that may influence the public preferences of various CESs based on social media and geographic data in Anshun, China. The findings are displayed below. The public preferences of various CESs are impacted by the landscape variables in different ways. Physical CESs are influenced by both natural and infrastructure elements, demonstrating that accessibility to restaurants, accommodation, and transit affects how people interact with plays in public. Experiential CESs are primarily influenced by sensory elements, particularly the visual senses, suggesting that when people visit such settings, they place more emphasis on sensory experiences. Intellectual CESs are mostly affected by sensory and natural elements, implying that intellectual CESs with a natural perception are more alluring to tourists. Inspirational CESs are mainly influenced by natural and infrastructure elements, people usually consider nature and convenience when they go to such scenic spots. From the standpoint of promoting people's wellbeing and boosting tourism appeal, the study's results can offer fresh perspectives and content additions for the tourism landscape planning and management in Anshun.

Keywords: social media; landscape preference; cultural ecosystem service; landscape planning

1. Introduction

With the advancement of urbanization and the improvement of people's living standards, society's requirements are gradually shifting from material requirements to non-material requirements, such as the pursuit of spiritual delight and the desire to be close to nature [1,2]. This shift is motivating an increasing number of individuals to travel to see nature up close and cherish the landscape's intangible benefits. Governments and society are therefore paying more attention to the inclusion of cultural ecosystem services (CES), which represent the intangible advantages of ecosystem services (ES), as a reference indicator for ecosystem valuation [3]. To inform the next stage of landscape planning and policy making, it is essential to establish communication between CES and landscape variables in order to collect scientifically sound information.

Culture Ecosystem Services (CES) refer to the non-material benefits that humans derive from ecosystems, providing them with services that include spiritual, aesthetic, educational and recreational dimensions [4]. CES is of great importance to promoting the harmonious development of man and nature and enhancing the well-being of mankind [5–7]. Previous research related to ecosystem services has focused on the ecological aspects of biophysical research and economic valuation [8,9]. Little is known about CES and there has been a failure to develop a unified system for understanding CES [10,11]. Moreover, the effective integration of CES into practical landscape planning and policy development to enhance the value of the non-material aspects of landscape services has been hampered by lack of information on how landscape variables affect CES, especially some sensory categories [12]. The quantification of landscape variables is often used as an important way to study the spatial distribution of CESs [13]. It has been repeatedly demonstrated in past studies that natural and infrastructural elements are fundamental landscape variables that primarily influence the spatial and temporal distribution of CESs [14]. However, other sensory experiences of sound, odour and landscape perception have not been included in many studies, possibly due to difficulties in obtaining and quantifying data, which should be addressed in future research.

CES is intangible, subjective and difficult to quantify; most studies on CES are mainly based on indirect evaluation methods such as qualitative description, monetised value assessment and non-monetised quantitative assessment [15]. Firstly, CES was described qualitatively using participatory mapping methods [3], smartphone location data [16], social media image data [17], and by asking for expertise or opinions [18]. Qualitative description methods are more detailed and incorporate the actual needs of different populations, but the credibility of the results may be questionable. Secondly, the monetised value of the CES is assessed by evaluating the economic value [19–21]. Monetary value data is widely available and easily quantifiable, but the method makes it difficult to capture the value of CES in terms of social relations, sense of place, access to inspiration, etc., through economic or monetary values [22,23]. Finally, the quantitative non-monetary assessment of CES is carried out using interviews [24], questionnaires [25] and indicator systems [26]. However, these assessment methods are often subject to semantic processing and conversion of questionnaire questions, which limit their applicability and lead to many uncertainties in practice [27].

In recent years, social media data have emerged as a new method for understanding CES due to their large data sample and the large range of people it can reach in recent years [17]. The geographic location, image content, text tags and keywords contained in social media data provide a wealth of data on the spatial distribution of human environmental activities for relevant research [17,28]. It resolved the research difficulties of intangibility and subjectivity. Many researchers have also tried to mine image information using deep learning models to indirectly predict the classification of CES from the classification results of image content [29,30]. Nevertheless, there are some drawbacks and controversies with using this approach, such as the complex composition of social users of social media and the fact that people in different regions have different preferences for the use of social platforms [31]. At the same time, a growing number of researchers believe that social media is likely to be increasingly valuable for research and management of nature-based tourism [32]. Therefore, social media data are suitable for this research to investigate how the landscape variables of the tourism landscape in Anshun area affect different CES.

To explore how landscape variables affect different CESs, an indicator that evaluates the results of that effect is also needed—landscape preferences. Landscapes are areas perceived by people, and the central component of landscape value is based on human perception [33]. Landscape preference usually reflects society's perceived preference for a particular landscape and is the result of a combination of perceptual activities, such as the public's emotional perception when confronted with the landscape [32]. Based on the attention restoration theory (ART) [34], different people perceive the same landscape differently, so landscape preferences are influenced by personal interests, differences in social and cultural backgrounds, and educational attainment [35]. Currently, some Chinese social media platforms, such as Ctrip, Tuniu and Where to go, can provide a large amount of data on tourists' reviews and ratings of landscape preferences [33], including their comments on the features, advantages and disadvantages of the attractions, as well as their subjective feelings (e.g., service quality, comfort, environmental quality, etc.), which can provide valuable data for assessing the public's landscape preferences. It can be a promising idea to understand CES by examining the public preferences using social media [36] and the landscape variables that influence them to each CES [37].

In this article, the tourism landscape of Anshun City is used as the research object, and this article primarily uses geographic and social media data to obtain the basic data. It classifies the landscape into different CESs based on comments made on social media, conducts regression analysis on the landscape variables (nature, infrastructure, sensory) that may influence the public preferences of CESs. Based on this, corresponding landscape planning recommendations are made for each type of CESs (physical, experiential, intellectual, inspirational) to guide landscape practices, strengthen the cultural service provision of ecosystems, enhance local tourism attractiveness and enhance people's well-being.

The aim of this study is to assess the dominant landscape variables of different CESs from physical, experiential, intellectual and inspirational aspects. This article focuses on the following two issues: (i) What are the main landscape variables that influence the public's landscape preferences in Anshun? (ii) Are there differences in the main landscape variables affecting the landscape preferences of different CESs? If differences exist, what are the main landscape variables that influence the landscape preferences of different CESs?

It is possible to gain an understanding of CES and contribute to incorporating CES as an important indicator in landscape planning and policy making to better meet the public's expectations by studying their preferences and the significant landscape variables, thereby exploring the various non-material benefits that different CESs provide to the public. The study is of great significance to the development of tourism in remote areas of China, which helps to clarify the future development direction of local tourist landscapes. The results and data of the study can help urban planners and managers to carry out landscape planning and practical policies in the Anshun area, in order to effectively improve tourists' satisfaction, and the well-being of local residents.

2. Materials and Methods

2.1. Study Area

Anshun, located in the mid-western part of Guizhou Province in southwestern China, is situated between 105°13′~106°34′ E and 25°21′~26°38′ N (Figure 1). It is an important watershed area between the Wujiang River Basin of the Yangtze River system and the Beipanjiang River Basin of the Pearl River system, with a total area of 9267 square kilometres and a total population of 3 million people. Anshun is part of the Qianzhong hill plain basin in the western plateau mountain region of Guizhou. The topography of the city is complex and varied, mainly dominated by karst landforms, accounting for 77.82%, which is a typical karst landform concentration distribution area in the world. Anshun belongs to the western part of the subtropical humid monsoon climate zone, there are often valley winds and summer winds in the territory, and there are many clouds and rain, cool, humid, low solar radiation, and the air quality is excellent at 99.8% all year round. The city's tourism resources account for 12% of the city's total area, much higher than the national average of 1% and Guizhou Province's 4.2% and is the earliest identified Class A tourism open city in China. Anshun is also a famous historical and cultural city in Guizhou, with unique historical and cultural heritage such as Tunbao, pierced cave and Yelang cultures. In recent years, to cope with the increased development of tourism, Anshun has made great efforts to improve infrastructure construction, with the Shanghai-Kunming Expressway running across the east and west, the Guiyang-Kunming Railway crossing the whole territory, and the Shanghai-Kunming High-speed Railway already having several stations in Anshun.

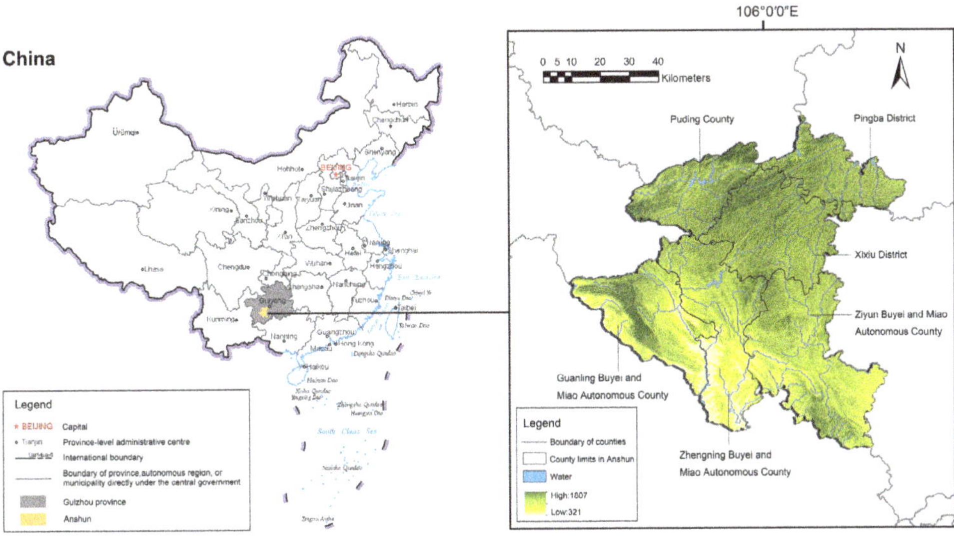

Figure 1. Study area.

Anshun has good natural conditions, a variety of beautiful landscape types, a rich historical and cultural heritage, and its infrastructure that is gradually being improved. However, because of the area's relatively slow economic development, the local investment in tourism development is sporadic and ungoverned, the attractions' attractiveness is insufficient, the region's tourism resources have not been used to their full potential, and the tourism industry has not developed as expected. Due to the wide-ranging practical implications for Anshun, this study investigates the cultural ecological services provided by the local tourism landscape in order to guide planning and actual policy formulation for the tourism landscape in the region, improve the cultural service supply of the local landscape, increase the local tourism attractiveness, and enhance the positive growth of the local tourism economy.

2.2. Research Frame

We investigate the dominant landscape variables that affect the landscape preferences of various CESs in Anshun using data from social media and geographic information. Based on the findings of the data analysis, we gain understanding of the landscape service values of various CESs (Figures 2 and 3).

The research methodology's specifics are as follows:

1. Crawl social media data on all scenic spots in Anshun from Ctrip (https://you.ctrip.com/place/anshun518.html (accessed on February 2023), excluding those with less than 10 reviews), including: names of scenic spots, tourist comments, and tourist ratings, excluding invalid data and integrating valid data.
2. Identify the text of tourist comments and classify the subword.
3. Analyse the word frequency of subword.
4. Based on the word frequency, experts were consulted to categorize all scenic spots into four different CESs (physical, experiential, intellectual, and inspirational) in accordance with the International Classification of Ecosystem Services (CICES).
5. Obtain the DEM and geographic information data on land cover, landform types, road networks, hydrology and POI, etc.
6. Process the geographic data in ArcGIS (the spatial resolution of these data is 12.5 m).
7. Determine the landscape variables from the natural, infrastructural and sensory perspectives.

8. Integrate the data obtained above on tourist ratings and landscape variables, random forest regression was conducted to obtain the ranking of importance of landscape variables affecting public preference for all CESs.
9. The same random forest regression was used to regress the data for different CESs separately to obtain the ranking of importance of landscape variables affecting public preference for different CESs.

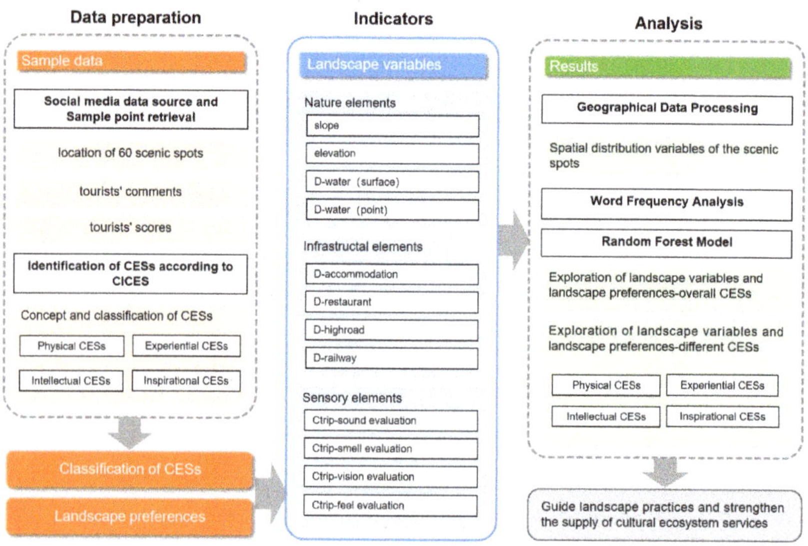

Figure 2. Research methods.

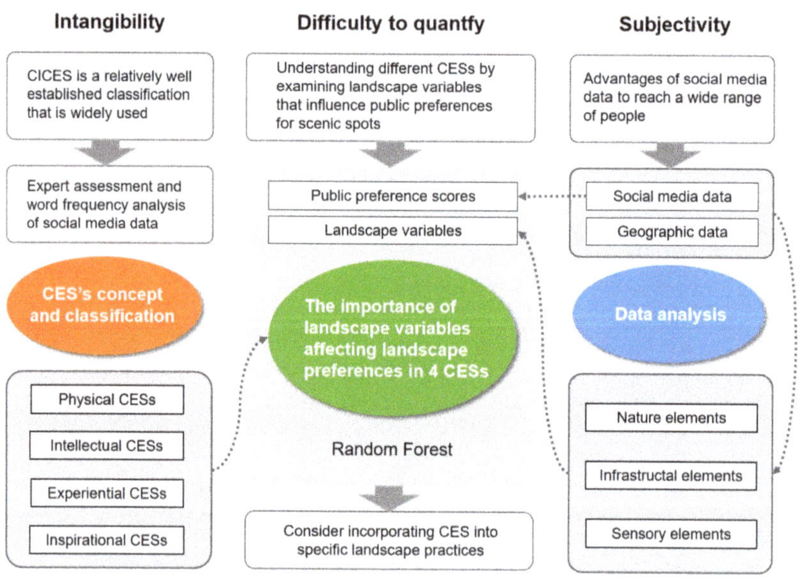

Figure 3. The treatment of the research difficulties in CES.

Specifically, the study period lasts from June 2022 to July 2023, and the date of crawling data is February 2023.

2.3. Social Media Data Source

2.3.1. Sample Point Retrieval

Ctrip is an important travel service app in China, offering a variety of services including tickets, accommodation and transportation, and its market size is one of the largest in China, and it has a sizable following. The sample data obtained from Ctrip is more representative due to its efficient services, authentic data and diverse products [38]. Ctrip includes basic information on most tourism products and their usage and evaluation data, including tourist ratings, reviews, number of comments, peak seasons and ticket prices for each attraction. These data can objectively reflect tourist preferences and behaviour [12], making it a reliable source of data for researching tourist attractions in China.

This study mainly used the Internet data collection software "Octopus Collector" (https://www.bazhuayu.com/) to obtain basic data from the Ctrip website. The sampling point retrieval process consisted of the following steps: cleaning the data and coordinating the conversion of the collected data; capturing the names of scenic spots, tourist comments, and tourist ratings in Anshun (Figure 4), excluding those with less than 10 reviews. Random and automatically generated comments have also been removed to ensure that the remaining comments are representative. Finally, sixty representative scenic spots were selected from all the attractions in Anshun City to be included in the analysis (Figure 5).

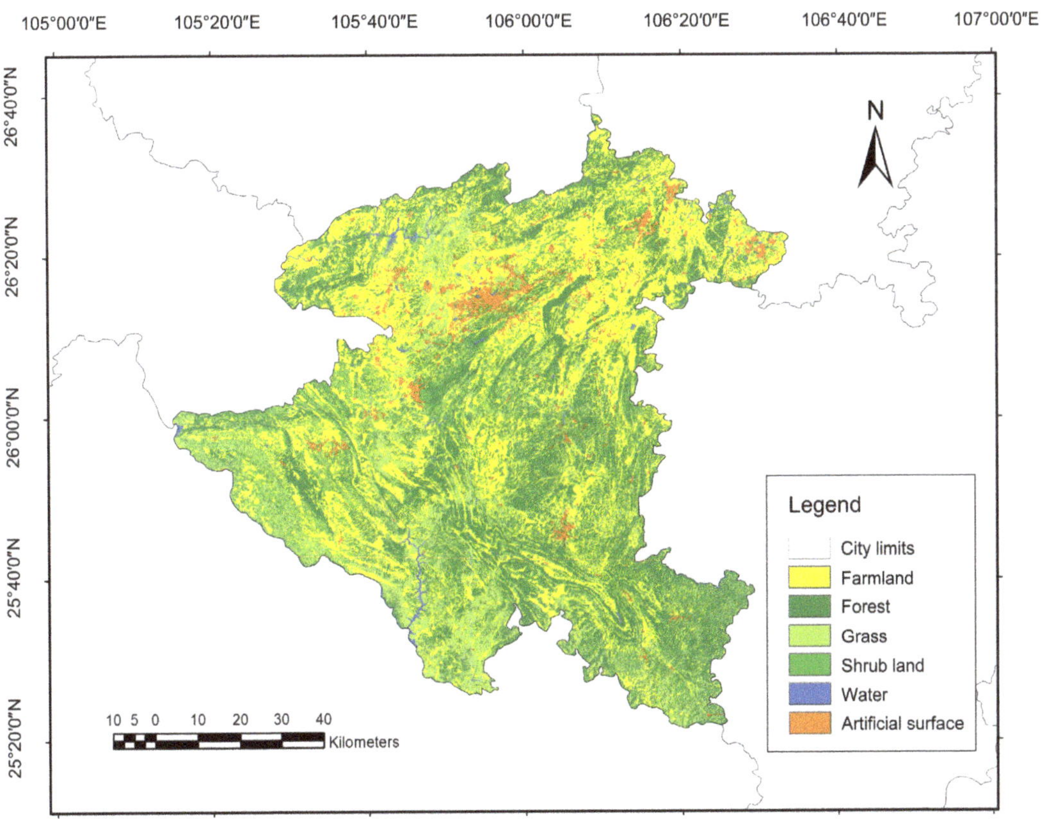

Figure 4. Land use data.

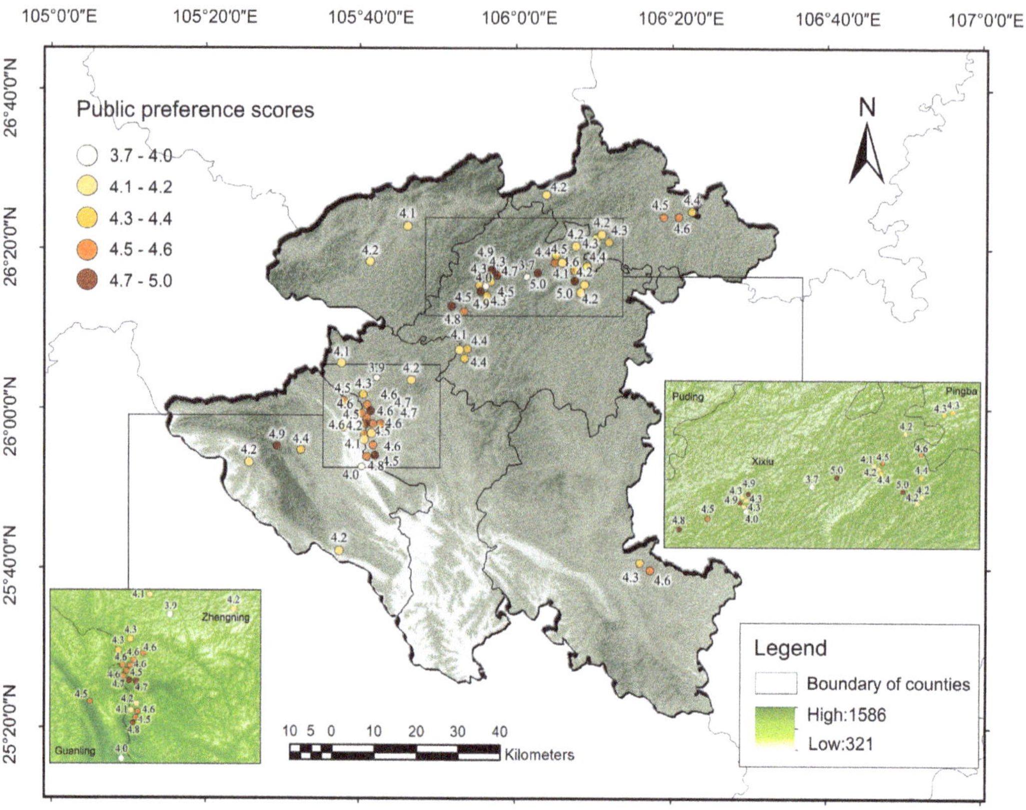

Figure 5. Public preference scores.

Among them, tourist's comments can reflect the subjective feelings of tourists, which are the source of data on the characteristics of the sensory category landscape and the main basis for the CES classification. Tourist's ratings represent the public's preference for the attraction, and tourist's ratings of the scenic spots on Ctrip are the comprehensive assessment result made by a large number of tourists who actually arrive and visit the attraction, and this score result can reflect to a certain extent the public's actual preference for the landscape.

2.3.2. Word Frequency Analysis

Using the word separation and classification search platform in the web data crawling tool "Jisouke", the comments of tourists on Ctrip were divided into phrases, and the key words of the comments were extracted through various filtering conditions such as word nature, word frequency and filtering, and repetitive words such as onomatopoeia were excluded to determine the effective word separation of the comments. In this study, the extracted words were divided into four categories (biological and natural landscape elements, cultural landscape elements, perceptual elements and human elements) and 20 sub-categories (Table 1).

Table 1. Different types of comment splitting.

Category	Sub-Category	Word Examples
Biological and natural landscape elements	Plant type	Bonsai, fruit trees, rape, cherry blossoms, petals, cherries, vegetation
	Land type	waterfalls, waterholes, caves, countryside, farms, gardens, stone forests, rivers
	Animal type	monkeys, birds
Cultural landscape elements	Building	architecture, ancient buildings, villages, villages
	Restaurant	buffet, dinner, lunch
	Homestay	hotel
	Tickets	buy tickets
	Price	cheap, expensive, cost-effective
	Entertainment	take pictures, perform, dance
	Distance	far, near
Perception elements	Sound	sound, loud
	Smell	air, fresh, refreshing
	Feel	shocking, comfortable, happy
	Vision	Spectacular, good-looking, clean and pleasant
	Weather	cool, cold
Sense of place	Traffic	walk, car, boat
	Service	commercial streets, tourists
	Mood	Pleased, comfortable
	Time	morning, evening
	People	tour guides, tourists, friends, children

The results of the word frequency analysis are an important basis for experts to classify CES and assess landscape variables.

2.4. Classification of CESs

CES provide social values to humans indirectly through subjective human intentions and feelings, such as aesthetics, spiritual healing, research and education, etc., and can be an important representation of the interaction between ecosystem services and human well-being. There are many proposals for the classification of CES in the academic community, including the Millennium Ecosystem Assessment [4], the Economics of Ecosystems and Biodiversity [39], the International Common Classification of Ecosystem Services (CICES) used by the EU initiative [39,40], the Nature's Contribution to People system used by the Intergovernmental Panel on Biodiversity and Ecosystem Services (IPBES) [41], and the classification system for final ecosystem goods and services (FEGS) proposed by the United States Environmental Protection Agency (USEPA) [42], etc. All of these classifications are intended to be general in nature, but they all derive from a specific context. Of these, CICES is widely used in the classification of ecosystem services (ES), particularly CES, which provides a relatively high level of detail in a nested hierarchy of 'taxonomic levels,' providing an appropriate structure for the assessment of ES [43], so CICES was chosen as the criterion for the CES classification in this study. According to the CICES definition of the CES classification, all CES were classified into four categories: physical, experiential, intellectual, and inspirational [44].

Previous research has found that the terms associated with landscape variables are similar within the same landscape type, so the public's comments can effectively distinguish between landscape types [24]. Thus, three professionals were invited to discriminate the lexical meaning of the 11,816 sub-words originating from the comments of the 60 scenic spots according to the CICES definition of CES classification, where sub-words related to CES were evaluated twice to represent different CES characteristics (Table 2), and finally the results of the word frequency analysis (Table 1) were combined to classify all attractions into four categories: physical, experiential, intellectual, and inspirational [45].

Table 2. Classification of CES.

CES Types Defined [44]	CES Category (Based on CICES)	Definition	Examples for Classification
Physical	Recreation	Resources provided for recreational activities in the ecosystem (its biological and non-biological elements).	Comments containing attractions for recreational activities (e.g., mountain climbing skiing and rafting).
Experiential	Aesthetics	Feelings provided by the aesthetic characteristic of natural and semi-natural landscapes and their biological and non-biological elements	Comments containing descriptions of the landscape and beauty (e.g., mountain and river).
Intellectual	Scientific and educational	Research or educational activities conducted through the natural environment of the ecosystem and its biological and non-biological components.	Comments containing attractions for educational training or research activities, (e.g., educational bases).
	Cultural heritage and identity	Value of the landscape, species or location to the local heritage and cultural heritage.	Comments containing cultural heritage or intangible cultural heritage (e.g., traditional buildings, local culture, cultural landscape and traditional practices).
Inspirational	Spiritual and religious	Landscapes, ecosystems and their elements that have religious or spiritual purposes.	Comments containing temples and religious attractions (e.g., churches, burning incense and worshiping buddha).
	Inspiration	Landscapes, ecosystems and their elements used in art architecture, advertising, local symbols, and folklore.	Comments containing attractions with art publicity and local symbols (e.g., art gallery and music)

To test the reliability of the word frequency classification procedure, three professionals from different professional backgrounds were invited separately and the professionals were asked to participate independently in the discrimination.

2.5. Landscape Variables

A set of key landscape variables mentioned in the literature that may influence the public's perception of the landscape were collected in order to later investigate how they affect the landscape preferences of CESs. The main landscape variables include three main categories: natural elements, infrastructural elements, and sensory elements (Table 3).

Table 3. Landscape variables.

Dimension	Code	Meaning
Natural elements	Slope	Slope of the scenic spots
	Elevation	Elevation of the scenic spots
	D-water (surface)	Distance to the nearest water bodies
	D-water (point)	Distance to the nearest water points
Infrastructural elements	D-highroad	Distance to the nearest highroad
	D-railway	Distance to the nearest railway
	D-accommodation	Distance to the nearest accommodation
	D-restaurant	Distance to the nearest restaurant
Sensory elements	Ctrip-sound	Sound evaluation word frequency in Ctrip
	Ctrip-smell	Smell evaluation word frequency in Ctrip
	Ctrip-vision	Vision evaluation word frequency in Ctrip
	Ctrip-feel	Touch or feel evaluation word frequency in Ctrip

(1) Natural elements

Anshun is located in the karst landscape region of southwest China, and it has been studied that the soil properties differ at different altitudes of the karst landscape region, with the increase in soil nutrients at lower altitudes being greater than at higher altitudes [46], soil nutrients directly influence vegetation recovery, so vegetation richness tends to be higher at lower elevations. At the same time, the slope has a greater influence on the redistribution of rainfall in the soil [47,48]. The lower the slope, the slower the soil loss, while the opposite will result in faster soil loss, degradation of vegetation, increased soil erosion, increased rock exposure and rock desertification. Therefore, both elevation and slope can be factors that contribute to different landscape perceptions in karst landscapes. In addition, water is one of the most important and attractive visual elements in a landscape and has for long been important to human perceptions of landscape quality and the quality of many outdoor recreational experiences [49].

To summarise, the most frequently selected natural element indicators were: elevation (elevation), slope (slope), distance to water surface (D-water (surface)) and distance to water system point (D-water (point)), which refer to the distance between the attraction and the nearest water surface and water system point, respectively. The first two were obtained through DEM data analysis of Anshun, the latter two were obtained in ArcGIS 10.8 using Euclidean distance and nearest neighbour distance analysis.

(2) Infrastructural elements

In terms of infrastructural elements, the distance of the attraction from the nearest highroad, railway, accommodation, restaurant was measured in ArcGIS 10.8 using Euclidean distance and nearest neighbour analysis, denoted as: distance to highroad (D-highroad), distance to railway (D-railway), distance to accommodation (D-accommodation), distance to restaurant (D-restaurant). It is worth mentioning that roads, railways and other infrastructures will give visitors easy access [50]; however, landscapes too close to roads can also bring greater noise to recreational activities and landscapes too close to roads can also bring louder noise to recreational activities and affect people's perception of the landscape.

(3) Sensory elements

Sensory elements at the emotional level guide people's perception of the landscape by triggering a sense of familiarity, attachment, connection and other emotions in the perceiving subject and are important factors influencing landscape perception [51]. In contrast to most of the above indicators, which can be directly quantified to describe socially physical characteristics, some sensory indicators, such as olfactory and tactile elements, are difficult to quantify and have rarely been considered in previous studies. In this study, the content of attraction reviews from social media data was used as a data source, which was judged by experts and relevant word frequencies were calculated to represent the corresponding sensory element: Ctrip-sound, Ctrip-smell, Ctrip-vision and Ctrip-feel.

2.6. Statistical Analysis

Random Forest (RF) was used to investigate the correlation between landscape preferences and landscape variables of different CESs. RF is a machine learning algorithm with a strong generalization ability [52], it will take a random sample of the original data set to form a number of different sample data sets, then build a number of different decision tree models based on these data sets, and finally, based on the average of these decision tree models or voting to obtain the final analysis results [53]. RF has many advantages, such as no need to preprocess the data, convenient and fast processing, and stable results, thus RF is a good fit for assessing the importance of each landscape characteristic.

The public rating data of 60 major scenic spots in Anshun and their landscape variables were composed into a sample data set, and 60 different decision tree models were built based on these datasets for random forest regression calculations. The public landscape preference scores are the output variables and the 12 landscape variables are the input variables.

The calculation formula is as follows,

$$V(Y_i) = \sum_{n=1}^{j} p(Y_i = X_j)(1 - p(Y_i = X_j)) = 1 - \sum_{n=1}^{j} p(Y_i = X_j)^2 \quad (1)$$

where $V(Y_i)$ denotes the public landscape preference score for Y_i, $p(Y_i = X_j)$ denotes the probability of the prediction set. There are two important parameters to optimise in the model: the number of spanning trees (N_{tree}) and the number of randomly selected variables at each node (M_{try}) [54].

During data analysis, the data were constantly re-ordered and re-segmented, and the training percentage was set to 0.7 for multiple training sessions to improve the accuracy of the training results.

The input variable importance in the model was ranked. It is defined as the cumulative contribution of the influence factor to the branch of the decision tree during the learning process. The larger the value, the more important is the variable's influence on the public preference. The contribution rate of each variable to the fitting accuracy was defined as the relative importance, with a sum of 1.

3. Results

3.1. Spatial Distribution Characteristics of Scenic Spots

Among all the scenic spots, most of the scenic spots in Anshun are located in the higher terrain in the north, where are Xixiu district, Zhenning Buyi and Miao Autonomous county and Guanling Buyi and Miao Autonomous county (Figure 6a).

According to the results of expert judgements and word frequency analysis, the 60 scenic spots in Anshun can be divided into four categories: physical CESs with eight scenic spots, experiential CESs with 25 scenic spots, intellectual CESs with 20 scenic spots, and inspirational CESs with seven scenic spots. The physical CESs are concentrated in the northern part of Xixiu District, with a few in Zhenning, Guanling and Ziyun Buyi Miao Autonomous Counties (Figure 6b). Experiential CESs are concentrated at the junction of Zhenning and Guanling Buyi Miao Autonomous Counties (Figure 6c). Intellectual CESs are concentrated in the northern part of Xixiu District, with a small distribution in Zhenning, Guanling and Buyi Miao Autonomous Counties (Figure 6d). Inspirational CESs are distributed at the junction of Zhenning and Guanling Buyi Miao Autonomous Counties, with only isolated distributions in northern Xixiu District and Ziyun Buyi Miao Autonomous County (Figure 6e).

The distribution of scenic spots clearly corresponds to the distribution of the highway network (Figure 7a). In the north of Anshun, the Hu Kun Expressway runs from northeast to northwest through the northern part of the city, and along the perimeter of the motorway are concentrated many of Anshun's well-known attractions. In contrast, the Ziyun Buyi Autonomous County, located in the southeast corner of Anshun, has fewer highways and is less accessible by car and less developed in terms of attractions. Compared to the highways, the railway network in Anshun is more evenly distributed, with the nearest railways in Anshun being relatively close to each other and less correlated with the distribution of scenic spots (Figure 7b).

As a tourist destination rich in natural landscape resources, water features have always been an important factor in attracting tourists to travel to Anshun, which is rich in water resources and has a relatively short distance from each scenic spot to the water (Figure 7c). Water resources are mainly concentrated in Zhenning Buyi Miao Autonomous County (Figure 7d), which has numerous scenic spots highlighted by water features and also concentrates on many of Anshun's famous scenic spots popular with tourists, such as Huangguoshu Waterfall, Steeple Pond Waterfall and Silver Chain Falling Pool Waterfall.

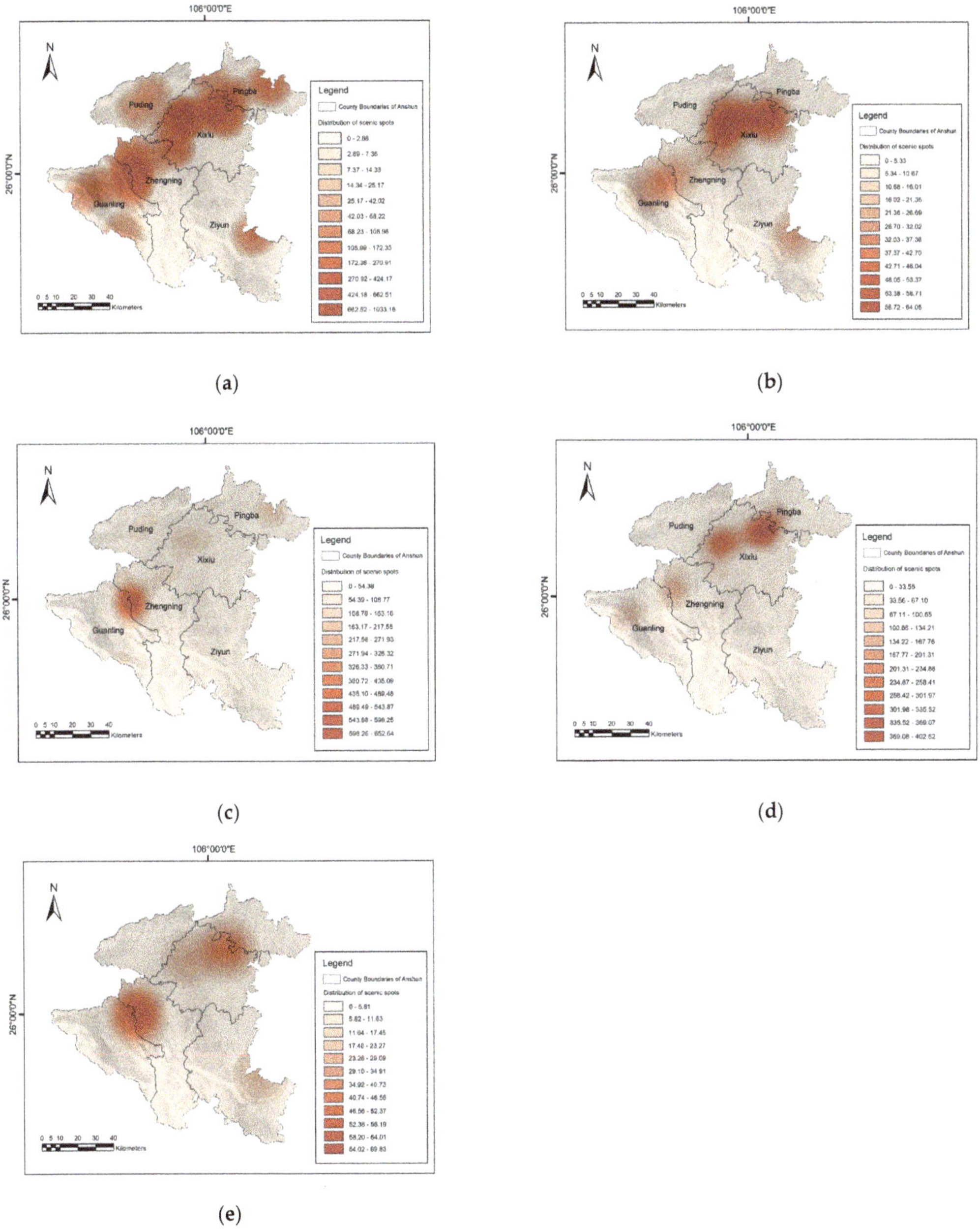

Figure 6. Distribution of scenic spots: (**a**) Distribution of scenic spots-overall CESs, in the legend the low value indicate 0 and max value is equal at 1033.16; (**b**) Distribution of scenic spots-Physical CESs, in the legend the low value indicate 0 and max value is equal at 64.05; (**c**) Distribution of scenic spots-Experiential CESs, in the legend the low value indicate 0 and max value is equal at 652.64; (**d**) Distribution of scenic spots-Intellectual CESs, in the legend the low value indicate 0 and max value is equal at 402.62; (**e**) Distribution of scenic spots-Inspirational CESs, in the legend the low value indicate 0 and max value is equal at 69.83.

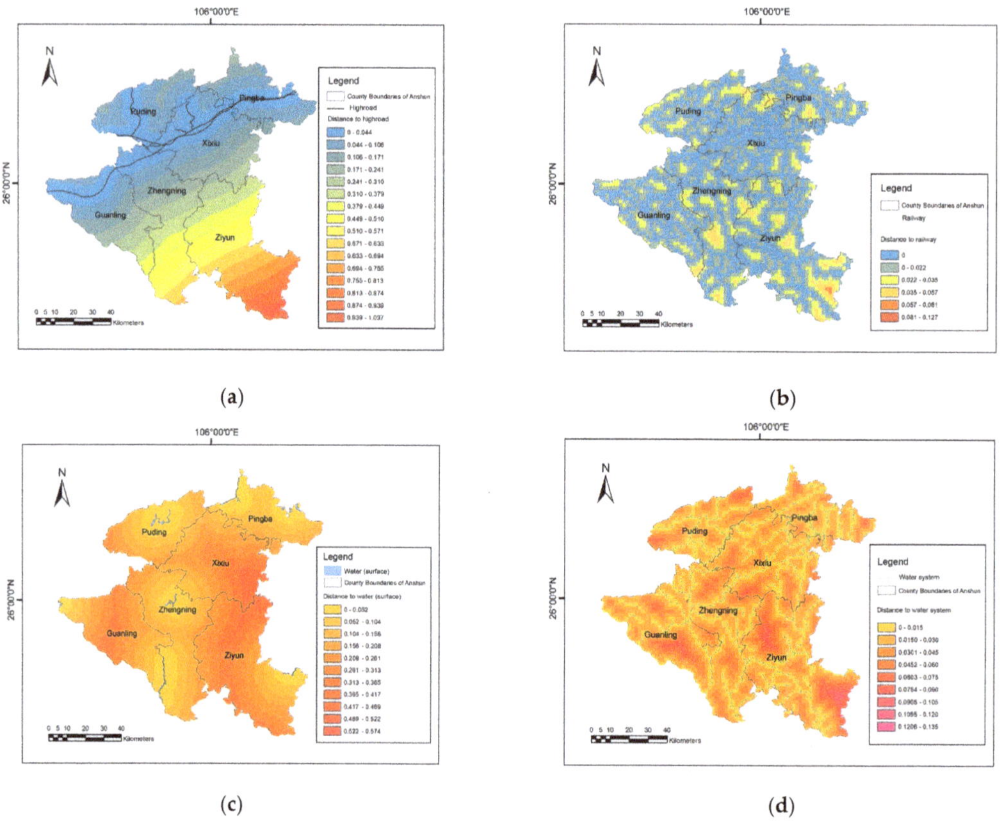

Figure 7. Spatial landscape characteristics: (**a**) Distance to the highroad; (**b**) Distance to the railway; (**c**) Distance to the water surface; (**d**) Distance to the water system.

3.2. Exploration of Landscape Variables and Landscape Preference

Twelve landscape character indicators were used as inputs to the Random forest to predict public landscape preferences for different CESs. Figures 8 and 9 illustrate the proportion of importance of each feature.

According to the predicted results of the overall CESs (Figure 8), it can be seen that natural elements, infrastructural elements and sensory elements all have different degrees of importance on the overall landscape preference, especially the natural elements have a greater influence on the overall landscape preference, with slope, elevation and D-water (surface) being the top three landscape variables affecting the overall landscape preference of Anshun, from which can be seen that natural scenery, such as rich topography and water features, are the main attraction for tourists to come to Anshun. For infrastructural elements, tourists are more concerned with practical needs such as D-restaurant and D-highroad. For sensory elements, there is a significant correlation between Ctrip-feel and Ctrip-vision on landscape preferences, which shows that public preferences are more influenced by sensory and visual sensory factors.

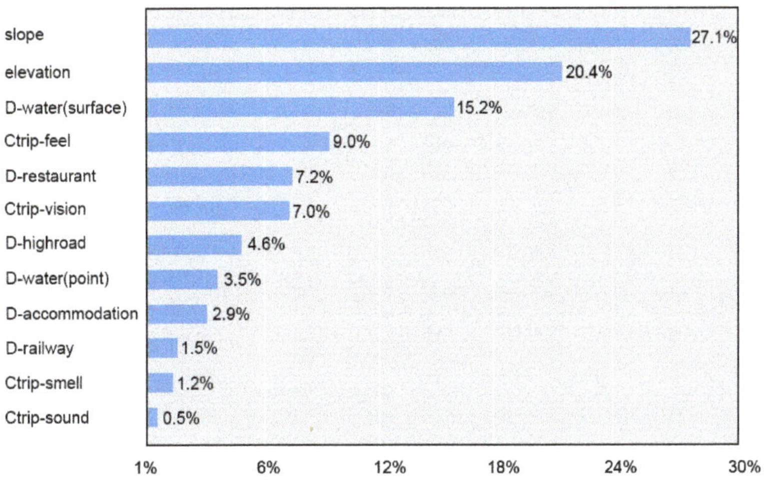

Figure 8. The relative importance of the indicators of the overall CESs.

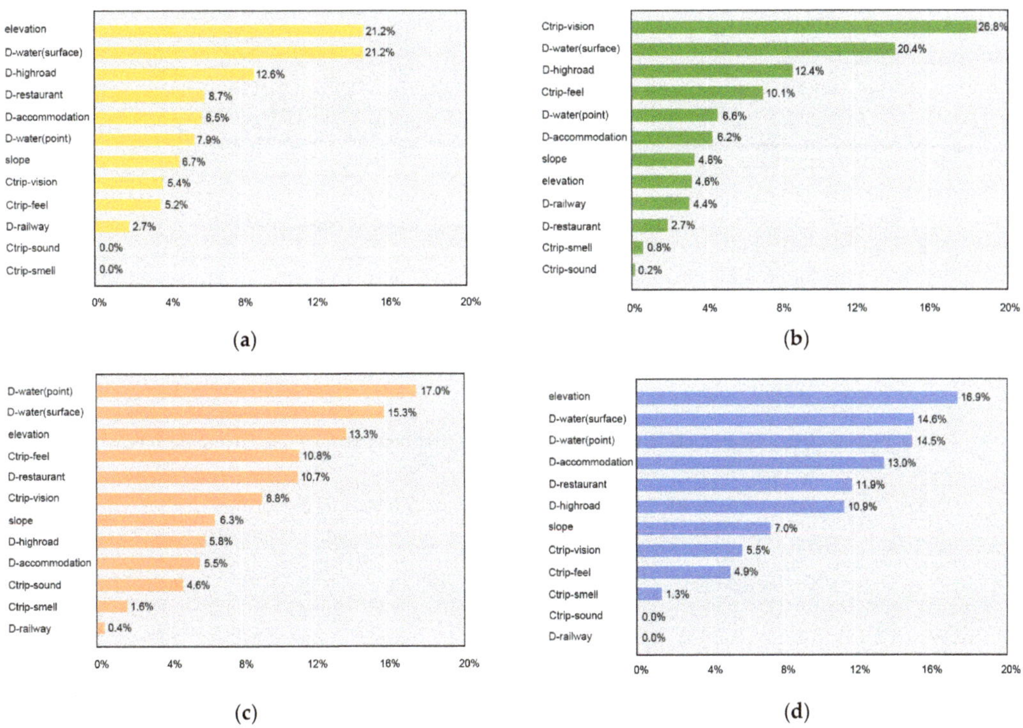

Figure 9. The relative importance of the indicators: (**a**) Physical CESs; (**b**) Experiential CESs; (**c**) Intellectual CESs; (**d**) Inspirational CESs.

The predictions were different for the different CESs (Figure 9):

For physical CESs, the top three factors influencing their landscape preferences are elevation, D-water (surface), and D-highroad. Physical CESs focus on the resources (both biotic and abiotic elements) that the ecosystem provides for recreation and focus on the variety of recreational activities that people will engage in such landscapes. The availability of natural conditions, such as good topography and water, is certainly popular with the public. At the same time, convenient infrastructure conditions such as transport, restaurants and accommodation are also the important factors for people to consider physical CESs for their excursions.

For experiential CESs, landscape preferences are mainly influenced by sensory elements, especially visual senses, with Ctrip-vision being the most significant factor influencing landscape preferences for this type of scenic spots. Experiential CESs focus on the aesthetic features of natural and semi-natural landscapes and the perceptions provided by their biotic and abiotic elements; therefore this result is not difficult to understand as these landscapes are more focused on bringing people a perceptual experience and visuals are the most direct sensory source of perceiving aesthetic features. In addition, the landscape preference of experiential CESs is also influenced by factors such as D-water (surface) and D-highroad.

For intellectual CESs, landscape preferences are mainly influenced by natural and sensory elements, with the main influencing indicators being D-water (point), D-water (surface), elevation, and Ctrip-feel. The specific content of intellectual CESs is divided into two aspects: science education and cultural heritage. These CESs focus on the study of educational activities through the natural environment of living and non-living factors in the ecosystem, and sites with landscape heritage and cultural heritage values usually belong to intellectual CESs. According to the analysis results, it is clear that natural ecological conditions are the basic conditions for conducting educational activities, on the basis of which people make good sensory perceptions of culturally valuable landscapes in order to obtain a better educational experience from them.

For inspirational CESs, landscape preferences are mainly influenced by elements of the natural and infrastructure, with the main influencing indicators being, in order, elevation, D-water (surface), D-water (point), D-accommodation, and D-restaurant. Inspirational CESs refer to landscapes, ecosystems and their elements that have religious or spiritual symbols or are used in art, architecture, advertising, local symbols and folklore. Such scenic spots are usually in good ecological base conditions, for example, Stone Cottage, Gaolaozhuang Scenic Area and Slippery Rock Wharf Scenic Area are built by water and have a quiet environment where people can enjoy the spiritual inspiration brought by the natural landscape, while the convenience of accommodation and restaurants can also influence visitors' preference for such scenic spots.

In the RF model, N_{tree} is set to the default value of 100, M_{try} is set to the square root of the number of input variable and a maximum depth of 10. Under these conditions, the model does a good job of predicting the landscape variables that influence public preferences across CESs, and the actual and estimated scores of public landscape preference show better agreement (Figure 10).

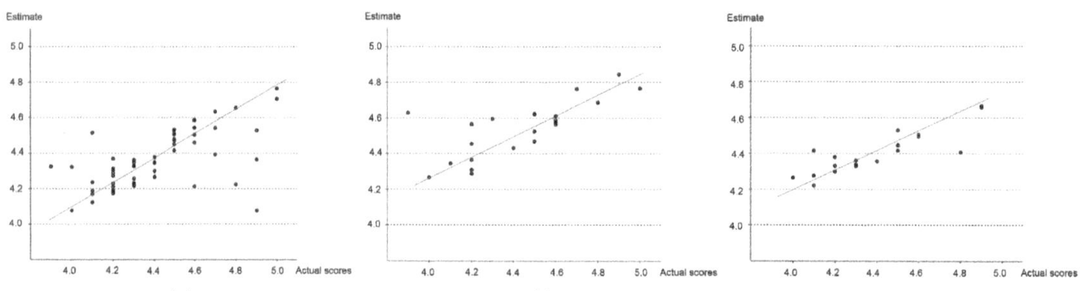

Figure 10. Scatter plot of predicted and true values: (**a**) Overall CESs; (**b**) Experiential CESs; (**c**) Intellectual CESs.

Table 4 shows the performance of the five models, where the smaller the values of MSE (mean square error), RMSE (root mean square error), MAE (mean absolute error), and MAPE (mean absolute percentage error) are, the more accurate the model is. R^2, the coefficient of determination, and the closer the result is to 1 the more accurate the model is. Since the R^2 for the test set was calculated using a nonlinear equation fit, there are some negative values that are not strictly R^2, and the R^2 is not informative.

Table 4. Performance of the RF Model.

Data Set		MSE	RMSE	MAE	MAPE	R^2
Overall CESs	Train set	0.009	0.096	0.072	1.641	0.859
	Test set	0.124	0.352	0.276	6.411	−0.283
Physical CESs	Train set	0.001	0.031	0.022	0.502	0.759
	Test set	0.234	0.484	0.413	9.644	−2.899
Experiential CESs	Train set	0.012	0.111	0.08	1.79	0.823
	Test set	0.129	0.359	0.304	6.669	−2.259
Intellectual CESs	Train set	0.021	0.145	0.124	2.811	0.721
	Test set	0.163	0.404	0.298	6.685	−0.38
Inspirational CESs	Train set	0.005	0.074	0.065	1.463	0.844
	Test set	0.075	0.275	0.218	4.865	−0.885

4. Discussion

4.1. Spatial Distribution Variables of the Scenic Spots

Most of the scenic spots in Anshun are located in the higher terrain in the north, which may be related to a number of economic and social reasons such as better natural scenery, better infrastructure, more concentrated population distribution, greater resource development and favourable policies in the north of Anshun. Xixiu District is the main urban area of Anshun City, with a more developed economy and a more concentrated population. It is the political, economic and cultural centre of the city and is rich in tourism resources, with three 4A-level tourist attractions, as well as a number of national and provincial key cultural heritage protection units. The resource in Xixiu District focus on entertainment and education, so the scenic spots in Xixiu District are mostly Physical CESs and Intellectual CESs. Zhenning buyi Miao Autonomous County has obvious karst geomorphological features, with a variety of caves, underground rivers, waterfalls, lakes and springs, and rich geothermal resources, making it "a karst kingdom". Due to the outstanding natural scenery of Zhenning County, known as "Silver City" and "Waterfall Township", the county's scenic spots are mostly based on natural sightseeing and inspiration, so the scenic spots mostly belong to Experiential CESs and Inspirational CESs. Meanwhile, the county's scenic spots in Guanling Buyi Miao Autonomous County are more Physical, Intellectual CESs, and a few Experiential, Intellectual CESs. The above three counties are rich in tourism resources, each with its own characteristics, and due to the early good development, has now developed into the city of Anshun tourism card. Moreover, the remaining counties of Puding, Pingba and Ziyun are not yet well developed because of the relative isolation of the traffic and the general natural and humanistic conditions, and even if there are a few scenic spots, they are scattered all over the place, and there has not been any centralised tourism development yet. The uneven distribution of scenic spots in Anshun is certainly affected by many factors such as nature, society and economy, but it is not conducive to the long-term development of Anshun. Therefore, the natural resources and human resources in the southern part of Anshun should be given enough attention by planners and managers, so that they can become the rising star of tourism development in Anshun and promote the balanced development of tourism economy.

The development of the tourism economy in the Anshun region can be approached in two ways. On the one hand, for the transport infrastructure, the distribution of scenic spots in Anshun is highly compatible with the distance from the scenic spots to the highway, which means that the distribution of major scenic spots in Anshun is clearly influenced by road traffic, and the road network can effectively drive the development of the tourism landscape along the route. Nowadays, with the gradual improvement of the road network in Anshun, self-drive tours have replaced the previously common train trips as the first choice for tourists travelling within Guizhou Province, which explains why the distribution of scenic spots is significantly correlated with their distance to the highway. Therefore, to develop the tourism economy in the south of Anshun, it is advisable to use the development of the highway network as an entry point to strengthen the construction of transport infrastructure in order to drive the development of tourism resources. On the other hand, for the natural landscape resources, tourists' preference for the landscape of Anshun is also closely related to the indicator of distance from scenic spots to water. Anshun benefits from its unique natural scenery of mountains and water, with numerous natural wonders within its borders, such as waterfalls and caves. These natural wonders are often made famous thanks to the good local water resources, which are just right to satisfy the pursuit of spiritual enjoyment and the desire to be close to nature for people who have lived in the city for a long time, so Anshun is increasingly becoming a tourist destination for urban tourists on short-term trips.

4.2. Landscape Variables That Influence Landscape Preferences of Different CESs

In the light of the rising standard of living, the strong demand for tourism and the increasing demand for non-material aspects, planners and managers of tourist attractions in Anshun should seize this opportunity in a timely manner, develop tourism resources in the territory in a scientific, rational and equitable manner, accelerate the improvement of related service infrastructure, incorporate the enhancement of cultural ecosystem service values into landscape planning and policy formulation, enhance the non-material aspects of tourists' landscape perceptions, and cater to the current expectations of tourists for attraction planning in order to increase the attractiveness of tourism in Anshun, enhance the well-being of local people and promote the good development of Anshun's tourism economy.

For all types of scenic spots, the most significant factors influencing their landscape preference were natural elements, with slope, elevation and D-water (surface) having a greater impact on overall landscape preference in Anshun. However, the results of the random forest regressions differed for different CESs of scenic spots.

Effectively enhancing the non-material aspects of tourists' landscape perceptions through rational landscape planning requires an understanding of the intrinsic correlation between public preferences and landscape variables in different CESs, and their application to concrete practice.

For physical CESs that rely mainly on natural conditions for recreational activities (e.g., mountain climbing, skiing and rafting), distance from infrastructure such as highways and restaurants needs to be considered when planning and formulating policies for their scenic areas in order to improve the accessibility and convenience of the scenic spots, and enabling visitors to more easily engage in rich recreational activities and have a better experience of physical-type cultural services. Obviously, physical CESs are strongly influenced by the accessibility of infrastructure and natural landscape features [13]. Good and convenient infrastructure not only provides the necessary conditions for mountaineering, skiing, rafting, and other related activities, but is also important for people to be able to rest and recover after the activity [55].

For experiential CESs that focus on aesthetic experiences with a focus on natural and semi-natural landscapes, planners and decision-makers should focus on the impact of sensory factors on the visitor experience, especially visual experiences, which can often bring the most intuitive and impactful aesthetic feelings to visitors. Several earlier studies have also demonstrated that people's perceptions of landscapes are mostly shaped by their visual environment. Since visual attention and landscape identification are closely intertwined, visitors' perceptions of landscapes are influenced by their visual perception, which in turn influences their encounters with various visual impacts [56].

For intellectual CESs, where science and research activities are carried out through the natural ecological environment, often rely on local traditional buildings, cultural heritage and cultural landscapes to bring educational perceptions to visitors. The natural landscape combined with physical perception can be used to bring better intellectual cultural services to visitors. Previous studies have shown that both natural and infrastructural elements were essential landscape characteristics for intellectual CESs, which are related to the diversity of science and cultural education [13]. The current study adds that sensory factors also have an influential factor on intellectual CESs.

For inspirational CESs with spiritual elements such as religion, folklore, local cultural symbols and art, which can bring new inspiration to visitors, they are usually located in places with good ecological base conditions. Needless to say, inspiration can come from ecological and natural sources [57]. Moreover, the planning of inspirational CESs also involves focusing on infrastructure, such as accommodation, restaurants and roads, as accessibility is an important factor in attracting visitors to CESs.

In summary, in the future, tourism landscape planning in Anshun should not only consider topographic, hydrological, economic, policy and other basic development conditions, but also try to evaluate and classify the cultural service content, purposefully considering the landscape variables that dominate the landscape preference of the scenic spots, and incorporate them into the indicators that guide the planning. This allows the landscape to supply a higher value of cultural services, tourists gain richer and more comprehensive perception of the landscape, attracting more tourists to Anshun for travel and tourism, and promoting the benign development of the local tourist attractions in Anshun.

4.3. Research Limitations

Based on geographical and social media data, this study investigates the correlation between landscape preferences and landscape variables of different CESs, and then gives reasonable suggestions for landscape planning and policy formulation in Anshun from the perspective of enhancing the non-material aspects of landscape perception. This helps us to gain a deeper understanding of the public's preferences for different CES types of tourist attractions, and to apply this information to the location of tourist attractions, the configuration of infrastructure, and the formulation of superordinate policies in Anshun, so as to effectively enhance tourists' experience of cultural ecosystem services. It is particularly important to note that the findings of this study are only applicable to guide the planning and policy formulation of scenic spots in Anshun due to the different natural ecological conditions, social-cultural background, economic conditions, policy context and major visitor segments in different regions. The article still has some unavoidable problems due to the limitations of insufficient sample size, difficulty in obtaining data, and time constraints of the study, as the research data mainly comes from the Internet. In addition, due to user privacy issues, the social media data used does not include information on social-demographic characteristics, such as user age, gender and education level; therefore, it is not possible to predict potential differences in preferences for landscape features among people from different backgrounds [58,59]. There are also many other influential factors on public preference besides the 12 variables mentioned in the article, such as: tourism resources of scenic spots, landscape richness, plant coverage, etc. In this paper, due to the factors of topic, research methodology, length and so on, it is not possible to consider all the influential factors into the regression model for statistical purposes, which can be supplemented in the future research. Furthermore, due to time restrictions, constrained circumstances, and other objective factors, we were unable to perform field interviews and attractiveness assessments in the local region for this article. By conducting field interviews and evaluating attractions, the researcher may gain a firsthand understanding of locals' perspectives on tourism landscapes and the cultural services offered by the sites. In subsequent related investigations, the variety of research methodologies may be enhanced and improved to increase the study's completeness and scientific validity.

5. Conclusions

The existing tourist attractions in Anshun are mostly located in the northern part of the city where the terrain is high and the highway network is dense, while the tourism resources in the south have not been vigorously developed. The natural landscape, history and humanities of the southern part of Anshun should be given sufficient attention by planners and managers in order to promote the long-term, balanced development of the tourism economy in Anshun. The development of tourism resources in the southern part of Anshun City can start from actively improving the highway network, restaurants, accommodation and other infrastructures, and choosing the more water-rich and potential tourism landscape along the road as the object of development, so as to attract more tourists to visit the area.

In order to make the planning and design of the tourism landscape in Anshun more responsive to the current expectations of tourists, planners can consider the non-material aspects of tourists' needs for spiritual enjoyment and closeness to nature when planning and designing the tourism landscape and enhance the experience of ecosystem cultural services for tourists. Managers can also consider incorporating indicators related to ecosystem cultural services when formulating relevant policies and regulations. For the future planning of the tourism landscape in Anshun, planners can try to use the findings of this study as a guide to target the landscape planning and design of each type of CESs from the perspective of enhancing the perceptual experience of the non-material aspects of the landscape. Physical CESs and inspirational CESs should pay attention to the convenience of visitors to the attractions. Experiential CESs should focus on the planning and design of the visual experience of the landscape. Intellectual CESs can be considered to enhance the sensory perception of visitors to the landscape on the basis of ensuring good natural ecology. Thus, the existing tourism landscape in Anshun City can play a correspondingly higher value in cultural services, so that the landscape experience of tourists is richer, strengthening the attractiveness and influence of tourism and driving the steady growth of the tourism economy in Anshun.

Author Contributions: Conceptualization, Y.Y. and W.D.; methodology, Y.Y.; software, Y.Y.; validation, W.D. and Y.Y.; formal analysis, Y.Y.; investigation, Y.Y.; resources, Y.Y.; data curation, Y.Y.; writing—original draft preparation, Y.Y.; writing—review and editing, Y.Y.; visualization, Y.Y.; supervision, W.D.; project administration, W.D.; funding acquisition, W.D. All authors have read and agreed to the published version of the manuscript.

Funding: This research received no external funding.

Institutional Review Board Statement: Ethical review and approval were waived for this study, due to involving no more than minimal risk.

Data Availability Statement: Data is contained within the article and Appendix A.

Acknowledgments: I would like to express my gratitude to Wei Duan and my peer HaoXian Cai for their support and assistance with this research work.

Conflicts of Interest: The authors declare no conflicts of interest.

Appendix A

	A	B	C	D	E	F	G	H	I	J	K	L	M	N	O	P	Q	R	S	T	U	V	W	X
1		1 Huangguoshu Waterfalls				2 The Dragon palace				3 Pingba Farms				4 Bailing Hot Springs				5 Pingba Cherry Blossom Garden				6 Huangguoshu Waterfalls Night Tour		
2	tag words	word frequency	file frequency	part of speech	tag words	word freq	file freq	part of speech	tag words	word freq	file freq	part of speech	tag words	word freq	file freq	part of speech	tag words	word freq	file freq	part of speech	tag words	word freq	file freq	part of speech
3	waterfall	260	69	noun	The Dragon	61	34	noun	cherry	31	18	noun	nice	17	14	adjective	cherry	40	18	noun	waterfall	38	23	noun
4	huangguoshu	106	50	noun	Scenic spo	52	28	noun	pingba	18	11	noun	hot sprin	13	13	noun	pingba	27	17	noun	fruit tre	23	18	noun
5	Scenic spot	103	41	noun	cave	30	28	noun	farm	17	10	noun	service	12	10	verb	farm	23	14	noun	night tou	19	14	noun
6	Celestial Stars	43	24	noun	waterfall	27	22	noun	have not	7	6	verb	environme	11	11	noun	can	17	9	verb	light	15	13	noun
7	steep slope	39	24	noun	Karst	19	14	noun	can	5	5	verb	whole	7	7	noun	garden	13	7	noun	act	13	12	verb
8	Attractions	33	21	noun	boat	16	13	verb	place	5	4	noun	have not	7	5	verb	Scenic sp	11	6	noun	experienc	11	10	verb
9	Curtain	32	23	noun	landform	14	11	noun	friend	5	5	noun	feel	6	6	noun	time	8	5	noun	same	11	10	adjective
10	guizhou	24	22	noun	orchard	14	13	noun	bloom	4	4	verb	restauran	6	2	noun	have not	8	5	verb	deserve	9	9	verb
11	spectacular	23	21	adjective	can	13	12	verb	scenery	4	4	noun	pool	5	3	noun	look	7	5	verb	perform	9	4	verb
12	escalator	18	9	noun	place	11	8	noun	air	3	3	noun	buffet	4	3	noun	times	6	3	noun	can	9	8	verb
13	play	17	13	verb	paly	11	7	verb	traffic	3	3	verb	have	4	4	verb	have	6	6	verb	shock	8	8	verb
14	china	16	16	noun	boating	11	11	noun	garden	3	2	noun	hotel	4	3	noun	need	6	3	verb	lighting	8	6	noun
15	anshun city	15	15	noun	pretty	11	11	adjective	season	3	3	noun	experienc	4	3	verb	parking	5	3	noun	spectacul	8	8	adjective
16	sighting	15	7	verb	special	11	8	adjective	guiyang	3	3	noun	place	3	3	noun	ferry	5	4	verb	recommend	8	6	verb
17	touring	14	10	verb	guizhou	10	10	noun	hour	3	3	noun	can	3	2	verb	guiyang	5	5	noun	Scenic sp	7	5	noun
18	queuing	14	10	verb	didn't	10	9	verb	bring	3	3	verb	comfortab	3	3	adjective	scenery	5	5	noun	nightscap	7	7	noun
19	time	13	8	noun	anshun	10	8	noun	drive	3	2	verb	anshun	3	2	noun	place	5	4	noun	feel	7	7	noun
20	Journey to the w	13	11	noun	scenery	10	8	noun	suit	3	3	verb	total	3	1	adjective	weather	5	3	noun	child	6	4	noun
21	walk	11	6	verb	have	8	8	verb	ticket	3	3	noun	child	3	3	noun	pretty	5	4	adjective	nature	6	6	adjective
22	shock	11	9	verb	visitor	8	8	noun	time	3	2	noun	mountain	13	3	noun	bloom	5	4	verb	guizhou	6	5	noun
23	anshun	10	8	noun	deserve	8	6	verb	traffic	3	3	noun	clean	3	3	adjective	burst	4	4	verb	feel	6	5	verb
24	world	10	7	noun	come	8	7	verb	estimate	3	3	verb	happy	3	3	adjective	deserve	4	4	verb	landscape	5	5	noun
25	foot	10	5	verb	Ventured	8	6	noun	deserve	3	3	verb	dinner	3	2	noun	attractio	4	4	noun	pretty	5	5	adjective
26	ticket	10	6	noun	landscape	8	6	noun	tourists	3	3	noun	speak	3	3	verb	ticket	4	4	noun	actor	5	4	noun
27	Physical	10	9	noun	nice	7	7	adjective	nice	3	3	adjective	fruit tre	3	3	noun	nice	4	3	adjective	look	5	5	verb
28	look	9	6	verb	enter	7	7	verb	see	2	2	verb	personnel	3	2	noun	traffic	4	4	noun	total	5	4	adjective
29	scenery	9	8	noun	ride	7	7	verb	lock	2	2	verb	come	2	2	verb	experienc	3	3	verb	vision	5	5	noun
30	recommend	9	8	verb	erjingong	7	7	noun	beautiful	2	2	adjective	strengthe	2	2	verb	feel	3	1	noun	nice	5	5	adjective
31	tributary	9	3	noun	Lighting	7	7	noun	spectacul	2	1	adjective	lunch	2	1	noun	beautiful	3	3	adjective	dance	4	2	noun
32	waterfalls	9	6	noun	service	6	4	verb	florescen	2	1	noun	cozy	2	2	adjective	beauties	3	3	noun	have	4	4	verb
33	convience	9	6	adjective	centre	6	6	noun	lovely	2	2	adjective	meat dish	2	1	noun	free	3	3	verb	play	3	3	verb
34	travel	8	8	verb	scenery	6	6	noun	viewing	2	2	verb	lies	2	1	verb	take	3	3	verb	nature	3	3	noun
35	guizhou province	8	8	noun	longritian	6	4	noun	anshun	2	2	noun	facility	2	2	noun	photograp	3	3	verb	TRUE	3	2	adjective
36	people	8	6	noun	form	6	6	verb	guizhou	2	2	noun	hall	2	2	noun	kilometre	3	2	noun	pretty	3	3	adjective
37	raincoat	8	8	noun	feel	6	4	noun	show	2	1	verb	fun	2	2	adjective	florescen	3	3	noun	mountains	3	3	noun
38	famous	8	8	adjective	Attraction	6	6	noun	come	2	2	verb	student	2	1	noun	recommend	3	3	verb	staff mem	3	2	noun
39	distance	8	8	noun	sight	6	4	verb	have	2	2	verb	have	2	2	verb	start	3	2	verb	visit	2	2	verb

Figure A1. The word frequency information crawled from Ctrip for landscape comments.

numbe	name	lng	lat	gra	slope	elevation	D-water (surfa	D-water (poir	D-accommodati	D-restaurant	D-highroad	D-railway	Ctrip-sour	Ctrip-small	Ctrip-vis	Ctrip-fer
10	Xiye Exquisite Campsite	106.129662	26.267442	5	19.29450035	1263	0.208614	0.169867	0.016086733	0.014388411	0.0651463	0	0	1	2	2
36	Lanshan Scenic Area	106.049259	26.283819	5	35.69810104	1328	0.194648	0.165638	0.025435358	0.032173466	0.00508707	0	0	0	2	2
28	Hongshan Lake	105.944938	26.262569	4.9	29.09250069	1390	0.182428	0.200519	0.007194206	0	0.011375	0.016	0	5	6	4
30	Anshun Wu Temple	105.939558	26.252098	4.9	4.575900005	1362	0.192666	0.200519	0	0	0.0143884	0	0	0	0	2
45	Huangguoshu Adventure Ridge	105.495813	25.924641	4.9	16.19129944	1608	0.244753	0.317048	0.057553645	0.128794357	0.0160867	0.016	0	0	2	3
12	Silver chain falling pool waterfall	105.682481	25.942813	4.8	17.29299927	921	0.121852	0.194532	0.046900481	0.041949067	0.135167	0.0357771	5	0	8	17
59	Shenjun Grand Theatre	105.86368	26.224192	4.8	5.828700211	1384	0.157582	0.153791	0.005087072	0	0.0254354	0.016	2	0	4	6
6	Huangguoshu Waterfall night	105.680318	25.982942	4.7	17.44330025	996	0.113137	0.165638	0.007194206	0.005087072	0.0950343	0.016	2	0	15	8
21	Talc Whistle	105.676895	25.983391	4.7	13.13710022	942	0.128996	0.165638	0.007194206	0.005087072	0.0950343	0	0	0	3	29
	Hongshan Park	105.944938	26.262569	4.7	29.09250069	1390	0.182428	0.200519	0.007194206	0	0.011375	0.016	0	0	8	8
3	Pingba farm	106.378704	26.411388	4.6			0.0226274	0.0687773	0.111915573	0.005087072	0.0160867	0	0	3	3	3
11	The steep pond waterfall	105.683558	26.003546	4.6	9.096130371	1029	0.113137	0.151466	0.005087072	0.010174143	0.077484	0	7	0	12	7
13	Water Curtain Cave	105.676146	25.995266	4.6	8.554940224	981	0.128	0.151466	0.005087072	0.010174143	0.0838981	0	0	0	12	5
19	Rhino pool	105.67445	25.994632	4.6	15.20779991	995	0.128	0.151466	0.005087072	0.010174143	0.0838981	0	4	0	13	14
29	Jiuxi Village	106.149327	26.310519	4.6	4.9198699	1304	0.16	0.145899	0.050101843	0.050101843	0.0296625	0.016	0	0	2	2
34	Potted landscape garden	105.672721	25.988023	4.6	23.85040092	934	0.128	0.165638	0	0.005087072	0.0888419	0	0	0	4	6
35	Maoshui Pool	105.683209	25.947525	4.6	23.55459976	979	0.121852	0.194532	0.041949067	0.037034441	0.130392	0.0357771	0	0	5	3
44	Zhong holo	106.304349	25.690612	4.6			0.0715542	0.141788	0.051878154	0.587661326	0.62573	0	0	0	1	2
56	Huangguoshu Stone Hall	105.69211	26.007142	4.6	27.5590992	1095	0.113137	0.131399	0	0	0.0804337	0	0	0	16	15
1	Huangguoshu waterfall	105.679142	25.985778	4.5	0.810760975	969	0.128	0.165638	0.007194206	0.005087072	0.0950343	0	4	0	23	23
8	Star Bridge Scenic Spot	105.68322	25.94822	4.5	23.85040092	959	0.121852	0.194532	0.041949067	0.037034441	0.130392	0.0357771	0	0	11	16
15	Baling River Bridge	105.632693	25.962297	4.5	25.19300079	809	0.163169	0.200519	0.018341698	0.015261214	0.0978518	0	0	0	11	4
22	Anshun Wen Temple	105.941905	26.255138	4.5	15.53759956	1375	0.192666	0.200519	0.005087072	0.005087072	0.0101741	0	0	0	9	11
37	Xiaohevan	106.335678	26.416574	4.5	32.51509857	1326	0.0357771	0.111961	0.066131927	0.032573152	0.0254354	0	0	0	3	0
38	Yunfeng eight villages	106.104182	26.299763	4.5	19.36730003	1382	0.176726	0.138412	0.018341698	0.039731298	0.0215826	0	3	0	0	0
49	Huangguoshu Catholic Church	105.673014	25.995309	4.5	24.10880089	1023	0.128	0.151466	0.005087072	0.010174143	0.0838981	0	1	0	2	3
55	Xingwei Stone Expo Park	105.897691	26.237488	4.5	21.7772007	1392	0.173066	0.173994	0.005087072	0	0	0	0	1	4	2
2	Dragon Palace	105.891836	26.112361	4.4	8.554940224	1143	0.115376	0.154558	0.096788138	0.066327296	0.0733668	0	0	0	11	6
5	Pingba Farm cherry garden	106.379837	26.4121	4.4			0.0226274	0.0687773	0.111915573	0.005087072	0.0160867	0	0	0	5	3
20	Jiuxian scenic spot in Guizhou Province	105.546762	25.916818	4.4	9.165719986	1204	0.268209	0.283576	0.046900481	0.080433659	0.0479913	0.016	0	0	6	6
24	Yunfeng Tunbao scenic spot	106.09834	26.293969	4.4	2.56230998	1288	0.176726	0.154558	0.010174143	0.036683396	0.0209745	0	8	7	12	21
33	Whirlpool scenic spot	105.890202	26.119461	4.4	19.4829998	1238	0.115376	0.154558	0.091708489	0.06209562	0.0591916	0	0	0	3	4
43	Langtang Village	106.150634	26.284391	4.4	13.68360043	1331	0.192	0.160562	0.025939077	0.032173466	0.0547894	0	0	0	0	3
4	Bailing hot spring	105.938828	26.26058	4.3	9.768239975	1368	0.182428	0.200519	0.005087072	0	0.0160867	0.016	0	0	3	3
7	Tianlong Tunbao	106.176864	26.356424	4.3	21.7276001	1338	0.116482	0.120114	0.010174143	0.010174143	0.011375	0.016	2	2	4	3
9	Ziyun Ge convex River scenic spot	106.270451	25.682331	4.3	19.99300003	1359	0.116482	0.120114	0.025435358	0.588739276	0.518806	0	0	2	5	14
16	Stone village	105.679753	26.026379	4.3	9.404959679	1062	0.131939	0.12399	0.015261214	0.016086733	0.0568752	0	0	2	3	12
23	Tiantai Mountain	106.186559	26.358178	4.3	7.646349907	1411	0.116482	0.108746	0.020348286	0.020348286	0.0183417	0.016	1	2	4	5
32	Anshun White Pagoda	105.940562	26.246631	4.3	3.339289904	1362	0.192666	0.209178	0	0.005087072	0.011375	0	0	0	6	6
46	Huangguo tree god Dragon Cave	105.665215	26.014153	4.3	4.12721014	1024	0.128996	0.145086	0.007194206	0.011375038	0.0514673	0	0	0	17	11
47	Rulin Road	105.940452	26.253707	4.3	4.047329903	1369	0.192666	0.200519	0	0.005087072	0.0152612	0	0	0	0	13
14	Jiuzhou ancient town	106.146934	26.255193	4.2	2.920900106	1240	0.224	0.190229	0	0.005087072	0.0804337	0	5	0	5	12
25	Guanling geochemical Group national geopark	105.434956	25.88984	4.2	18.04509926	1142	0.193329	0.355053	0.127176791	0.196363598	0.0406966	0.016	0	0	4	7
31	Huajiang Grand Canyon	105.63066	25.704182	4.2	32.64899826	1212	0.16	0.443867	0.234226361	0.194974914	0.273238	0	0	0	5	4
39	Sila River	106.071758	26.443665	4.2	4.9198699	1336	0.0357771	0.0307581	0.129695386	0.11711318	0.111103	0	0	1	2	3
48	Qianshanxiushui scenic spot	105.697101	26.301843	4.2	11.8803997	1242	0.0659697	0.0984741	0.035609502	0.040696573	0.00719421	0.016	0	0	2	4
51	Tiantai Mountain Wulong Temple	106.09834	26.293969	4.2	2.56230998	1288	0.176726	0.154558	0.010174143	0.036683396	0.0209745	0	0	0	0	1
52	Jiangnan Mountain tourism and leisure resort	106.151262	26.259515	4.2	6.940750122	1288	0.208	0.190229	0.005087072	0.011375038	0.077484	0.016	0	0	5	9
54	Shushengbu	105.685536	25.953038	4.2	15.04959965	920	0.116482	0.190229	0.037034441	0.032173466	0.125641	0.0226274	0	0	0	2
57	Baojiatun Village	106.132318	26.334709	4.2	4.9198699	1287	0.144886	0.111961	0.041013282	0.041013282	0	0	0	0	2	3
58	Rhinoceros cave	105.784754	26.060247	4.2	31.34840012	1293	0.0576888	0.0896746	0.010174143	0.007194206	0.0661319	0.016	0	0	4	1
17	Gaodang millennium Buyi ancient village cultural tourism scenic spot	105.698319	26.079378	4.1	2.56230998	1145	0.124964	0.0768953	0.043760642	0.071400449	0.0160867	0.016	0	0	0	4
26	Gao Lao Zhuang	105.683182	25.951068	4.1	23.73060036	858	0.121852	0.194532	0.041949067	0.037034441	0.130392	0.0357771	0	0	6	4
41	Guanyin Cave	105.887873	26.118019	4.1	23.28689957	1267	0.107331	0.154558	0.091567285	0.059324939	0.0663273	0	0	0	0	3
42	Yunjiu Mountain	106.094927	26.29566	4.1	2.292109966	1292	0.176	0.154558	0.010174143	0.036683396	0.0209745	0	0	0	2	2
53	Yelang Lake	105.773177	26.372713	4.1	0	1109	0	0	0.066131927	0.022750076	0.0935247	0.016	0	0	3	3
50	Yuantong Temple	105.940259	26.247459	4	4.575900005	1364	0.192666	0.209178	0	0	0.0143884	0	1	0	2	1

Figure A2. Summary of relevant analytical data for each scenic spot.

References

1. Emborg, J.; Gamborg, C. Land Use Policy A wild controversy: Cooperation and competition among landowners, hunters, and other outdoor recreational land-users in Denmark. *Land Use Policy* **2016**, *59*, 197–206. [CrossRef]
2. Guo, Z.; Zhang, L.; Li, Y. Increased dependence of humans on ecosystem services and biodiversity. *PLoS ONE* **2010**, *5*, e13113. [CrossRef] [PubMed]
3. Gajardo, L.J.; Sumeldan, J.; Sajorne, R.; Madarcos, J.R.; Goh, H.C.; Culhane, F.; Langmead, O.; Creencia, L. Cultural values of ecosystem services from coastal marine areas: Case of Taytay Bay, Palawan, Philippines. *Environ. Sci. Policy* **2023**, *142*, 12–20. [CrossRef]
4. Millennium Ecosystem Assessment. *Ecosystems and Human Well-Being: Synthesis Reports*; Millennium Ecosystem Assessment: Washington, DC, USA, 2005.
5. Gret-Regamey, A.; Weibel, B. Global assessment of mountain ecosystem services using earth observation data. *Ecosyst. Serv.* **2020**, *46*, 101213. [CrossRef]
6. Lakshmi, A. Coastal ecosystem services & human wellbeing. *Indian J. Med. Res.* **2021**, *153*, 382–387. [PubMed]
7. World Resources Institute. *Ecosystems and Human Well-Being: A Framework for Assessment*, 2nd ed.; Island Press: Washington, DC, USA, 2003.
8. Daniel, T.C.; Muhar, A.; Arnberger, A.; Aznar, O.; Boyd, J.W.; Chan, K.M.A.; Costanza, R.; Elmqvist, T.; Flint, C.G.; Gobster, P.H.; et al. Contributions of cultural services to the ecosystem services agenda. *Proc. Natl. Acad. Sci. USA* **2012**, *109*, 8812–8819. [CrossRef] [PubMed]
9. Plieninger, T.; Dijks, S.; Oteros-Rozas, E.; Bieling, C. Assessing, mapping, and quantifying cultural ecosystem services at community level. *Land Use Policy* **2013**, *33*, 118–129. [CrossRef]
10. Hattam, C.; Broszeit, S.; Langmead, O.; Praptiwi, R.A.; Lim, V.C.; Creencia, L.A.; Tran, D.H.; Maharja, C.; Mitra Setia, T.; Wulandari, P.; et al. A matrix approach to tropical marine ecosystem service assessments in South east Asia. *Ecosyst. Serv.* **2021**, *51*, 101346. [CrossRef]
11. Liquete, C.; Piroddi, C.; Drakou, E.G.; Gurney, L.; Katsanevakis, S.; Charef, A.; Egoh, B. Current status and future prospects for the assessment of marine and coastal ecosystem services: A systematic review. *PLoS ONE* **2013**, *8*, e67737. [CrossRef]
12. Bai, Y.; Chen, Y.; Alatalo, J.M.; Yang, Z.; Jiang, B. Scale effects on the relationships between land characteristics and ecosystem services- a case study in Taihu Lake Basin, China. *Sci. Total Environ.* **2020**, *716*, 137083. [CrossRef]
13. Li, Y.; Xie, L.; Zhang, L.; Huang, L.; Lin, Y.; Su, Y.; AmirReza, S.; He, S.; Zhu, C.; Li, S.; et al. Understanding different cultural ecosystem services: An exploration of rural landscape preferences based on geographic and social media data. *J. Environ. Manag.* **2022**, *317*, 115487. [CrossRef] [PubMed]
14. Zhang, H.; Huang, R.; Zhang, Y.; Buhalis, D. Cultural ecosystem services evaluation using geolocated social media data: A review. *Tour. Geogr.* **2020**, *24*, 646–668. [CrossRef]
15. Dong, L.; Zhu, W.; Gao, Y.; Li, S. Progress in the study of ecosystem cultural services. *J. Peking Univ. Nat. Sci. Ed.* **2014**, *50*, 1155–1162.
16. Wang, Y.; Hayashi, K. Methodological development of cultural ecosystem services evaluation using location data. *J. Clean. Prod.* **2023**, *396*, 136523. [CrossRef]
17. Richards, D.R.; Tunçer, B. Using image recognition to automate assessment of cultural ecosystem services from social media photographs. *Ecosyst. Serv.* **2018**, *31*, 318–325. [CrossRef]
18. Balázsi, Á.; Danhardt, J.; Collins, S.; Schweiger, O.; Settele, J.; Hartel, T. Understanding cultural ecosystem services related to farmlands: Expert survey in Europe. *Land Use Policy* **2021**, *100*, 104900. [CrossRef]
19. Ginsburgh, V. Contingent valuation, willingness to pay, and willingness to accept. In *Economic Ideas You Should Forget*; Springer: Cham, Switzerland, 2017; pp. 65–66.
20. Zhao, N.; Wang, H.; Zhong, J.; Sun, D. Assessment of recreational and cultural ecosystem services value of islands. *Land* **2022**, *11*, 205. [CrossRef]
21. Sumarga, E.; Hein, L.; Edens, B.; Suwarno, A. Mapping monetary values of ecosystem services in support of developing ecosystem accounts. *Ecosyst. Serv.* **2015**, *12*, 71–83. [CrossRef]
22. Zheng, S.; Ma, M.; Li, H.; Wang, Q.; Zhang, X. Evaluation of Ecospatial Cultural Services in Urban Centres from the Perspective of Residents' Well-being—A Case Study of Beijing. *Urban Dev. Res.* **2021**, *28*, 21–27.
23. Christie, M.; Fazey, I.; Cooper, R.; Hyde, T.; Kenter, J.O. An evaluation of monetary and non-monetary techniques for assessing the importance of biodiversity and ecosystem services to people in countries with developing economies. *Ecol. Econ.* **2012**, *83*, 67–78. [CrossRef]
24. Wartmann, F.M.; Purves, R.S. Investigating sense of place as a cultural ecosystem service in different landscapes through the lens of language. *Landsc. Urban Plann.* **2018**, *175*, 169–183. [CrossRef]
25. Riechers, M.; Barkmann, J.; Tscharntke, T. Diverging perceptions by social groups on cultural ecosystem services provided by urban green. *Landsc. Urban Plann.* **2018**, *175*, 161–168. [CrossRef]
26. Liu, Z.; Huang, Q.; Yang, H. Supply-demand spatial patterns of park cultural services in megalopolis area of Shenzhen, China. *Ecol. Indic.* **2021**, *121*, 107066. [CrossRef]
27. Buchel, S.; Frantzeskaki, N. Citizens' voice: A case study about perceived ecosystem services by urban park users in Rotterdam, the Netherlands. *Ecosyst. Serv.* **2015**, *12*, 169–177. [CrossRef]

28. Ghermandi, A.; Sinclair, M. Passive crowdsourcing of social media in environmental research: A systematic map. *Glob. Environ. Chang.* **2019**, *55*, 36–47. [CrossRef]
29. Gosal, A.S.; Ziv, G. Landscape aesthetics: Spatial modelling and mapping using social media images and machine learning. *Ecol. Indicat.* **2020**, *117*, 106638. [CrossRef]
30. Jean-Christophe, F.; Jens, I.; Nicolas, B. Coupling crowd-sourced imagery and visibility modelling to identify landscape preferences at the panorama level. *Landsc. Urban Plann.* **2020**, *197*, 103756.
31. Van Zanten, B.T.; Verburg, P.H.; Koetse, M.J.; Van Beukering, P.J.H. Preferences for European agrarian landscapes: A meta-analysis of case studies. *Landsc. Urban Plann.* **2014**, *132*, 89–101. [CrossRef]
32. da Mota, V.T. Catherine Pickering, Using social media to assess nature-based tourism: Current research and future trends. *J. Outdoor Recreat. Tour.* **2020**, *30*, 100295. [CrossRef]
33. Ren, X. Consensus in factors affecting landscape preference: A case study based on a cross-cultural comparison. *J. Environ. Manag.* **2019**, *252*, 109622. [CrossRef]
34. Cai, K.; Huang, W.; Lin, G. Bridging landscape preference and landscape design: A study on the preference and optimal combination of landscape elements based on conjoint analysis. *Urban For. Urban Green.* **2022**, *73*, 127615. [CrossRef]
35. Kaplan, R.; Kaplan, S.; Brown, T. Environmental preference: A comparison of four domains of predictors. *Environ. Behav.* **1989**, *21*, 509–530. [CrossRef]
36. Kenter, J.O.; Jobstvogt, N.; Watson, V.; Irvine, K.N.; Christie, M.; Bryce, R. The impact of information, value-deliberation and group-based decision-making on values for ecosystem services: Integrating deliberative monetary valuation and storytelling. *Ecosyst. Serv.* **2016**, *21*, 270–290. [CrossRef]
37. Czembrowski, P.; Kronenberg, J.; Czepkiewicz, M. Integrating non-monetary and monetary valuation methods–SoftGIS and hedonic pricing. *Ecol. Econ.* **2016**, *130*, 166–175. [CrossRef]
38. Yang, B.; Zhang, J. A study on tourism image perception of Tianmu Mountain based on network text analysis--Taking Ctrip.com travelogue and review as an example. *Fujian For. Sci. Technol.* **2017**, *44*, 118–125. [CrossRef]
39. Haines-Young, R.; Potschin, M. Common international classification of ecosystem services (CICES, Version 4.1). *Eur. Environ. Agency* **2012**, *33*, 107.
40. Potschin, M.; Haines-Young, R. Defining and measuring ecosystem services. *Routledge Handb. Ecosyst. Serv.* **2016**, *1*, 25–44.
41. Pascual, U.; Balvanera, P.; Díaz, S.; Pataki, G.; Roth, E.; Stenseke, M.; Watson, R.T.; Dessane, E.B.; Islar, M.; Kelemen, E.; et al. Valuing nature's contributions to people: The IPBES approach. *Curr. Opin. Environ. Sustain.* **2017**, *26*, 7–16. [CrossRef]
42. Landers, D.H.; Nahlik, A.M.; Rhodes, C.R. *The Beneficiary Perspective: Benefits and Beyond*; Routledge Handbook of Ecosystem Services; Routledge: London, UK, 2016; pp. 74–87.
43. Czúcz, B.; Arany, I.; Potschin-Young, M.; Bereczki, K.; Kertész, M.; Kiss, M.; Aszalós, R.; Haines-Young, R. Where concepts meet the real world: A systematic review of ecosystem service indicators and their classification using CICES. *Ecosyst. Serv.* **2018**, *29*, 145–157. [CrossRef]
44. Clemente, P.; Calvache, M.; Antunes, P.; Santos, R.; Cerdeira, J.O.; Martins, M.J. Combining social media photographs and species distribution models to map cultural ecosystem services: The case of a Natural Park in Portugal. *Ecol. Indic.* **2019**, *96*, 59–68. [CrossRef]
45. Haines-Young, R.; Potschin, M. *Common International Classification of Ecosystem Services (CICES): Consultation on Version 4*; European Environment Agency: Nottingham, UK, 2013.
46. Zhang, J.; Chen, H.; Fu, Z.; Wang, K. Effects of vegetation restoration on soil properties along an elevation gradient in the karst region of southwest China. *Agric. Ecosyst. Environ.* **2021**, *320*, 107572. [CrossRef]
47. Fan, C.; Zhao, L.; Hou, R.; Fang, Q.; Zhang, J. Quantitative analysis of rainwater redistribution and soil loss at the surface and belowground on karst slopes at the microplot scale. *Catena* **2023**, *227*, 107113. [CrossRef]
48. Fang, Q.; Zhao, L.; Hou, R.; Fan, C.; Zhang, J. Rainwater transformation to runoff and soil loss at the surface and belowground on soil-mantled karst slopes under rainfall simulation experiments. *Catena* **2022**, *215*, 106316. [CrossRef]
49. Burmil, S.; Daniel, T.C.; Hetherington, J.D. Human values and perceptions of water in arid landscapes. *Landsc. Urban Plan.* **1999**, *44*, 99–109. [CrossRef]
50. Mühl, S. Landscape and infrastructure preferences of recreational rowers in Germany. *J. Outdoor Recreat. Tour.* **2020**, *29*, 100271. [CrossRef]
51. Lee, C.-H. Understanding rural landscape for better resident-led management: Residents' perceptions on rural landscape as everyday landscapes. *Land Use Policy* **2020**, *94*, 104565. [CrossRef]
52. El Islem Karabadji, N.; Korba, A.A.; Assi, A.; Seridi, H.; Aridhi, S.; Dhifli, W. Accuracy and diversity-aware multi-objective approach for random forest construction. *Expert Syst. Appl.* **2023**, *225*, 120138. [CrossRef]
53. Wang, Y.; Chen, X.; Gao, M.; Dong, J. The use of random forest to identify climate and human interference on vegetation coverage changes in southwest China. *Ecol. Indic.* **2022**, *144*, 109463. [CrossRef]
54. Belgiu, M.; Drăguţ, L. Random forest in remote sensing: A review of applications and future directions. *ISPRS J. Photogramm. Remote Sens.* **2016**, *114*, 24–31. [CrossRef]
55. Gidlow, C.; Cerin, E.; Sugiyama, T.; Adams, M.A.; Mitas, J.; Akram, M.; Reis, R.S.; Davey, R.; Troelsen, J.; Schofield, G.; et al. Objectively measured access to recreational destinations and leisure-time physical activity: Associations and demographic moderators in a six-country study. *Health Place* **2019**, *59*, 102196. [CrossRef]

56. Yuan, G.; Wang, H.; Wang, M.; Lin, M. Visual attention and ethnic landscape perception: A case of three cities in the Guangdong–Hong Kong–Macao greater bay area. *Appl. Geogr.* **2022**, *147*, 102781. [CrossRef]
57. Dai, P.; Zhang, S.; Gong, Y.; Yang, Y.; Hou, H. Assessing the inspirational value of cultural ecosystem services based on the Chinese poetry. *Acta Ecol. Sin.* **2022**, *42*, 467–475. [CrossRef]
58. Kalivoda, O.; Vojar, J.; Skřivanová, Z.; Zahradník, D. Consensus in landscape preference judgments: The effects of landscape visual aesthetic quality and respondents' characteristics. *J. Environ. Manag.* **2014**, *137*, 36–44. [CrossRef] [PubMed]
59. Sharafatmandrad, M.; Mashizi, A.K. Visual value of rangeland landscapes: A study based on structural equation modeling. *Ecol. Eng.* **2020**, *146*, 105742. [CrossRef]

land

MDPI

Article

Coupling Coordination Relationship and Driving Force Analysis between Gross Ecosystem Product and Regional Economic System in the Qinling Mountains, China

Pengtao Wang [1], Yuxuan Chen [2,*], Kang Liu [3], Xupu Li [4], Liwei Zhang [4], Le Chen [1], Tianjie Shao [4], Peilin Li [1], Guoqing Yang [1], Hui Wang [5], Shang Gao [6] and Junping Yan [4]

1 School of Tourism, Research Institute of Human Geography, Xi'an International Studies University, Xi'an 710128, China; wnpengtao@xisu.edu.cn (P.W.); 107242018100023@xisu.edu.cn (L.C.); peilinli@xisu.edu.cn (P.L.); yangguoqing@xisu.edu.cn (G.Y.)
2 Xi'an Environmental Sanitation Science Research Institute, Xi'an 710075, China
3 College of Urban and Environmental Sciences, Northwest University, Xi'an 710127, China; 20132381@nwu.edu.cn
4 School of Geography and Tourism, Shaanxi Normal University, Xi'an 710119, China; xupuli@snnu.edu.cn (X.L.); zlw@snnu.edu.cn (L.Z.); tjshao@snnu.edu.cn (T.S.); yanjp@snnu.edu.cn (J.Y.)
5 School of Geographical Sciences, China West Normal University, Nanchong 637009, China; hegwanghui@cwnu.edu.cn
6 School of Tourism Management, Henan Finance University, Zhengzhou 451464, China; 33200031@hafu.edu.cn
* Correspondence: yuxuanchen1995@yeah.net

Abstract: As a new concept for systematically evaluating ecosystem services, Gross Ecosystem Product (GEP) provides an effective means to comprehensively reveal the overall status of the ecosystem, the impact of economic activities on the ecological environment, and the effectiveness of ecological protection efforts. GEP accounting has been conducted in various regions; however, GEP's application in natural reserves still requires further exploration. Taking the Qinling Mountains as the research area, this paper aims to assess the relationship between GEP and economic development on the basis of the GEP accounting system. The results indicated that: (1) From 2010 to 2020, GEP tended to increase continuously and exhibited a distribution pattern with high value regions in the east and west, and low value regions in the north and south. (2) Over the years, the coupling coordination degree between GEP and GDP was in a consistent upward trend. In 2020, a good coupling coordination state between GEP and GDP was achieved in most districts and counties. (3) With the relative development between GEP and GDP, the social economy of most districts and counties lagged behind GEP in 2010. The number of districts and counties lagging in GEP in 2020 increased, while the number of regions with a balanced development of GEP and GDP was still relatively discouraging. (4) In general, elevation, contagion, temperature, population density, and precipitation were the main drivers of coupling coordination degree between GEP and GDP. If the relationship between economic development and ecological environmental protection can be reasonably balanced, it will further promote the sustainable development of nature reserves, and provide a scientific basis for sustainable policy-making in other similar areas.

Keywords: gross ecosystem product (GEP); ecosystem services; Qinling mountains; Shaanxi province; coupling coordination degree (CCD) model; geographical detector

1. Introduction

Ecosystem service (ES) is an important quantitative index to measure the degree of support services and contributions of natural ecosystem to human society [1–4], and has become a hot topic in global change ecology, ecological economics, environmental science, and geography [5–10], and an important foundation for regional ecological management and social economic development [11–16]. At present, the assessment of ecosystem services can be mainly divided into two types: physical amount assessment and value assessment [17–19]. The first method is to simulate the ecosystem process and mechanism of ES formation with geochemical models, biophysical models, and other mechanism models, and to quantitatively simulate the quality of ecosystem services with comparative accuracy [20–22], but it often fails to reflect the economic value of the whole ES. Conversely, the second method can simply and intuitively reflect the economic value of the whole ES. In this method, an equivalence factor can directly reflect the potential contributions of different ecosystems to mankind so that people can intuitively understand the remarkable contributions and great value of ecosystems and dedicate themselves to the sustainable development of society and economy to a higher extent [23,24]. However, it could hardly reveal the ecosystem change process of ESs.

It was the impossible mission in the past research on the value assessment of ESs to integrate the ecological significance and economic value. In 2012, on the basis of Constanza et al. [7] and Daily et al. [13], referring to the concept of GDP, Ouyang et al. introduced the concept of GEP and established a complete GEP accounting system, thereby providing improved methods for the valuation of ESs [25–27].

First of all, GEP can be defined as the total value of the final products and services provided by the ecosystem for human society [26]. In its calculating process, it includes both the quality evaluation based on a biogeographic model and the monetization value evaluation of ESs. Therefore, it can effectively integrate the advantages of the quality evaluation and value evaluation, and become an effective breakthrough in ES evaluation [28].

Secondly, the ecosystem and the economic system are the vital powers in the survival and development of human society [29]. GEP and GDP are parallel indicators in these two systems, respectively. GDP can solely summarize the development of the entire social economy, while GEP can summarize the status of the ecosystem on its own. The accounting of GEP goes beyond GDP, which is an important supplement and improvement in the statistics and management of the social economy. It can measure the economic value of the ecosystem to human society [25], and provide a quantitative scientific reference for ecological protection and ecological compensation.

In addition, achieving sustainable development is an urgent need for human society, as it aims to maintain economic growth and guarantee the stability of ecological environment quality and the health of the ecosystem [30,31]. Based on these coupling coordination models, scholars have tried to introduce GEP indicators to evaluate the relationship between economic development and ecosystem status, and deeply explore the interaction of economic system and ecosystem. Typical studies include the following examples: Zang et al. used the four quadrant analysis method to evaluate the synergistic states and evolution process of GEP and GDP in mainland China from 2000 to 2015, and found that GEP–GDP synergy continued to increase [30]; Xie et al. quantified the coupling coordination degree between GEP and economic system in Jiangxi Province, China, from 2010 to 2020, and found that the coupling coordination degree of Jiangxi province was continuously improved, but the development of GEP lagged behind economic development [29]; Guan et al. adopted the Tapio decoupling model to assess the coupling relationship between GEP and GDP in Hubei Province during 2010–2019, and the results showed that the decoupling coefficient between GEP and economic growth gradually decreased, and the decoupling relationship changed from weak decoupling to strong decoupling [32]. In summary, with GEP accounting, people can better use monetary value to measure the ecological value provided by the ecosystem for economic development. This allows for a more effective integration of economic accounting and ecological asset accounting in the coupling coordination analysis of GDP and GEP, providing support for evaluating the interaction of economic development and ecosystem reasonably and making decisions regarding regional sustainable development.

As for GEP research and practice, scholars have conducted extensive research globally, nationally, and provincially [26,29,30,33]. The majority of GEP calculations occur mainly in urbanized areas; however, fewer cases can be associated with physical geographical units such as mountain areas, river basins, nature reserves, and ecological fragile zones [27,28]. An urbanization area differs from a mountain area in that it is characterized by a large population density, fast economic development, a tense relationship between man and land, and greater pressure on ecological environmental protection. It is imperative to address the conflict between economic development and ecological protection in urban areas. A mountainous area features a sparsely distributed population, slow economic development, and less pressure on the environment. It boasts well-preserved ecological conditions with abundant nature reserves and designated ecological functional areas. There is an urgent need to transform ecological advantages into economic assets. Therefore, research on the coordinated development between GDP and GEP in urban and mountain areas has practical significance and theoretical value for the sustainable development of the whole social economy. However, there are few studies on GEP assessment, as well as coordination development assessment, of GEP and GDP in mountainous areas and nature reserves.

The Qinling Mountains (QMs) are a huge mountain system that runs from east to west in central China and serve as the geographical demarcation between north and south of China, playing a crucial role as important ecological barriers in China. Meanwhile, this region is a typical concentrated poverty-stricken area [34]. It has become a common and significant challenge for nature reserves to effectively manage the harmonious interaction between economic development and ecological protection, fully leverage the advantages of ecological environment and assets, and facilitate the conversion of ecological benefits into economic benefits.

Therefore, taking the QMs as a case study, based on the GEP accounting system and Coupling Coordination Degree model, the purpose of this paper was to: (1) quantify the ESs and measure the GEP with a monetary value; (2) investigate the spatio-temporal variations of GEP from 2010 to 2020; (3) explore the interaction of GEP and GDP in the QMs. This study will raise policy suggestions for ecological management and social economic development of nature reserves, aiming to protect the good ecological background, give full play to ecological advantages, and enhance sustainable development of social economy.

2. Materials and Methods

2.1. Study Area

The construction of ecological civilization is a key strategic priority for China, and the QMs have an important strategic position in this endeavor (Figure 1). First of all, the QMs are the central mountains of China, and the key components of the geographical demarcation between north and south of China [35]. In terms of climate, this demarcation lies within the transitional zone between subtropical monsoon and temperate monsoon climate. Hence, the QMs serve as an ecological transition zone, delineating the interface between subtropical evergreen broad-leaved forests and temperate deciduous broad-leaved forests, as well as demarcating the boundary separating semi-arid and humid regions [36]. At the same time, the QMs are the watershed and important water conservation areas of the Yangtze River and the Yellow River (the two largest rivers in China), and an important water source for China's North-South transfer project and other water conservancy projects [22,31]; therefore, the QMs have a very important geographical demarcation function and are abundant in water resources [36,37]. Moreover, the QMs have important ecological advantages and are important ecological barriers in China. The diverse geographical environment provides a solid foundation for animal and plant growth, ensuring the stability of ecosystems and the preservation of biodiversity and earning the recognition as the "Kingdom of animals and plants", "National Central Park", and "gene bank" [37].

Notes: In the QMs, Xi'an has six districts and counties: Baqiao District (BQ), Lintong District (LT), Chang'an District (CA), Huyi District (HUY), Zhouzhi County (ZZ), and Lantian County (LAT); Baoji has six districts and counties: Weibin District (WB), Chencang District (CC), Qishan County (QS), Meixian County (MX), Taibai County (TB), and Fengxian County (FX); Weinan has four districts and counties: Linwei District (LW), Huazhou District (HZ), Huayin City (HY), and Tongguan County (TG); Hanzhong has nine districts and counties: Hantai District (HT), Chenggu County (CG), Yangxian County (YX), Xixiang County (XX), Mianxian County (MIX), Ningqiang County (NQ), Lueyang County (LY), Liuba County (LB), and Foping County (FP); Ankang City has seven districts and counties: Hanbin District (HB), Hanyin County (HAY), Shiquan County (SQ), Ningshan County (NS), Ziyang County (ZY), Langao County (LG), and Xunyang County (XY); Shangluo City has seven districts and counties: Shangzhou District (SZ), Luonan County (LN), Danfeng County (DF), Shangnan County (SN), Shanyang County (SY), Zhen'an County (ZA), and Zhashui County (ZS).

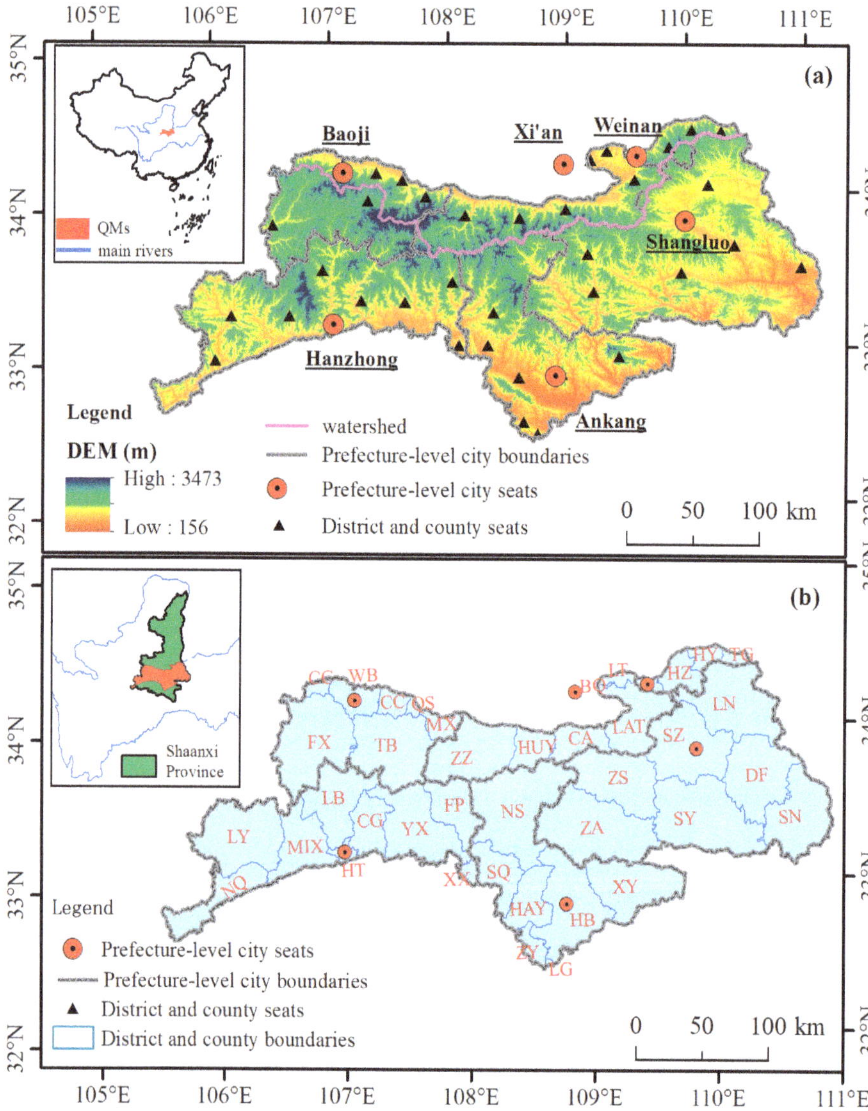

Figure 1. Overview of the Qinling Mountains: (**a**) geographical range, elevation, district and county seats, and prefecture-level city seats in the QMs; (**b**) administrative division of prefecture-level cities and districts and counties in the QMs.

The QMs have a broad sense and a narrow sense; the narrow sense of the QMs mainly refers to the Qinling Mountains in Shaanxi Province, which is the core area of the Qinling Mountains embracing 12 national nature reserves [38]. The QMs (105°29′18″–111°01′54″ E, 32°28′53″–34°32′33″ N) cover an area of 58,800 km^2, accounting for 28.59% of the total area of Shaanxi Province. From north to south, this province can be divided into the Loess Plateau of Northern Shaanxi Province, the Guanzhong basin zone, and the QMs [38,39]. The Guanzhong basin zone is at the heart of economic and cultural development, which is clustered by the provincial capital city Xi'an and its significant neighbor cities. The QMs is closely adjacent to the Guanzhong basin zone, spreading to the southernmost part of Shaanxi Province.

The QMs are divided into a south slope and a north slope by the watershed [36] and contain six prefecture-level cities, namely Xi'an, Weinan, Baoji, Hanzhong, Ankang, and Shangluo, and 39 districts and counties [40]. Baqiao District of Xi'an and Langao County of Ankang have relatively small land areas, neither exceeding 30 km^2, while other districts and counties in the region have larger land areas, all exceeding 100 km^2. Geographically, Weinan and Xi'an are located in the north slope, Hanzhong, Ankang and Shangluo are located in the south, while the districts and counties of Baoji are distributed on both north and south sides around the watershed.

Due to the tough mountainous conditions, inconvenient traffic, and insufficient cultivated land, social and economic development here is relatively backward. Especially, the QMs are on the list of the 14 centralized contiguously poor areas in China. In addition, in recent years, China has attached increasing importance to the ecological environmental protection of the QMs, and prohibits destruction of the natural environment. The urbanization and industrial development processes here are restricted, and higher requirements are put forward for the high-quality development of the local social economy. Therefore, the social economic development in the north slope of the QMs is relatively fast, while the social economic development in the south slope is relatively slow.

2.2. Research Framework

The research specifically includes the following three processes (Figure 2).

Firstly, ES types were first divided into three ES categories, Provisioning ecosystem services, Regulating ecosystem services, and Cultural ecosystem services, and the main ecosystem types in the QMs were selected. Within the framework of GEP accounting, remote sensing data, meteorological observation data, social economic data, and quantitative models such as RULSE and raster calculation methods were used to quantitatively evaluate the material quality of the ESs. Then, the monetary value of each ESs was quantitatively assessed through market value and other methods, so as to obtain the value of different ecosystem types and the total GEP value in the QMs.

Secondly, within the framework of the Coupling Coordination Degree, the three methods of Coupling Degree, Coupling Coordination Degree, and Relative Development Degree were comprehensively adopted to evaluate the interaction between GEP and GDP.

Thirdly, the geographic detector model was introduced, the driving factors of the relationship between GEP and GDP were selected, and the two methods of factor detection and interactive factor detection were used to quantitatively evaluate the contribution degree of different factors to the spatial differentiation of the relationship between GEP and GDP, and the main influencing factors that affect the development relationship were identified. In light of the need to maintain a balance between ecological environment and social economic development of nature reserves, corresponding countermeasures for sustainable development has been proposed.

Notes: GEP represents Gross Ecosystem Product, PES represents Provisioning ecosystem services, RES represents Regulating ecosystem services, CES represents Cultural ecosystem services, CD represents Coupling Degree, and CCD represents Coupling Coordination Degree.

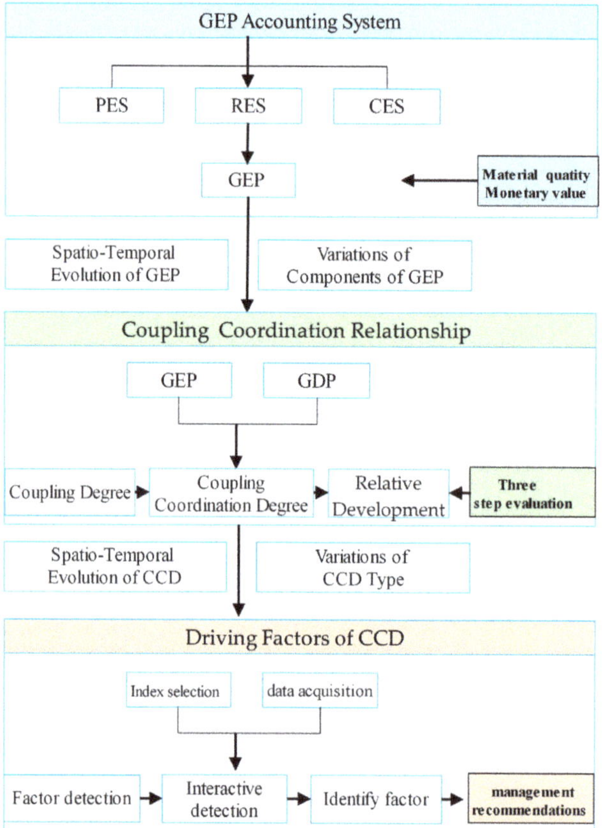

Figure 2. Research framework for the study.

2.3. Data Sources

We collect historical period data from multiple sources to carry out GEP accounting in the QMs. Specifically, the following types of data are introduced in Table 1:

(1) Daily climate data of national meteorological stations in the QMs were downloaded from the China Meteorological Data Service Centre, and the meteorological elements included precipitation, temperature, etc. Then, with the meteorological elements, the local runoff and evaporation data were obtained using the hydrological formula;

(2) Environmental data included air quality and water environment quality data, collected from Shaanxi Province air quality monitoring station and Shaanxi Provincial Department of Ecology and Environment, respectively;

(3) Remote sensing data mainly included land use, soil, DEM, NPP, and NDVI. Since there were many types of remote sensing data involved, and the spatial resolutions of different data were significantly different, the remote sensing data were resampled to a resolution of 1000 m in the data processing process, and the data output resolution of ES and GEP assessment results were also set to 1000 m;

(4) Social economic data, including GDP, population, water price, agricultural product price, tourist income, etc., were obtained from Shaanxi provincial statistical yearbook, Shaanxi provincial tourism development statistical bulletin, Shaanxi water conservancy statistical yearbook, etc. The statistical scale of social economic data is district-county scale, so the data output resolution of ES value evaluation results is

also unified to district-county scale. Therefore, the total GEP results are also read with the same scale. In view of the fact that some districts and counties are not fully included in the scope of the QMs [40], the social economic statistics of these districts and counties are assigned within the coverage area by an area-weighted method.

Table 1. Data source information.

Category	Index	Time Resolution	Spatial Resolution	Data Sources
Climate data	Precipitation, Temperature, etc.	2010–2020	Meteorological station	China Meteorological Data Service Centre (http://data.cma.cn, accessed on 20 June 2021)
	Runoff Evaporation			Calculated from meteorological data
Environmental data	Air quality		Observation station	Shaanxi Province air quality monitoring station, accessed on 20 September 2021
	Water environment quality			Shaanxi Provincial Department of Ecology and Environment, accessed on 20 September 2021
Remote sensing data	Land use	2010, 2015, 2020	30 m	China Multi-Period Land Use Remote Sensing Monitoring Dataset (CNLUCC) (https://www.resdc.cn, accessed on 6 September 2021)
	Soil	2017	250 m	Global gridded soil information (https://www.isric.org/explore/soilgrids, accessed on 17 August 2021)
	DEM	2011	12.5 m	NASA EARTHDATA Advanced Land Observing Satellite data (https://search.earthdata.nasa.gov/search, accessed on 15 August 2021)
	NPP	2010, 2015, 2020	500 m	NASA's Land Processes Distributed Active Archive Center (https://e4ftl01.cr.usgs.gov/, accessed on 23 October 2021)
	NDVI	2010, 2015, 2020	1000 m	MODIS/Terra Vegetation Indices Monthly L3 Global 1 km SIN Grid V006, NASA EOSDIS Land Processes DAAC (https://search.earthdata.nasa.gov/search, accessed on 28 April 2022)
Social economic data	Density of road network	2019	1000 m	A dataset of 1 km Grid Road network density in China (2019) (https://cstr.cn/31253.11.sciencedb.02938, accessed on 12 October 2023) [41]
	GDP, Population, Water price, Agricultural product price, Tourist income, etc.	2011, 2016, 2021	District-county scale	Shaanxi provincial statistical yearbook, Shaanxi provincial tourism development statistical Bulletin, Shaanxi water conservancy statistical yearbook

2.4. Methods

2.4.1. GEP Accounting Methodology

This paper is based on the first provincial GEP accounting standard in China [42], where methods and models are chosen to calculate the quality and value of most ES types. Additionally, some popular models from previous studies are employed to assess remaining ESs, thereby establishing a more comprehensive GEP accounting framework in the QMs (Table 2) [20,27,33,43,44]. The specific models and algorithms employed in this study are detailed in the supplementary materials (Tables S1–S3).

First of all, the Millennium Ecosystem Assessment divides ESs into four ES categories: Provisioning ecosystem services (PES), Regulating ecosystem services (RES), Cultural ecosystem services (CES), and Supporting ecosystem services (SES) [12]. However, GEP accounts for the final goods and services provided by the ecosystem. Counting SES would result in double counting, leading to inflated results, so GEP only accounts for the final economic value of the first three ES categories [25,26,43]. Therefore, the types of ES are divided into PES, RES, and CES in the QMs (Table 2). Among them, PES includes five indicators such as agricultural products, RES includes eight indicators such as water conservation service, and CES is evaluated by ecological tourism service value.

Secondly, the biophysical value of the three types of ES is evaluated by the material quality method. Since most of these data are remote sensing data, the ES quality results are mainly raster data.

Thirdly, the monetary value of 14 indicators in three categories of ES with social economic data at district-county scale is evaluated.

Finally, the GEP of all districts and counties in the QMs is obtained by summing up the monetary values of the three ES types using the following formula:

$$GEP = V_{PES} + V_{RES} + V_{CES} \quad (1)$$

where GEP represents Gross Ecosystem Product in the QMs, V_{PES} represents the value of PES, V_{RES} represents the value of RES, V_{CES} represents the value of CES [29]. The value of each ecosystem service is estimated by monetary value, and the unit is billion of the Chinese currency (CNY).

In the process of calculation, the *GEP* value is determined by utilizing climate and environmental data, soil erosion factor, vegetation cover factor, and social economic statistics specific to the region. Ultimately, the obtained results are compared and validated against relevant findings from previous studies [33,45].

Table 2. Accounting scheme for Gross Ecosystem Product of the QMs.

Categoriy	Accounting ES	Material Quantity Method	Monetary Value Method
Provisioning ecosystem service	Agricultural products Forestry products Animal husbandry products Fishery products Water resources	Statistical survey method	market value
Regulating ecosystem service	Water conservation service Water purification service Flood regulating service Carbon sequestration service Oxygen release service Air purification service Climate regulating service Soil conservation service	water balance equation empirical method empirical method Vegetation photosynthesis model empirical method Ecosystem Evapotranspiration Model RUSLE model	shadow project replacement cost shadow project market value replacement cost replacement cost replacement cost
Cultural ecosystem service	Ecological tourism service	Statistical survey method	replacement cost
Gross Ecosystem Product	Total value of 14 ES types	-	accumulation

2.4.2. Coupling Coordination Degree (CCD) Model

As a physical concept, coupling coordination is a common method used to evaluate the degree of interaction between different systems [29,46], which is widely used in the fields of economic development, urbanization, land use, ecological environment, and ecosystem services, etc. In this paper, the Coupling Coordination Degree (CCD) Model is used to study the dynamic interaction between GEP and GDP.

1. Data standardization processing

Before the CCD analysis, in order to eliminate the differences in the dimensions of the indicators for the later calculation and comparative analysis, it is necessary to use the extreme value standardization method to normalize the GEP and GDP data to ensure that the index values of the two data are within the range of [0, 1], which is shown as follows [46]:

$$G = \frac{G_{ij} - G_{min}}{G_{max} - G_{min}} \quad (2)$$

where G refers to the standardized value of GEP or GDP, G_{ij} refers to the original value of GEP or GDP, G_{max} and G_{min} refer to the maximum and minimum of the original value of GEP or GDP, respectively.

2. Coupling Degree (CD) model

Coupling Degree refers to the relationship between two or more systems that interact and influence each other [46,47]. Therefore, we calculate the Coupling Degree of GEP and GDP for judging the relationship between the two systems:

$$C = \frac{2\sqrt{G_1 \times G_2}}{G_1 + G_2} \quad (3)$$

where G_1 and G_2 refer to normalized data of GEP and GDP, respectively, C refers to the coupling degree between GEP and GDP, with a value range of [0, 1]. An increasing value of C indicates a benign relationship between the two systems, and suggests that the development of the two systems tends to follow an orderly pattern; a decreasing value of C indicates a weaker coupling state between the two systems, and the two systems tend to develop in a disordered manner [48]. According to previous research, this paper divides the coupling degree of the two systems into four states ranging from high to low based on the C value [49]: low-level coupling (LC, $0 < C \leq 0.3$), antagonistic development (AD, $0.3 < C \leq 0.6$), running-in development (RD, $0.5 < C \leq 0.8$), and high-level coupling (HLC, $0.8 < C \leq 1$).

3. Coupling Coordination Degree (CCD) model

Coupling Degree can effectively characterize the interaction between multiple systems; however, it solely represents a resonance relationship, which fails to adequately reflect the overall synergy effect and development level between multiple systems [46]. On the basis of Coupling Degree, the Coupling Coordination Degree can provide a deeper reflection of the development of two systems [47]. The calculation formula is as follows:

$$T = \alpha G_1 + \beta G_2 \quad (4)$$

$$D = \sqrt{C \times T} \quad (5)$$

where T represents the comprehensive coordination index of these two subsystems, with a value range of [0, 1], α and β represent the contribution rates of the GEP and GDP subsystems in the eco-economic composite system, respectively, that is, the relative importance of them in the eco-economic system. Social economic development is as important as the ecological environment for a region, so both values are set at 0.5.

D refers to the coupling coordination degree between GEP and GDP, with a value range of [0, 1], which can be obtained by combining the coupling degree and comprehensive coordination index. A high value of D indicates that GEP and GDP are mutually promoting at a high level in the eco-economic system, while a low value of D indicates that they are mutually restricting in the overall system. According to previous studies [29] and the local social development, the coupling coordination degree between GEP and GDP is divided into five levels from high to low (Table 3): severe unbalance (SU, $0 < D \leq 0.2$), moderate unbalance (MU, $0.2 < D \leq 0.4$), slight coordination (SC, $0.4 < D \leq 0.6$), moderate coordination (MC, $0.6 < D \leq 0.8$), and high coordination (HC, $0.8 < D \leq 1$).

4. Relative Development Degree (RDD) model

Though CCD can better reflect the mutual relationship between GEP and GDP in the eco-economic system, however, there is no consideration of the development rate or the development gap between them [39]. Therefore, on the basis of Coupling Coordination Degree results, it is necessary to introduce Relative Development Degree, which calculates the ratio of normalized GEP to GDP, in order to continue measuring relative development status between GEP and GDP in the eco-economic system. It is easy to know whether the regional development status of the ecosystem is ahead of or behind the development of the economy relatively with the following calculation formula:

$$R = \frac{G_1}{G_2} \quad (6)$$

where G_1 and G_2 refer to normalized GEP and GDP, respectively, R represents the relative development status of these two subsystems in the eco-economic system. Based on the relevant research results and the quotient of normalized GEP and GDP, the RDD in the QMs is divided into three types: GEP lag, GDP lag, and Balance [29,39]. Among them, GEP lag indicates that ecology lags behind economic development, GDP lag indicates that the economy lags behind the ecological status, and Balance indicates that ecology and the economy have achieved synchronous and balanced development.

Finally, combined with CD, CCD, and RDD, this paper constructs a comprehensive coupling coordination analysis framework between GEP and GDP in the QMs (Table 3).

Table 3. Coupling coordination levels and types of GEP and GDP.

Level	Classification	CCD	Relative Development Degree (RDD)	CCD Features	Type
1	Severe unbalance	$0 < D \leq 0.2$	$0 < RDD \leq 0.9$ $0.9 < RDD \leq 1.1$ $1.1 < RDD$	Severe unbalance—GEP lag Severe unbalance Severe unbalance—GDP lag	SU—GEP lag SU—Balance SU—GDP lag
2	Moderate unbalance	$0.2 < D \leq 0.4$	$0 < RDD \leq 0.9$ $0.9 < RDD \leq 1.1$ $1.1 < RDD$	Moderate unbalance—GEP lag Moderate unbalance Moderate unbalance—GDP lag	MU—GEP lag MU—Balance MU—GDP lag
3	Slight coordination	$0.4 < D \leq 0.6$	$0 < RDD \leq 0.9$ $0.9 < RDD \leq 1.1$ $1.1 < RDD$	Slight coordination—GEP lag Slight coordination Slight coordination—GDP lag	SC—GEP lag SC—Balance SC—GDP lag
4	Moderate coordination	$0.6 < D \leq 0.8$	$0 < RDD \leq 0.9$ $0.9 < RDD \leq 1.1$ $1.1 < RDD$	Moderate coordination—GEP lag Moderate coordination Moderate coordination—GDP lag	MC—GEP lag MC—Balance MC—GDP lag
5	High coordination	$0.8 < D \leq 1$	$0 < RDD \leq 0.9$ $0.9 < RDD \leq 1.1$ $1.1 < RDD$	High coordination—GEP lag High coordination High coordination—GDP lag	HC—GEP lag HC—Balance HC—GDP lag

2.4.3. Geographic Detector Model

The geographic detector model can better detect the spatial heterogeneity of geographical phenomena, effectively explore the contribution of different driving factors to geographical differentiation, and analyze the impact of interaction between variables on the driving process, which has become an important tool for driving analysis in spatial statistics [50].

1. Factor detection

In this study, the contribution of each single factor (X) to the spatial variation of dependent variable CCD (Y) was investigated by using the factor detection method. The calculation method is as follows:

$$q = 1 - \frac{1}{N\sigma^2} \sum_{h=1}^{L} N_h {\sigma_h}^2 \quad (7)$$

where q represents the contribution degree of a factor (X) to the spatial variation of CCD (Y), h represents the stratification or partitioning of Y or X, N and N_h represent the sample size of the whole region and the sample size of layer h, ${\sigma_h}^2$ and σ^2 represent the variance of the whole region and the intra-layer variance of layer h, respectively. The value of q is between 0 and 1, and the larger the value of q, the higher the contribution of the factor to spatial variation.

2. Interaction detector

Interactive detection can detect the interpretation degree of the spatial variation of CCD when two factors (X1 and X2) act simultaneously.

Firstly, the q values of X1 and X2 contributing to the change of Y, namely q(X1) and q(X2), are calculated, and then the interaction q value (q(X1∩X2)) is calculated. The interaction relationship between the two factors is determined by judging the value of q(X1), q(X2), and q(X1∩X2). For the specific calculation principle and process, refer to the literature [50].

On this basis, the interaction of the two factors can be divided into five types: nonlinear weakening, single-factor nonlinear weakening, two-factor enhancement, independent, and nonlinear enhancement. When q(X1∩X2) > max (q(X1), q(X2)), the relationship between X1 and X2 is two-factor enhanced; when q(X1∩X2) > q(X1) + q(X2), the relationship is nonlinear enhanced [50].

3. The selection and processing of indicators

The spatial differentiation of regional CCD is influenced by natural environment conditions and local social economic development. Therefore, CCD is taken as the explained variable (Y) of this study. Based on the natural environment characteristics, regional development status of the QMs, and relevant studies [51,52], we select temperature (TEM, X1), precipitation (PRE, X2), altitude (X3), slope (X4), NDVI (X5) as key natural environmental factors, Aggregation index (AGI, X6), Landscape shape index (LSI, X7), Shannon diversity index (SDI, X8), Shannon evenness index (SEI, X9), Contagion index (CAI, X10) as Landscape pattern indices, and density of road network (RND, X11), and density of population (POD, X12) as factors which reflect social economic development (Table 4).

Table 4. Driving factors for the CCD between GEP and GDP.

Type	Factor	Abbreviation	Code	Unit
Geographical conditions	Temperature	TEM	X1	°C
	Precipitation	PRE	X2	mm
	Altitude	Altitude	X3	m
	Slope	Slope	X4	°
	NDVI	NDVI	X5	-
Landscape features	Aggregation index	AGI	X6	%
	Landscape shape index	LSI	X7	-
	Shannon diversity index	SDI	X8	-
	Shannon evenness index	SEI	X9	-
	Contagion index	CAI	X10	%
Social economic features	Road network density	RND	X11	km/100 km^2
	Population density	POD	X12	persons/km^2

In data processing, the vector layers of six landscape pattern indices from 2010 to 2020 were calculated using the moving window method in Fragstats 4.2, based on the land use data. At the same time, since geographic detectors are effective at processing type variables, this study first processed variables X1–X12 as categorical variables using the natural split point method in the ArcGIS 10.6 platform. Then, according to the boundaries of different districts and counties, the type variables of different indicators in different regions are obtained using the tool Zonal Statistics as Table in the ArcGIS platform, so that the driving force analysis can be carried out with the geographic detector.

3. Results

3.1. Spatial Variations of ESs and GEP

3.1.1. Variations of PES from 2010–2020

From 2010 to 2020, PES in the QMs had obvious spatio-temporal changes (Figure 3). In 2010, PES took on a spatial pattern of high in the east and low in the west. The lowest

PES values were scattered in Langao County, Tongguan County, Baqiao District, and Foping County, with PES lower than 0.01 billion CNY, while the highest PES values were distributed in Zhouzhi County and Weibin District, with PES higher than 2.00 billion CNY. In 2015, the PES of most districts and counties showed a certain increasing tendency, and only Weibin District, Linwei District, and Lintong District had a certain decrease in PES. In 2020, the increase rate of PES in various districts and counties was growing faster; only the PES of Chang'an decreased, while the increasing amount of Weibin exceeded 4 billion CNY. A total of eight districts and counties in the region exceeded 2.00 billion CNY.

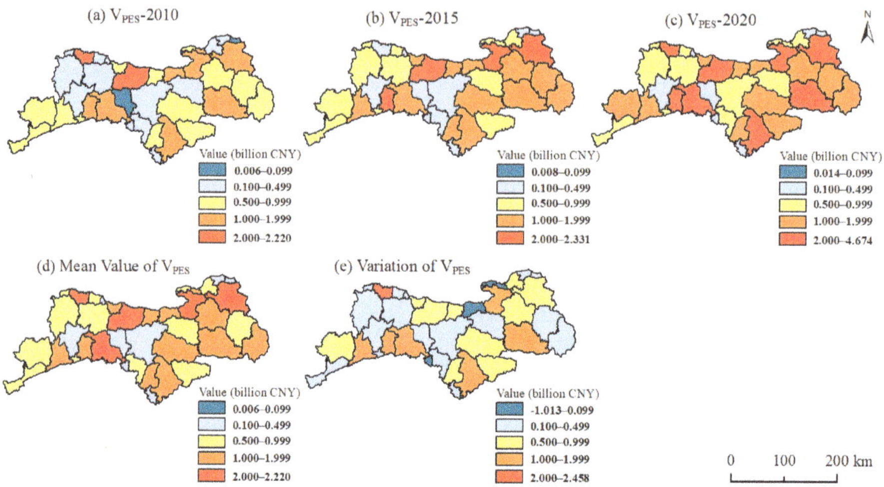

Figure 3. Spatial distribution of the monetary value of PES in the QMs from 2010 to 2020.

In the whole QMs, the cumulative PES values of all districts and counties in 2010, 2015, and 2020 were 29.28 billion, 35.86 billion, and 51.30 billion CNY, respectively. Over the years, PES in the whole area showed a continuous upward tendency, with a growth rate of 75.17%. In terms of stage change, the growth rate of PES in the first five years was 22.44%, while the greater growth rate in the second five years reached 43.06%.

3.1.2. Variations of RES during 2010–2020

From 2010 to 2020, the spatial pattern of RES showed the distribution characteristics of high values in the east and west, and low values in the north and south (Figure 4). In 2010, the highest values were found in Ningshan County, Zhen'an County, and Shanyang County in the eastern and central QMs, with RES values exceeding 20 billion CNY. In 2015, a total of 28 districts and counties saw a decrease in RES compared with that in 2010, among which the reduction in RES of Shangzhou District and Shanyang County was more than 2.0 billion CNY. In 2020, RES in the whole region had a relatively obvious downward tendency, and a total of 35 districts and counties had an increase in RES compared with 2015, including six districts and counties with an increase of more than 2.0 billion CNY. From 2010 to 2020, a total of 15 districts and counties in the region showed a downward trend in RES, and almost all these districts and counties were distributed in the eastern region. Therefore, it was seen that although the RES value in the eastern region was larger, the downward trend was also more obvious.

In the whole region, the cumulative RES values of all districts and counties in 2010, 2015, and 2020 were 373.22 billion, 355.60 billion, and 390.05 billion CNY, respectively. RES in the whole region of the QMs had a fluctuation trend with a growth rate of 4.5% in the decade. In terms of stage change, the first five years showed a downward trend with a decline rate of 4.7%, and the second five years showed an upward trend with a growth rate of 9.7%. However, from the perspective of the magnitude of change, the changes in each county and district in this decade were not large.

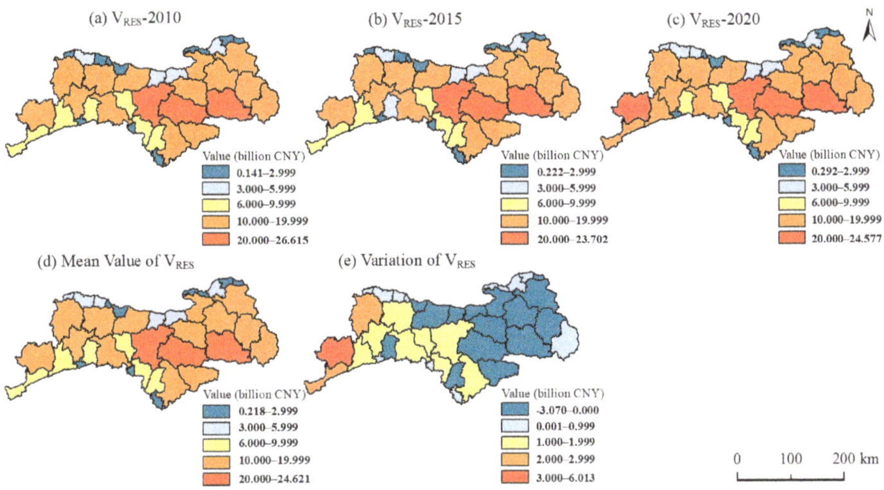

Figure 4. Spatial distribution of the monetary value of RES in the QMs from 2010 to 2020.

3.1.3. Variations of CES during 2010–2020

The spatial pattern of CES underwent significant changes from 2010 to 2020 (Figure 5). In 2010, CES was generally higher in the east and lower in the west. There were five districts and counties with a value less than 0.10 billion CNY and five districts and counties with a value more than 2.00 billion CNY. In 2015, CES in the whole region increased significantly, and only Luonan County and Hanyin County showed a slight decline in CES. In 2020, there was a downward trend in eight districts and counties and an obvious increase in other regions in CES. In particular, Linwei District exceeded 20.00 billion CNY. In terms of total volume, seven districts and counties in the region were above 5 billion CNY in CES, and only nine districts and counties were below 1 billion CNY.

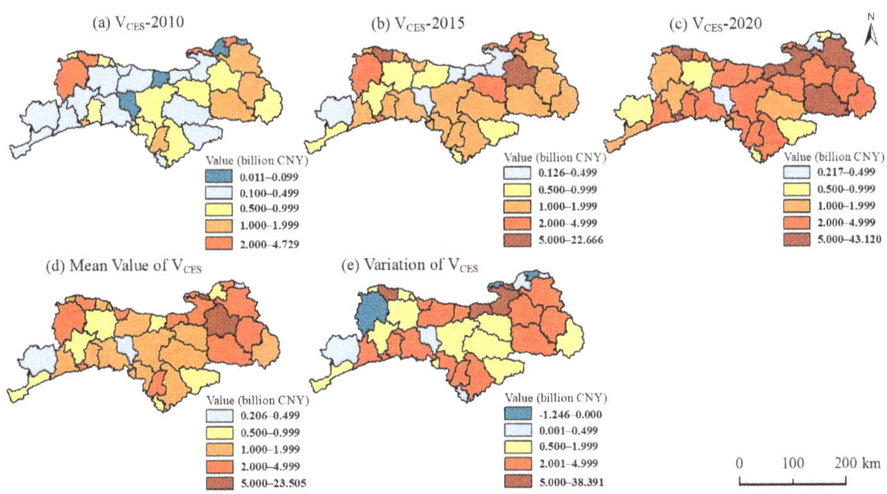

Figure 5. Spatial distribution of the monetary value of CES in the QMs from 2010 to 2020.

For whole region, in 2010, 2015, and 2020, the cumulative CES values were 32.88 billion, 89.62 billion, and 148.24 billion CNY, respectively. The CES of the whole QMs showed a very obvious upward trend with the growth rate of 350.9% in the past decade. In terms of phase change, the growth rate of the first five years was 172.6%, and the growth rate of the second five years was 65.4%.

3.1.4. Variations of GEP during 2010–2020

From 2010 to 2020, the spatial pattern of GEP exhibited relatively similar spatial distribution characteristics to RES, namely high values in the east and west, and low values in the north and south (Figure 6). In 2010, the GEP values of all districts and counties were generally low, within the range from 0.45 to 29.46 billion CNY, and a total of 13 districts and counties had less than 6 billion CNY in GEP. The GEP values of Shangzhou District, Luonan County, Ningshan County, Zhen'an County, and Shanyang County were the highest, all exceeding 20 billion CNY. In 2015, only 11 districts and counties experienced a decrease in GEP, while the remaining districts and counties witnessed a significant increase in GEP. Notably, Linwei District's GEP surpassed 10 billion CNY. In 2020, a decrease in GEP was observed in a total of five districts and counties, while the remaining areas exhibited a substantial increase in GEP. Specially, Linwei District witnessed an impressive surge of over 20 billion CNY in its GEP. A total of eight districts and counties had a GEP less than 6 billion CNY and two districts and counties achieved a GEP greater than 30 billion CNY. During 2010–2020, the average GEP of 12 districts and counties was lower than 6 billion CNY, while eight districts and counties had an average GEP of more than 20 billion CNY. The GEP of Huyi District and Lintong District exhibited a declining trend from 2010 to 2020, while Hanbin District, Lintong District, Lueyang County, Chang'an District, Weibin District, and Linwei District experienced prosperity with an increasing GEP. Particularly, the added value of GEP in Linwei District surpassed 40 billion CNY over the past decade.

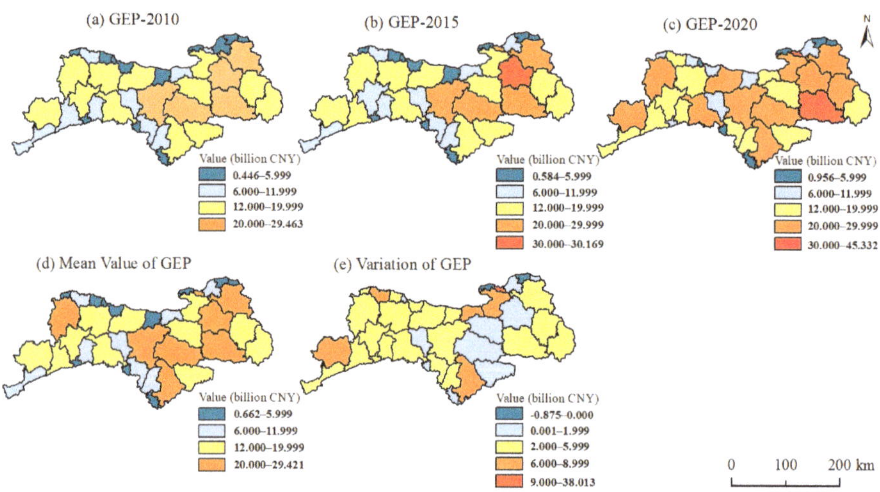

Figure 6. Spatial distribution of GEP in the QMs from 2010 to 2020: (**a**–**c**) GEP Value, (**d**) mean value of GEP during 2010–2020, (**e**) variations of GEP during 2010–2020.

In 2010, 2015, and 2020, the cumulative GEP values of all districts and counties were 435.38 billion, 481.07 billion, and 589.58 billion CNY, respectively. In the past ten years, the GEP of the QMs showed a very continuous upward trend, with a growth rate of 35.4%. Meanwhile, the growth rate of the first five years was 10.5%, while the greater growth rate of the last five years reached to 22.6%.

The GEP of prefecture-level cities in the northern QMs was generally lower, while the GEP of prefecture-level cities in the southern QMs was generally higher (Figure 7). The list of cities ranked from lowest to highest in GEP were the following: Weinnan City, Xi'an City, Baoji City, Ankang City, Hanzhong City, and Shangluo City.

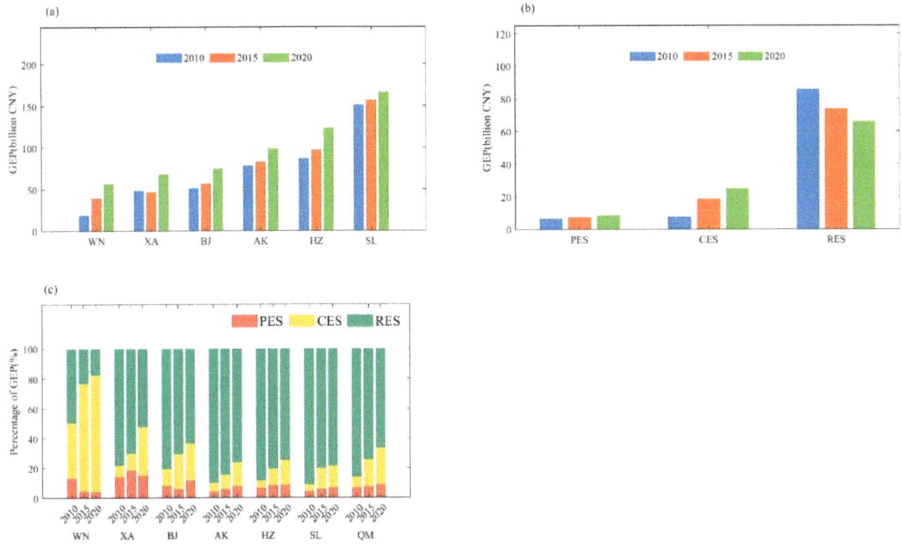

Figure 7. Temporal variations of GEP in prefecture-level cities (**a**), ESs in the QMs (**b**), and in prefecture-level cities (**c**) from 2010–2020.

The economic value of RES constituted a significant proportion of the composition of GEP in the QMs. This was mainly because the QMs were key protection areas with well-protected ecological environments and high levels of biodiversity, allowing for effective utilization of ESs such as climate regulation. From 2010 to 2020, the proportion of RES in GEP in the QMs showed a continuous decreasing trend, while the proportion of PES and CES showed a continuous increasing trend. This was because in recent years the social economy, production activities, culture, and tourism industries in the QMs had developed, and the proportion of PES and CES in GEP increased significantly.

The GEP composition of these cities also reflected a relatively similar pattern of evolution. From 2010 to 2020, the proportion of CES in each city showed a continuous upward trend, and the proportion of RES showed a downward trend. The PES proportion in the southern QMs showed a certain increasing trend, while the proportion in the northern QMs showed an inter-annual fluctuation.

From a regional perspective, the proportion of CES in GEP is generally higher in the northern QMs. This can be attributed to their strategic location in the culturally rich and historically significant Guanzhong region, which boasts abundant cultural relics, scenic spots, and tourism resources that attract a relatively high level of tourism income. Among these regions, the contribution of CES to GEP in Weinnan City exceeded 60%.

3.2. Spatiotemporal Variations of the Coupling Relationship between GEP and GDP

3.2.1. Evolution Characteristics between CD and CCD

In 2010, the Coupling Degree (CD) between GEP and GDP was relatively low among districts and counties in the QMs, and the spatial distribution difference was significant (Figure 8). The districts and counties with a spatial coupling degree less than 0.6 (belonging to low-level coupling and antagonistic development states) accounted for 58.97%, mainly located in the central QMs. There were 18 districts and counties in antagonistic development state, and the CD values of Baqiao District and Langao County were the lowest. The districts and counties with high CD values accounted for 41.03%, mainly located in the northern and southern edge of QMs. The CD values reached the high-level coupling state in nine districts and counties, mainly distributed in the area to the north of the watershed, in which the Chang'an District and Weibin District had the highest CD values.

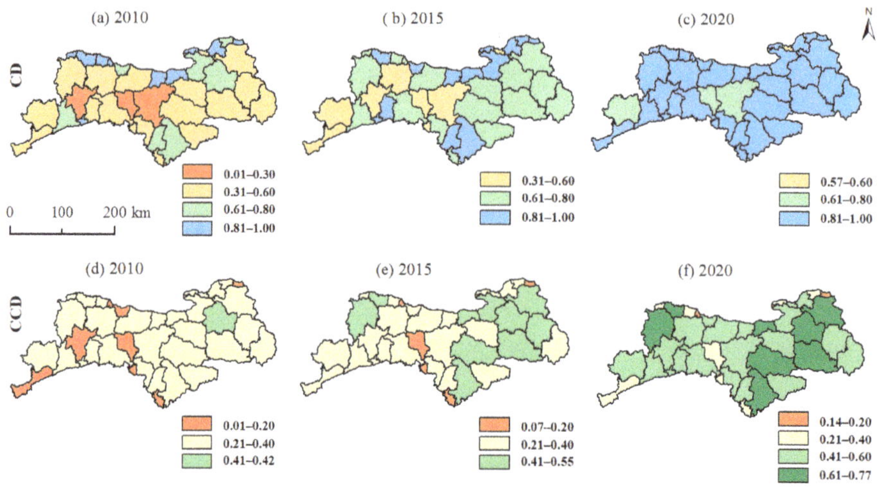

Figure 8. Spatial distribution of CD and CCD between GEP and GDP in the QMs from 2010–2020.

In 2015, the coupling degree between GEP and GDP of the districts and counties was greatly improved. In space, the CD values of districts and counties were greater than 0.3, so there was no area of low-level coupling. The CD values of 10 districts and counties were in an antagonistic development state and these areas were mainly distributed in the western QMs. In the east, the CD values were generally higher, and the districts and counties with running-in development and high-level coupling state accounted for 16 and 13, respectively.

In 2020, the coupling degree between GEP and GDP improved significantly. The districts and counties with a high-level coupling state accounted for 84.62% of the whole region. During these years, there had been a significant improvement in the CD values across all regions, leading to an attainment of a relatively ideal state of the coupling relationship between GEP and GDP in 2020.

As for the Coupling Coordination Degree (CCD) between GEP and GDP in the QMs, this was relatively low in 2010, among which Baqiao District and Langao County had the lowest CCD values. There were 10 districts and counties in a severely unbalanced state, 28 districts and counties in a moderately unbalanced state, and only Shangzhou District in a slight coordination state.

In 2015, CCD in the QMs improved, and CCD in the eastern QMs improved significantly. The highest values of CCD were detected in Weibin District in the west and Shangzhou District in the east. In the central QMs, CCD was generally low in unbalanced development (severely unbalanced and moderately unbalanced state). The lowest CCD values of the whole region were distributed in Langao County and Baqiao District.

In 2020, CCD in the region increased significantly, and the scope of balanced development between GEP and GDP expanded. There were only 11 districts and counties with unbalanced CCD, which were mainly found in the northern and southern edge of the QMs. The CCD level in the eastern QMs was generally high, with six districts and counties' CCD reaching a high coordination level. The CCD level in the central and western QMs was also relatively high. The highest values of CCD were seen in Hanbin District in the west and Weibin District in the east. Therefore, over these years, the CCD between GEP and GDP of all districts and counties showed a continuous upward trend, and the number of districts and counties reaching the coupling coordination state continued to increase. In 2020, most of the counties and counties in the QMs obtained a good coupling and coordination between GEP and GDP.

3.2.2. Evolution Characteristics of RDD and CDD Type

According to the relative development degree (RDD) of GEP and GDP, in 2010, the GDP of almost all districts and counties lagged behind the development of GEP (Figure 9). Only Weibin District, Baqiao District and Chang'an District were GEP lag areas. In 2015, the regional economy had a certain development in comparison with the ecological background, and there was a district, Huyi District, where the GEP and GDP reached a balanced development state. At the same time, Hantai District and Huayin City changed into GEP lag regions. In 2020, the spatial difference of the relative development of regional GEP and GDP was more obvious. In the eastern QMs, Lintong District, Shangzhou District, Luonan County, and Xunyang County developed into GEP–GDP balanced regions. The distribution range of the GEP lag area was remarkably large, with a total of 19 districts and counties mainly distributed in the northern fringe and the southeastern QMs. There were 16 districts and counties with GDP lag, mainly distributed in the central QMs.

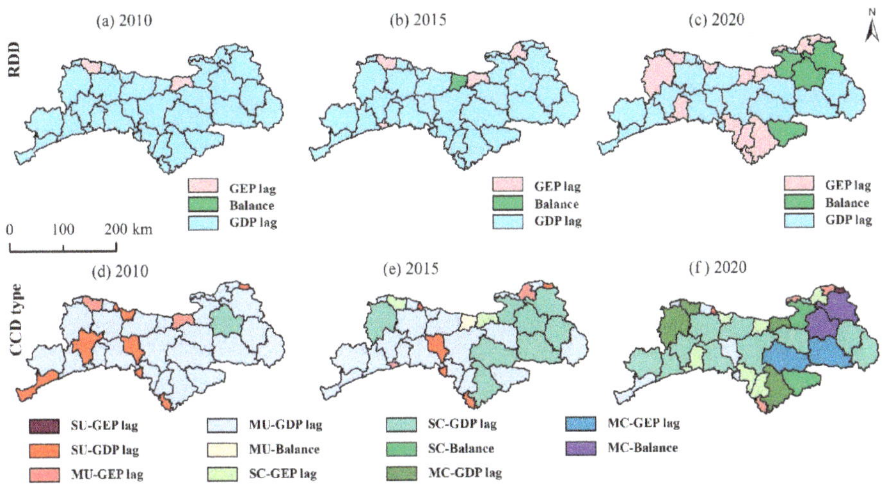

Figure 9. Spatial distribution of RDD and CCD between GEP and GDP in the QMs from 2010–2020.

Based on relative development degree, combined with the corresponding CCD development status of each district and county, different development types were distinguished, and the coupling coordination relationship between GEP and GDP of each district and county was further analyzed. It could be seen that there were 11 types of coupling coordination relationship between GEP and GDP in all districts and counties of the QMs from 2010 to 2020.

Specifically, in 2010, only the interaction between GEP and GDP in Shangzhou District achieved slight coordination, so this region was in the state of GDP lag and coupling coordination development. MU–GDP lag regions were widely distributed in the QMs, accounting for 26 in total, which indicated that most regions of the QMs were in moderate unbalance and GDP lagged state. The regional distribution of SU-GDP lag exhibited a relatively large scale, accounting for nine districts and counties, mainly distributed in the western QMs.

In 2015, the relative development degree and CCD between GEP and GDP of several districts and counties changed greatly, which led to certain changes in the CDD types of 15 districts and counties in the QMs. Spatially, the distribution of MU–GDP lag was still relatively wide, but the total number reduced to 19. At the same time, three districts and counties changed from SU–GDP lag to MU–GDP lag, two districts and counties changed from MU–GDP lag to SC–GDP lag, and seven districts and counties changed from MU–GDP lag to SU–GDP lag. It indicated that, although the relative development degree of GEP and GDP in these regions did not change, the CDD type did change due to the change of CDD level.

In 2020, there were more CDD type changes in the whole region, reaching 33, and unchanged types were only found in six districts and counties. In total, two districts and counties changed from MU–GEP lag to SC–GEP lag, nine districts and counties changed from MU–GDP lag to SC–GDP lag, two districts and counties changed from SC–GEP lag to MC–GEP lag, and two districts and counties changed from SC–GDP lag to MC–GDP lag. Spatially, a total of four districts and counties reached simultaneous coupling coordination and balance, all of which were distributed in the eastern region, namely Luonan County, Shangzhou District, Lintong District, and Xunyang County. It showed that the development of the social economy and the protection of ecological environment in these regions was coordinated well and maintained in a good balance.

3.3. *Driving Factors Affecting the Coupling Coordination Degree between GEP and GDP*

3.3.1. Factor Detection Analysis

From 2010 to 2020, there were significant differences in various driving factors influencing the CCD between GEP and GDP in the whole region (Figure 10). In 2010, the change of CCD was mainly influenced by elevation with the explanatory power q value surpassing 0.5. The other driving factors, such as PRE, CAI, TEM, and POD, exceeded 0.3 at a moderate intensity, indicating that these factors also had a great influence on the change of CCD. In 2015, the change of CCD was mainly influenced by elevation and TEM, both of which had q values above 0.4, but not above 0.5. At the same time, it also reflected that the influence of TEM on CCD had been enhanced, while the influence of elevation had a certain downward trend. In addition, CAI, PRE, NDVI, SEI, and other factors had great influence on CCD, and their q values were all higher than 0.3. It can be seen that the influence of PRE and ROD on CCD waned, while the influence of NDVI on CCD significantly enhanced. In 2020, CAI had significantly enhanced its influence on CCD ($q > 0.4$) and became the main driving factor of regional CCD change; SDI's q value also increased greatly, and it became the driving factor of CCD change along with TEM, elevation, POD, and NDVI ($q > 0.4$), while the influence of PRE on CCD decreased ($q < 0.2$).

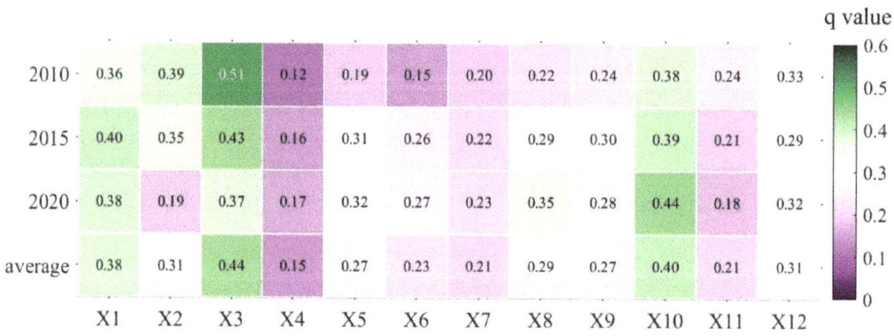

Figure 10. Contribution rates of driving factors of CCD between GEP and GDP from 2000 to 2020.

From the perspective of multi-year average state, q values of elevation (X3), CAI (X10), TEM (X1), POD (X12), and PRE (X2) were relatively large, which indicated that natural factors and social economic development had a great impact on the coupling interaction between GEP and GDP in the QMs. However, LSI (X7), ROD (X11), slope (X3), and other factors had limited influence on the coupling relationship between GEP and GDP. In addition, the influence of different factors on CCD had a large inter-year fluctuation, which also indicated that CCD was more sensitive to changes in the social economy and natural environment.

3.3.2. Interaction Detection Analysis

By employing interaction detection, the impact of driving factors on the spatial differentiation of CCD in different years can be further investigated. These years, under the pairwise interaction of influencing factors, the contribution degree q to the spatial variation of CCD was greatly enhanced (Figure 11). These results indicated that these factors had a good synergistic effect on the spatial variation of CCD in the QMs and could promote or inhibit the spatial variation of CCD at the same time. From the perspective of the category, these interaction types were mainly manifested as nonlinear enhancement, and a small number of factors were manifested as bifactorial enhancement.

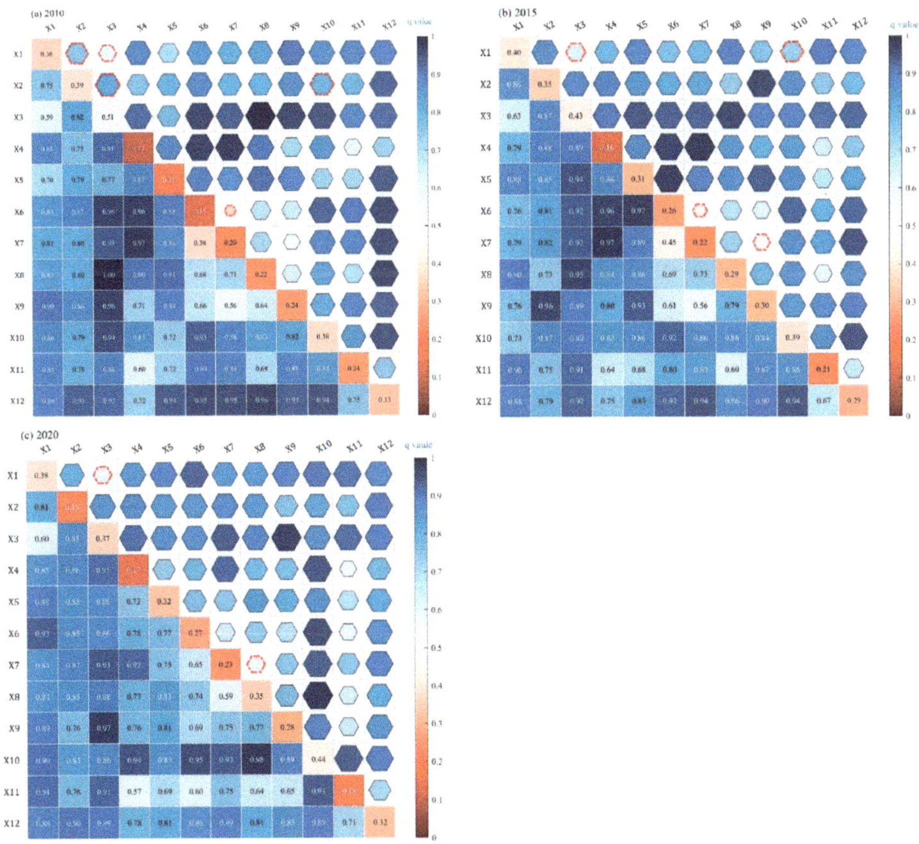

Figure 11. Interaction detection results of driving factors of CCD between GEP and GDP from 2000 to 2020. Hexagons with red dashed edges represent the interaction type as bivariate enhancement, while the rest indicate the interaction type as nonlinear enhancement.

In 2010, the interaction between elevation and SDI was strong, and the strongest interaction occurred between elevation and SDI (q = 0.995). In addition, the interactions between TEM and SEI, and slope and SEI were also strong ($q > 0.9$). On the whole, ROD had the most significant influence in the interactions. The q values of its interactions with seven factors exceeded 0.9.

In 2015, the interaction of the two factors was relatively weaker than that of 2010. A total of eight groups of factors had a significant interactive impact ($q > 0.9$). The strongest interaction was between slope and LSI ($q = 0.938$). In addition, three groups had an interaction effect value of more than 0.9 with POD, indicating that population density still played a key role in the change of CCD at this stage, and had a greater impact on the change of CCD through the interactions with other indicators.

In 2020, there were still eight groups of factors with a significant interactive impact ($q > 0.9$), and four groups of CAI and other factors had a significant interactive impact ($q > 0.9$). The strongest interaction was between CAI and SDI ($q = 0.977$). In addition, over the course of three years, the interaction effect of population density on CCD showed a consistent decreasing trend.

4. Discussion

4.1. Implications of GEP Assessment

The comprehensive evaluation of regional ecosystem function and the integrated accounting of ecosystem value are multifaceted processes [19,26,53]. In the previous calculation process, the quality assessment makes accurate calculations of different ESs on the basis of different localized parameters and models, which required a large workload. However, the types of ESs vary widely, so it was difficult to directly sum up the total value of ESs. In contrast, the value assessment sets the ESs proper value coefficients, so as to estimate their value at each unit area simply and intuitively [24,54]. However, ignoring the heterogeneity of ecosystems in different spatial ranges would bring about difficulties in characterizing the dynamic change process of ecosystem quality, which would obstruct the application of this method.

GEP accounting integrates the advantages of the two methods and overcomes these research gaps well. First, the biological mechanism model in GEP accounting can comprehensively evaluate the quantity of ESs, taking of the heterogeneity of ESs fully into consideration. Subsequently, GEP adopts the market value method to further calculate the economic value of each ES and the total value of the regional ecosystem.

From 2010 to 2020, GEP in the QMs showed a trend of continuous growth, which was mutually confirmed with the continuous construction of ecological projects and the continuous improvement of ecological environment quality in the QMs. The north slope was close to the urban agglomeration of Guanzhong Plain, and the ecosystem was greatly disturbed by urbanization, so the GEP was relatively small. The southern slope was the core area of the QMs with better ecological environment quality. As for GEP composition, RES played a leading role. The stability and health of the ecosystem in the southern QMs ensured the continuous operation of RES and maintained the stable development of the local GEP. Secondly, the proportion of RES in GEP had consistently decreased from 2010 to 2020, while the proportion of PES and CES continued to rise, which indicated that the social economy of the QMs had achieved significant development. The relationship between the social economy and GEP, as well as the impact of social economic development on the ecological environment, deserves more attention in the QMs.

Over the years, the concepts of natural capital, ecological assets, ESs, GEP, and the contribution of nature to human beings have been put forward successively, and the supporting role of the ecosystem to human social economic development has been analyzed deeply. GEP accounting effectively integrates physical quantity and ecological asset value, and can systematically evaluate the overall benefits of the ecosystem to human society. The framework can carry out value accounting for forest, grassland, desert, and urban ecosystems, and can integrate detailed accounting indicators and localized parameters on the basis of more accurate ecosystem assessment at the national or regional level. Therefore, it can be widely used in the assessment of the overall status of the regional ecosystem and ecological protection effectiveness, the accounting of ecological assets, the implementation of inter-regional ecological compensation, and the evaluation of social sustainable development.

4.2. Interaction between GEP and GDP

The relationship between the ecosystem and economic development is characterized by mutual constraints and reciprocal influences. Only by promoting the benign interaction and coupling coordination of ecology and the economy can we gradually realize the sustainable development of society. The preservation of a sound ecological environment and the provision of high-quality ecosystem services are the primary objectives for regional development in the QMs. However, this region is also a typical poverty-stricken area, so it is equally important to improve the life and well-being of local residents. Therefore, it is imperative to effectively harmonize social economic development with the ecosystem and appropriately foster social economy while ensuring the ecological environment's quality and ecosystem health. This is pivotal for achieving sustainable and high-quality development in local society.

In this study, the coupling degree and the coupling coordination degree between GEP and GDP in the QMs showed a continuous improvement from 2010 to 2020. In terms of CD, most districts and counties were in a state of high-level coupling between GEP and GDP. In terms of CCD, similarly, in 2020, most districts and counties in the QMs showed slight and moderate coordination between GEP and GDP, but there was no highly coordinated development between GEP and GDP in the region. There was a significant difference between CD and CCD relationship in this region. It can be seen that, on the one hand, the two methods can fairly consistently represent the regions where GEP and GDP have a close relationship and develop in an orderly manner. On the other hand, CCD can further reflect the synergic development effect of the two subsystems as compared with CD. Therefore, CCD is a more accurate way to characterize the interaction between GEP and GDP.

Secondly, combined with the relative development degree, the development difference of GEP and GDP in the QMs can be further evaluated. Due to the ecological background and development orientation of nature reserves, most districts and counties exhibited a lag in GDP development compared to GEP. Therefore, in 2020, a certain number of districts and counties showed a balance between GEP and GDP, and many districts and counties showed the GEP lag development. Generally, due to the stability of the ecosystem structure of a region, GEP will remain in a relatively stable condition without the influence of external factors or human activities, and the situation of slow growth or stagnation may occur. At the same time, the growth rate of GDP may be significantly faster than that of GEP. Therefore, although GEP in the QMs showed a trend of continuous increase over the years, the GDP growth rate was more rapid, resulting in GEP lagging behind GDP development in many districts and counties. Combined with the relative development degree and CCD model, the coordinated development status and mutual relationship between GEP and GDP of various districts and counties in the QMs can be more clearly understood, providing a reference for the formulation of policies related to regional ecosystem management and social sustainable development.

From the single-factor-driven analysis process of CCD, both natural factors and social and economic development have great impacts on the CCD between GEP and GDP. In addition, the influence of different factors on CCD had a large inter-year fluctuation, which also indicates that CCD is more sensitive to changes in the social economy and natural environment. Elevation, CAI, temperature, population density, and precipitation are the main driving factors of CCD in the QMs. According to the interaction analysis, the influencing factors have a good nonlinear enhancement effect on the spatial variation of CCD. Among these years, elevation and SDI, slope and LSI, and CAI and SDI have the strongest interaction influences, which indicates that landscape ecological pattern interaction plays an important driving role in the spatial variation of CCD in the QMs. The POD had an obvious enhancement trend with other factors. Population density represents the impact of human activities on the natural environment and social economic conditions, so the population density plays a key role in the change of CCD. The alteration in population density will induce the modifications in other natural factors or social and human factors, thus intensifying their impacts on CCD. Therefore, in the process of sustainable development of the QMs, we need to pay more attention to the change of population density.

4.3. Policy Recommendations

The unique geographical location and mountain features have created diverse natural environment features and various ecosystems in the QMs. There are 12 national nature reserves and 21 provincial nature reserves in the QMs, so the QMs are important ecological barriers in central China and key national ecological functional areas. Therefore, in such nature reserves, the red line of ecological protection should be strictly observed according to the national main function zoning and regional ecological civilization construction. It would ensure that development, construction, and economic and industrial activities are not carried out within the red line, effectively protecting biodiversity, ecosystem stability and ecological environment quality. It would provide basic support for sustainable social development and ecological civilization construction in the region and surrounding areas.

Second, from the perspective of land use, the development of regional production and life activities will have certain encroachment or influence on regional ecological space [55–58]. However, in ecological functional areas, economic development and people's well-being are also important contents of social development. Therefore, scientific and comprehensive research on the relationship between GEP and GDP in ecological functional areas should be carried out and reasonable economic development goals should be set to ensure that the economy operates within a controllable interval or threshold. There will be a good prospect through adjusting industrial structure, changing the mode of economic development, reducing the environmental impact of industrial development, and effectively promoting the balanced development of regional ecology and economy.

Thirdly, the ecological environment and ESs are both regional and external. As a nature reserve, apart from fulfilling the fundamental needs of local residents for ESs, it will also provide enhanced support for the maintenance of the ecological environment and social economic development of the surrounding region. Therefore, it is urgent to deepen the market value evaluation system of ecological products and ESs, and carry out an applicable GEP evaluation in the QMs. Only by fully understanding the real situation and evolutionary law of the local ecosystem, and quantitatively assessing the ecological relationship between different ESs types and the surrounding beneficiary areas in the QMs, can we determine the contribution degree of the ecosystem to local and regional socio-economic development and human welfare. On the basis of the ecological compensation system, economic compensation should be imposed for the loss of regions due to their inputs to protect ecosystems and the environment, and forgone opportunities for social economic development. Only in this way can we effectively ensure better coordination and sustainable development between regional ecology and economy for some ecological functional areas that are unable develop their economy on a large scale.

Fourth, the relationship between the ecosystem and the social economy is complicated. In the early stage of development, the rapid increase in GDP was obtained through the massive consumption of natural resources, which led to great destruction of the ecological environment, and there was a great contradiction between the two. After reaching a turning point in economic development, the regional industrial structure has undergone an upgrade, leading to advancements in environmental protection technology and scientific research. Consequently, the ecological environment quality has been enhanced by economic progress. In this study, after ten years of development, the coupling coordination relationship between GEP and GDP in most areas is in a benign state, which indicates that while the social economy of various districts and counties is developing rapidly, the ecological benefits and welfare provided by the ecosystem for the development of human society are also increasing steadily. However, from the perspective of relative development, until 2020, the development of GEP in many areas had lagged behind the development of the social economy. This shows that although the social economy and GEP in these places show synergistic growth, the growth rate of GEP is relatively slow, and the development of both has not yet attained an optimal equilibrium. Therefore, it is necessary to combine the coupling coordination degree with relative development degree to deeply study the relationship between GEP and GDP in different regions, delineate different types of the relationship between the two spatially, and conduct spatial management by classification. We should focus on the districts and counties whose GDP lags behind the development of the ecological environment. While pursuing economic growth, we should fully consider the possible impact of development on the ecosystem, rationally utilize natural resources, and gradually realize the harmonious development of ecology and economy.

4.4. Limitations and Future Research

In different regions, there are great differences in ES types. In the study of GEP in the QMs, typical and important ESs were selected for ecological value accounting according to regional characteristics. Some regulating services, such as pest control, wind prevention, and sand fixation, etc., were not taken into account. In addition, in the accounting of ecosystem cultural functions, most studies take the tourism resource value of natural landscape as representative. Therefore, in the current GEP accounting studies, due to the variations in the selection of ES types and inherent limitation of capturing all local ES types, the obtained results will be smaller than the true GEP value.

In addition, the selected model and specific parameters will also have an impact on the calculation results. First of all, for the same ES, the selection of different models and different parameters will inevitably have an impact on the final result. Second, the calculation of GEP involves not only quantitative data such as physical geography, but also a lot of social economy data. The latter type of data are often statistical data, which are difficult to characterize in a spatial raster, so there is a problem of scale mismatch between the GEP and raster data in model calculation. Therefore, the calculation results of some ES types can only be counted in districts and counties. This affects the accuracy of calculation results. Third, due to data limitations, it is difficult to make long-term time series or year-by-year ecosystem assessments.

Of course, as a relatively innovative assessment framework of ES value, GEP can comprehensively assess the ecological value of the final products and services provided by the ecosystem for human survival, and effectively incorporate environmental and ecological benefits into the evaluation of social economic development. It provides scientific reference for ecological asset assessment and ecological civilization construction. From the perspective of practice, although GEP accounting standards have been preliminarily established at the national, provincial, and regional levels, there are still differences in accounting standards and ES types, which need to be further deepened and improved. Meanwhile, due to the heterogeneity of geographical environment and regional differences of social economic factors, the applicability of different models and the localization of parameters in ES assessment still need to be strengthened. The accuracy and accessibility of data is a necessary condition for conducting research. If the grid processing of social economic data and the collection density of social statistics data can be improved, the accuracy of ES value assessment will be enhanced.

5. Conclusions

In this paper, the QMs region was taken as a remarkable example, and multi-source data including meteorological data, environmental data, remote sensing data, and social economic data were used to quantitatively assess the spatio-temporal evolution of ecosystem services and GEP, and the coupling coordination relationship between GEP and GDP.

The results show that: (1) From 2010 to 2020, the GEP in the whole QMs showed a continuous upward trend, and the spatial pattern of GEP exhibited the distribution characteristics of high values in the east and west, and low values in the north and south. (2) The GEP composition in the QMs included RES, PES, and CES, of which RES accounted for a relatively large proportion but has been declining in recent years. (3) The degree of coupling coordination between GEP and GDP of all districts and counties had shown a continuous upward trend, and the number of districts and counties reaching the state of coupling coordination continues to increase. By 2020, most of the counties and counties had achieved a satisfactory coupling coordination state between GEP and GDP. (4) The social economy of most districts and counties lagged behind GEP in 2010. However, this spatial pattern has been changing over time, with an increase in the number of districts and counties lagging in GEP in 2020. This indicates significant social economic development in the region. (5) Elevation, contagion, temperature, population density, and precipitation are the main driving factors influencing coupling coordination state between GEP and GDP. In the future, further research is needed to understand the relationship between the social economic development and GEP in these regions, such as nature reserves. Corresponding countermeasures should be taken to ensure the coordinated and sustainable development of the regional social economy and ecological environment.

Supplementary Materials: The following supporting information can be downloaded at: https://www.mdpi.com/article/10.3390/land13020234/s1, Table S1: Detailed accounting methods of Provisioning ecosystem service; Table S2: Detailed accounting methods of Regulating ecosystem service; Table S3: Detailed accounting methods of Cultural ecosystem service.

Author Contributions: Conceptualization, P.W. and K.L.; methodology, Y.C. and P.W.; software, Y.C. and P.W.; validation, Y.C., K.L. and P.W.; formal analysis, P.W. and Y.C.; investigation, Y.C., X.L. and L.Z.; resources, P.W., K.L. and L.C.; data curation, Y.C., P.W. and L.C.; writing—original draft preparation, P.W., Y.C. and L.C.; writing—review and editing, P.W., X.L., L.Z., L.C. and T.S.; visualization, P.L. and P.W.; supervision, G.Y., H.W., S.G. and J.Y.; project administration, J.Y. and K.L.; funding acquisition, K.L., P.W. and T.S. All authors have read and agreed to the published version of the manuscript.

Funding: This research was funded by the Natural Science Basic Research Program of Shaanxi Province of China (2021JQ-768), the Scientific Research Project of Shaanxi Provincial Education Department (21JK0306), National Natural Science Foundation of China (42001132), the Key Forestry Science and Technology Innovation Projects in Shaanxi Province (SXLK2023-02-4), the Social Science Foundation project of Shaanxi Province (2023H012), and the Social Science Planning Research Project of Shandong Province (22CGLJ40).

Data Availability Statement: All data and materials are available upon request.

Conflicts of Interest: The authors declare no conflicts of interest.

References

1. Chaplin-Kramer, R.; Sharp, R.P.; Weil, C.; Bennett, E.M.; Pascual, U.; Arkema, K.K.; Brauman, K.A.; Bryant, B.P.; Guerry, A.D.; Haddad, N.M.; et al. Global modeling of nature's contributions to people. *Science* **2019**, *366*, 255–258. [CrossRef]
2. Díaz, S.; Pascual, U.; Stenseke, M.; Martín-López, B.; Watson, R.T.; Molnár, Z.; Hill, R.; Chan, K.M.A.; Baste, I.A.; Brauman, K.A.; et al. Assessing nature's contributions to people. *Science* **2018**, *359*, 270–272. [CrossRef] [PubMed]
3. Liu, Y.; Fu, B.; Wang, S.; Rhodes, J.R.; Li, Y.; Zhao, W.; Li, C.; Zhou, S.; Wang, C. Global assessment of nature's contributions to people. *Sci. Bull.* **2023**, *68*, 424–435. [CrossRef] [PubMed]
4. Pascual, U.; Balvanera, P.; Díaz, S.; Pataki, G.; Roth, E.; Stenseke, M.; Watson, R.T.; Başak Dessane, E.; Islar, M.; Kelemen, E.; et al. Valuing nature's contributions to people: The IPBES approach. *Curr. Opin. Environ. Sustain.* **2017**, *26–27*, 7–16. [CrossRef]
5. Peng, J.; Xia, P.; Liu, Y.; Xu, Z.; Zheng, H.; Lan, T.; Yu, S. Ecosystem services research: From golden era to next crossing. *Trans. Earth Environ. Sustain.* **2023**, *1*, 9–19. [CrossRef]
6. Liu, Q.; Qiao, J.; Li, M.; Huang, M. Spatiotemporal heterogeneity of ecosystem service interactions and their drivers at different spatial scales in the Yellow River Basin. *Sci. Total Environ.* **2024**, *908*, 168486. [CrossRef]
7. Costanza, R.; dArge, R.; deGroot, R.; Farber, S.; Grasso, M.; Hannon, B.; Limburg, K.; Naeem, S.; Oneill, R.V.; Paruelo, J.; et al. The value of the world's ecosystem services and natural capital. *Nature* **1997**, *387*, 253–260. [CrossRef]

8. Bullock, J.M.; Aronson, J.; Newton, A.C.; Pywell, R.F.; Rey-Benayas, J.M. Restoration of ecosystem services and biodiversity: Conflicts and opportunities. *Trends Ecol. Evol.* **2011**, *26*, 541–549. [CrossRef]
9. Watson, L.; Straatsma, M.W.; Wanders, N.; Verstegen, J.A.; de Jong, S.M.; Karssenberg, D. Global ecosystem service values in climate class transitions. *Environ. Res. Lett.* **2020**, *15*, 024008. [CrossRef]
10. Helseth, E.V.; Vedeld, P.; Vatn, A.; Gómez-Baggethun, E. Value asymmetries in Norwegian forest governance: The role of institutions and power dynamics. *Ecol. Econ.* **2023**, *214*, 107973. [CrossRef]
11. Costanza, R.; de Groot, R.; Braat, L.; Kubiszewski, I.; Fioramonti, L.; Sutton, P.; Farber, S.; Grasso, M. Twenty years of ecosystem services: How far have we come and how far do we still need to go? *Ecosyst. Serv.* **2017**, *28*, 1–16. [CrossRef]
12. Millennium Ecosystem Assessment. *Ecosystems and Human Well-Being: Synthesis*; Island Press: Washington, DC, USA, 2005.
13. Daily, G.C. *Nature's Services: Societal Dependence on Natural Ecosystems*; Island Press: Washington, DC, USA, 1997.
14. Tan, P.Y.; Zhang, J.; Masoudi, M.; Alemu, J.B.; Edwards, P.J.; Gret-Regamey, A.; Richards, D.R.; Saunders, J.; Song, X.P.; Wong, L.W. A conceptual framework to untangle the concept of urban ecosystem services. *Landsc. Urban Plan.* **2020**, *200*, 103837. [CrossRef] [PubMed]
15. Zhang, Z.; Liu, Y.; Wang, Y.; Liu, Y.; Zhang, Y.; Zhang, Y. What factors affect the synergy and tradeoff between ecosystem services, and how, from a geospatial perspective? *J. Clean. Prod.* **2020**, *257*, 120454. [CrossRef]
16. Jones, L.; Norton, L.; Austin, Z.; Browne, A.L.; Donovan, D.; Emmett, B.A.; Grabowski, Z.J.; Howard, D.C.; Jones, J.P.G.; Kenter, J.O.; et al. Stocks and flows of natural and human-derived capital in ecosystem services. *Land Use Policy* **2016**, *52*, 151–162. [CrossRef]
17. Jiang, W.; Lü, Y.; Liu, Y.; Gao, W. Ecosystem service value of the Qinghai-Tibet Plateau significantly increased during 25 years. *Ecosyst. Serv.* **2020**, *44*, 101146. [CrossRef]
18. Daily, G.C.; Söderqvist, T.; Aniyar, S.; Arrow, K.; Dasgupta, P.; Ehrlich, P.R.; Folke, C.; Jansson, A.; Jansson, B.-O.; Kautsky, N.; et al. The Value of Nature and the Nature of Value. *Science* **2000**, *289*, 395–396. [CrossRef] [PubMed]
19. de Groot, R.; Brander, L.; van der Ploeg, S.; Costanza, R.; Bernard, F.; Braat, L.; Christie, M.; Crossman, N.; Ghermandi, A.; Hein, L.; et al. Global estimates of the value of ecosystems and their services in monetary units. *Ecosyst. Serv.* **2012**, *1*, 50–61. [CrossRef]
20. Ouyang, Z.; Zheng, H.; Xiao, Y.; Polasky, S.; Liu, J.; Xu, W.; Wang, Q.; Zhang, L.; Xiao, Y.; Rao, E.M.; et al. Improvements in ecosystem services from investments in natural capital. *Science* **2016**, *352*, 1455–1459. [CrossRef]
21. Yu, Y.; Li, J.; Han, L.; Zhang, S. Research on ecological compensation based on the supply and demand of ecosystem services in the Qinling-Daba Mountains. *Ecol. Indic.* **2023**, *154*, 110687. [CrossRef]
22. Wang, P.; Zhang, L.; Li, Y.; Jiao, L.; Wang, H.; Yan, J.; Lü, Y.; Fu, B. Spatio-temporal variations of the flood mitigation service of ecosystem under different climate scenarios in the Upper Reaches of Hanjiang River Basin, China. *J. Geogr. Sci.* **2018**, *28*, 1385–1398. [CrossRef]
23. Sannigrahi, S.; Chakraborti, S.; Joshi, P.K.; Keesstra, S.; Sen, S.; Paul, S.K.; Kreuter, U.; Sutton, P.C.; Jha, S.; Dang, K.B. Ecosystem service value assessment of a natural reserve region for strengthening protection and conservation. *J. Environ. Manag.* **2019**, *244*, 208–227. [CrossRef] [PubMed]
24. Xie, G.; Zhang, C.; Zhen, L.; Zhang, L. Dynamic changes in the value of China's ecosystem services. *Ecosyst. Serv.* **2017**, *26*, 146–154. [CrossRef]
25. Ouyang, Z.; Zhu, C.; Yang, G.; Xu, W.; Zheng, H.; Zhang, Y.; Xiao, Y. Gross ecosystem product: Concept, accounting framework and case study. *Acta Ecol. Sin.* **2013**, *33*, 6747–6761. [CrossRef]
26. Ouyang, Z.; Song, C.; Zheng, H.; Polasky, S.; Xiao, Y.; Bateman, I.J.; Liu, J.; Ruckelshaus, M.; Shi, F.; Xiao, Y.; et al. Using gross ecosystem product (GEP) to value nature in decision making. *Proc. Natl. Acad. Sci. USA* **2020**, *117*, 14593–14601. [CrossRef] [PubMed]
27. Zou, Z.; Wu, T.; Xiao, Y.; Song, C.; Wang, K.; Ouyang, Z. Valuing natural capital amidst rapid urbanization: Assessing the gross ecosystem product (GEP) of China's 'Chang-Zhu-Tan' megacity. *Environ. Res. Lett.* **2020**, *15*, 124019. [CrossRef]
28. Wang, W.; Xu, C.; Li, Y. Priority areas and benefits of ecosystem restoration in Beijing. *Environ. Sci. Pollut. Res. Int.* **2023**, *30*, 83600–83614. [CrossRef]
29. Xie, H.; Li, Z.; Xu, Y. Study on the Coupling and Coordination Relationship between Gross Ecosystem Product (GEP) and Regional Economic System: A Case Study of Jiangxi Province. *Land* **2022**, *11*, 1540. [CrossRef]
30. Zang, Z.; Zhang, Y.; Xi, X. Analysis of the Gross Ecosystem Product—Gross Domestic Product Synergistic States, Evolutionary Process, and Their Regional Contribution to the Chinese Mainland. *Land* **2022**, *11*, 732. [CrossRef]
31. Ma, Q.; Zhou, M.; Liu, J.; Zhao, J.; Xi, M. Coupling Relationship between Ecosystem Service Value and Socioeconomic Development in the Qinba Mountains, China. *Diversity* **2022**, *14*, 1105. [CrossRef]
32. Guan, S.; Liao, Q.; Wu, W.; Yi, C.; Gao, Y. Revealing the Coupling Relationship between the Gross Ecosystem Product and Economic Growth: A Case Study of Hubei Province. *Sustainability* **2022**, *14*, 7546. [CrossRef]
33. Jiang, H.; Wu, W.; Wang, J.; Yang, W.; Gao, Y.; Duan, Y.; Ma, G.; Wu, C.; Shao, J. Mapping global value of terrestrial ecosystem services by countries. *Ecosyst. Serv.* **2021**, *52*, 101361. [CrossRef]
34. Ge, Y.; Hu, S.; Ren, Z.; Jia, Y.; Wang, J.; Liu, M.; Zhang, D.; Zhao, W.; Luo, Y.; Fu, Y.; et al. Mapping annual land use changes in China's poverty-stricken areas from 2013 to 2018. *Remote Sens. Environ.* **2019**, *232*, 111285. [CrossRef]
35. Sun, X.; Lu, H.; Wang, S.; Xu, X.; Zeng, Q.; Lu, X.; Lu, C.; Zhang, W.; Zhang, X.; Dennell, R. Hominin distribution in glacial-interglacial environmental changes in the Qinling Mountains range, central China. *Quat. Sci. Rev.* **2018**, *198*, 37–55. [CrossRef]

36. Qi, G.; Bai, H.; Zhao, T.; Meng, Q.; Zhang, S. Sensitivity and areal differentiation of vegetation responses to hydrothermal dynamics on the northern and southern slopes of the Qinling Mountains in Shaanxi province. *J. Geogr. Sci.* **2021**, *31*, 785–801. [CrossRef]
37. Zhao, T.; Bai, H.; Yuan, Y.; Deng, C.; Qi, G.; Zhai, D. Spatio-temporal differentiation of climate warming (1959–2016) in the middle Qinling Mountains of China. *J. Geogr. Sci.* **2020**, *30*, 657–668. [CrossRef]
38. Shang, X.; He, Z.; Chen, W.; He, L.; Yang, H. Changes and response mechanisms of leaf area index and evapotranspiration in the typical natural landscapes of the Loess Plateau in northern Shaanxi of China under the human intervention. *Ecol. Indic.* **2023**, *154*, 110517. [CrossRef]
39. Zou, C.; Zhu, J.; Lou, K.; Yang, L. Coupling coordination and spatiotemporal heterogeneity between urbanization and ecological environment in Shaanxi Province, China. *Ecol. Indic.* **2022**, *141*, 109152. [CrossRef]
40. Taibai County People's Government. Notice of General Office of Shaanxi Provincial People's Government on Issuing the Overall Plan for Ecological and Environmental Protection of Qinling Mountains. Available online: https://www.taibai.gov.cn/art/2021/2027/2013/art_7770_2821.html (accessed on 15 December 2021).
41. Zhang, D.; Zhang, X.; Bai, X.; Peng, J. A Dataset of 1km Grid Road Network Density in China (2019). Available online: https://cstr.cn/31253.11.sciencedb.02938 (accessed on 12 October 2023).
42. Zhejiang Provincial Market Supervision and Administration Bureau. Technical Specification for Accounting Gross Ecosystem Product (GEP)—Terrestrial Ecosystems. *DB33/T 2274-2020*. Available online: http://zjamr.zj.gov.cn/art/2020/9/29/art_1229047334_58814039.html (accessed on 12 October 2023).
43. Ni, R.; Wang, F.; Yu, J. Spatiotemporal changes in sustainable development and its driving force in the Yangtze River Delta region, China. *J. Clean. Prod.* **2022**, *379*, 134751. [CrossRef]
44. Wen, Y.; Sun, Q.; Yan, Y.; Xiao, M.; Song, W.; Yang, J. Impacts of the terrestrial ecosystem changes on the carbon fixation and oxygen release services in the Guangdong-Hong Kong-Macao Greater Bay Area. *Acta Ecol. Sin.* **2020**, *40*, 8482–8493.
45. Ma, G.; Yu, F.; Wang, J.; Zhou, X.; Yuan, J.; Mou, X.; Zhou, Y.; Yang, W.; Peng, F. Measuring gross ecosystem product (GEP) of 2015 for terrestrial ecosystems in China. *China Environ. Sci.* **2017**, *37*, 1474–1482.
46. Guo, X.; Fang, C.; Mu, X.; Chen, D. Coupling and coordination analysis of urbanization and ecosystem service value in Beijing-Tianjin-Hebei urban agglomeration. *Ecol. Indic.* **2022**, *137*, 108782. [CrossRef]
47. Sun, Y.; Liu, S.; Dong, Y.; An, Y.; Shi, F.; Dong, S.; Liu, G. Spatio-temporal evolution scenarios and the coupling analysis of ecosystem services with land use change in China. *Sci. Total Environ.* **2019**, *681*, 211–225. [CrossRef] [PubMed]
48. Ding, T.; Chen, J.; Fang, Z.; Chen, J. Assessment of coordinative relationship between comprehensive ecosystem service and urbanization: A case study of Yangtze River Delta urban Agglomerations, China. *Ecol. Indic.* **2021**, *133*, 108454. [CrossRef]
49. Yang, Y.; Bao, W.; Liu, Y. Coupling coordination analysis of rural production-living-ecological space in the Beijing-Tianjin-Hebei region. *Ecol. Indic.* **2020**, *117*, 106512. [CrossRef]
50. Zhang, Y.; He, Y.; Li, Y.; Jia, L. Spatiotemporal variation and driving forces of NDVI from 1982 to 2015 in the Qinba Mountains, China. *Environ. Sci. Pollut. Res. Int.* **2022**, *29*, 52277–52288. [CrossRef] [PubMed]
51. Li, L.; Fan, Z.; Feng, W.; Yuxin, C.; Keyu, Q. Coupling coordination degree spatial analysis and driving factor between socio-economic and eco-environment in northern China. *Ecol. Indic.* **2022**, *135*, 108555. [CrossRef]
52. Hu, Z.; Gong, J.; Li, J.; Li, R.; Zhang, Z.; Zhong, F.; Wen, C. Valuing the coordinated development of urbanization and ecosystem service value in border counties. *J. Clean. Prod.* **2023**, *415*, 137799. [CrossRef]
53. Weiskopf, S.R.; Rubenstein, M.A.; Crozier, L.G.; Gaichas, S.; Griffis, R.; Halofsky, J.E.; Hyde, K.J.W.; Morelli, T.L.; Morisette, J.T.; Munoz, R.C.; et al. Climate change effects on biodiversity, ecosystems, ecosystem services, and natural resource management in the United States. *Sci. Total Environ.* **2020**, *733*, 137782. [CrossRef] [PubMed]
54. Costanza, R.; de Groot, R.; Sutton, P.; van der Ploeg, S.; Anderson, S.J.; Kubiszewski, I.; Farber, S.; Turner, R.K. Changes in the global value of ecosystem services. *Glob. Environ. Chang.* **2014**, *26*, 152–158. [CrossRef]
55. Li, W.; Cai, Z.; Jin, L. Spatiotemporal characteristics and influencing factors of the coupling coordinated development of production-living-ecology system in China. *Ecol. Indic.* **2022**, *145*, 109738. [CrossRef]
56. Fu, J.; Bu, Z.; Jiang, D.; Lin, G.; Li, X. Sustainable land use diagnosis based on the perspective of production–living–ecological spaces in China. *Land Use Policy* **2022**, *122*, 106386. [CrossRef]
57. Liu, X.; Wang, X.; Chen, K.; Li, D. Simulation and prediction of multi-scenario evolution of ecological space based on FLUS model: A case study of the Yangtze River Economic Belt, China. *J. Geogr. Sci.* **2023**, *33*, 373–391. [CrossRef]
58. Wang, A.; Liao, X.; Tong, Z.; Du, W.; Zhang, J.; Liu, X.; Liu, M. Spatial-temporal dynamic evaluation of the ecosystem service value from the perspective of "production-living-ecological" spaces: A case study in Dongliao River Basin, China. *J. Clean. Prod.* **2022**, *333*, 130218. [CrossRef]

land

MDPI

Article

Evaluating the Landscape and Ecological Aspects of Urban Planning in Byblos: A Multi-Faceted Approach to Assessing Urban Forests

Mira Hobeika [1], Victoria Dawalibi [1], Georgio Kallas [1] and Alessio Russo [2,*]

1 Landscape and Territory Planning Department, Lebanese University, Beirut P.O. Box 6573, Lebanon; mira.hobeika@st.ul.edu.lb (M.H.); victoria.dawalibi@ul.edu.lb (V.D.); georgio.kallas@ul.edu.lb (G.K.)

2 School of Architecture and Built Environment, Faculty of Engineering, Queensland University of Technology, Brisbane, QLD 4000, Australia

* Correspondence: alessio.russo@qut.edu.au

Abstract: Byblos, designated as a UNESCO World Heritage site, stands as one of Lebanon's most ancient urban centers, known for its expansive green spaces. However, ongoing urbanization threatens these valuable areas. This study uses a multi-faceted approach to evaluate the structure and landscape attributes of Byblos' urban and peri-urban forests (UPFs). Landscape canopy cover, diversity indices, forest structure, and a silhouette perceptual test were assessed across 24 streets in the city center, residential zones, and areas with heavy vehicular traffic. Findings reveal that 28% of Byblos' canopy cover is concentrated mostly in the northeastern region. Native tree species account for 30% of the total, and a notable variation in tree diversity exists among different land-use types (Shannon diversity index (H) was 1.02 for the city center, 1.35 for residential streets, and 0.64 for vehicular areas). Additionally, a normal J-shaped distribution of tree diameters was identified across all street types. This study highlights a correlation between tree silhouettes and visual preferences, with densely spreading canopies being favored. Residential trees demonstrate the highest structural diversity and varied blossoming seasons. This research represents the first investigation into the current state of urban forestry in Byblos and offers recommendations for sustainable management and planning strategies.

Keywords: urban forestry; ecological indicators; class distribution; silhouette; perceptual testing

1. Introduction

Nowadays, more than half of the world's population already lives in cities, and this proportion is anticipated to rise to 70% by 2050 [1]. Rural–urban migration resulted in rapid urbanization, and people began settling in the cities and their surroundings. This increased the number of surfaces made of asphalt and other hard materials [2], which had a severe impact on natural ecosystems and might have even led to a radical loss in these ecosystems [1,3]. Although urban spaces cover only 2% of the surface of the earth, they emit considerable amounts of greenhouse gases and more than 70% of global CO_2 emissions; this is due to the heavy use of fossil fuels for transportation, heating, etc. [1,4].

To effectively address urban challenges, urban and peri-urban forests (UPF) can play a vital role in promoting the sustainability and livability of future cities. Specifically, preserving and advancing urban and peri-urban forests, as well as fostering the sustainable provision of ecosystem services (ES) from UPFs, are critical to ensuring the well-being of present and future generations [5]. In this context, the preservation and enhancement of urban nature are critical for ensuring urban dwellers' health and improving their general quality of life [5,6], as well as ensuring the integrated and systematic management of trees in urban and peri-urban areas, maximizing their contribution to the physiological, sociological, and economic well-being of urban society [7]. Rooted in North America with a rich history spanning over four decades, the concept of UPFs gained momentum in the United States during the 1970s as environmental concerns intensified [8].

The advantages of UPFs are well-documented, with studies demonstrating their ability to improve air quality, regulate temperature, and mitigate noise pollution [9–11]. These benefits translate into tangible improvements in human health, reducing respiratory problems, cardiovascular risks, and psychophysiological stress levels [12]. Moreover, UPF and afforestation projects enhance the overall quality of life by creating aesthetically pleasing spaces that foster social interaction, recreation, and a sense of community [13,14].

The economic benefits of UPFs are equally compelling. Studies have shown that UPFs can increase residential property values, boost tourism, and attract businesses, leading to job creation and economic growth [14]. Additionally, UPFs can reduce an area's overall energy demand by providing shade and insulation, and lowering heating and cooling demands [15,16].

Despite the wealth of evidence supporting the benefits of UPFs, most studies have focused on cities in the Global North [17]. In the Global South, particularly in the Middle East and North Africa (MENA) region, where climate change is exacerbating existing challenges such as rising temperatures, prolonged droughts, and increasing urbanization, UPF research is relatively limited.

The MENA region is particularly vulnerable to the impacts of climate change, with scarce water resources facing additional strain from population growth, socioeconomic development, and urbanization [18]. These factors pose serious risks to lives, biodiversity, financial stability, and human security. It is expected that the challenges will worsen in the future, especially since the MENA region is one of the regions most vulnerable to climate change impacts [19].

Recognizing the urgency of the situation, countries in the MENA region have initiated several UPF projects, drawing inspiration from successful initiatives elsewhere. A 2016 report by The Nature Conservancy, a prominent US environmental NGO, highlighted the role of urban tree planting in reducing air pollution and heat stress. This report has spurred several UPF projects in the MENA region, with one of the most ambitious being Saudi Arabia's Riyadh Green Plan. This visionary initiative, aligned with Vision 2030, aims to plant 7.5 million trees in the capital, transforming it into a sustainable and livable city [20].

Apart from various urban forestry projects in the MENA/NENA region, Lebanon stands out as a country with a limited number of studies on urban forestry planning [21]. According to a UN-Habitat report [22], Lebanon's cities are experiencing a population surge, leading to haphazard and unregulated urban sprawl. This has resulted in additional environmental issues, including shrinking green spaces, water contamination, car-centric transportation, and reliance on non-renewable energy sources. Additionally, climate change has yet to receive the attention it deserves from Lebanon's national authorities. While its effects may seem mild, Lebanon is witnessing longer, hotter summers; decreased winter snowfall; and dwindling water resources—clear indications of climate change's social and environmental ramifications [23]. To address these concerns, a sustainable environmental solution is imperative. While a few UPF projects have been implemented in Beirut, such as the Beirut River Forest, they have not been fully completed.

Several cities need to enhance their urban forestry planning in this context. For instance, Byblos, one of Lebanon's most captivating cities and the world's oldest continuously inhabited city listed on the UNESCO World Heritage List, faces numerous challenges. This mesmerizing town boasts a rich tapestry of cultural and natural features, including coastlines, beaches, agricultural lands, and orchards, contributing significantly to its allure and defining its landmarks [24]. However, Byblos's topography has undergone significant alterations over time, with the city expanding eastward up the hills in an uncontrolled manner, neglecting the precious green spaces that once dotted its landscape [24]. Consequently, only one large green area in the north of the city has been preserved. This neglect has led to a loss of biodiversity and the emergence of pressing issues such as waste and sewage contamination of rivers and aquifers, flooding, and wastewater.

Furthermore, Byblos faces microclimate challenges due to its growing tourism sector and economic importance. Despite a doubling in visitor numbers over the last decade, the city lacks adequate public spaces for rest and exploration, leaving visitors uncomfortable during hot days and pushing residents to rely on cars for transportation, thereby severing their connection with nature.

Recognizing the scarcity of studies in this area, this research aims to fill the gap by focusing on the following objectives:

1. Understand the ecological characteristics that contribute to the diversity and resilience of Byblos's ecology.
2. Analyze the landscape components that influence the visual guidance and aesthetic value of Byblos.
3. Develop recommendations for the implementation of a UPF project in Byblos that aligns with urban forestry principles and the city's identity.

2. Materials and Methods

2.1. Site Analysis

2.1.1. Overview of Byblos

The city is part of the Byblos district, which consists of 84 towns, and is surrounded by Hboub, Edde, Amchit, Qartaboun, and Blat. The city's area is around 8 km^2. Since Byblos is situated next to the sea, its altitude ranges between 0 m and 350 m (Figure 1).

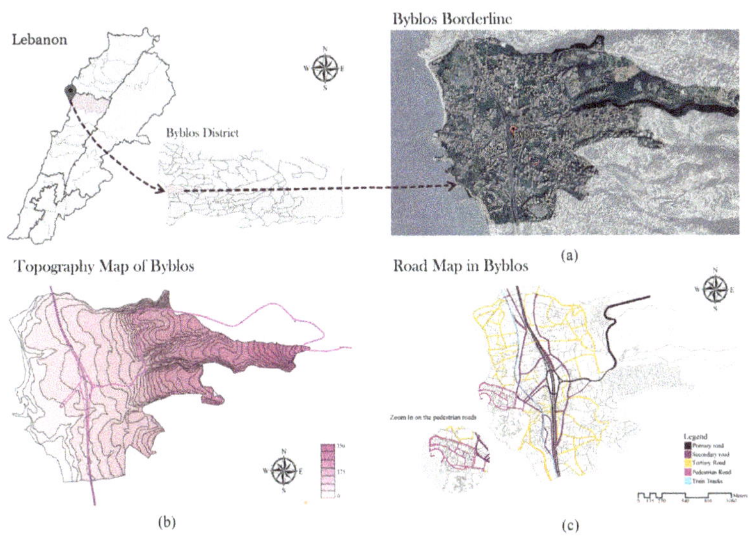

Figure 1. (**a**) The location of Byblos, (**b**) the topography of Byblos, retrieved from National Council for Scientific Research–Lebanon (CNRS) 2022, (**c**) The circulation in Byblos, made by the author in 2024.

As one of the oldest continuously inhabited villages in the East, dating back more than 7000 years, Byblos has many hidden layers that offer great knowledge about its historical story and the footprint of every civilization that lived there—whether Romans, Greeks, Persians, or others [25]. The cultural heritage embedded in this city not only made it a pole of attraction in Lebanon but also contributed to it being registered on the UNESCO World Heritage List in 1984, as it fits three of their criteria: Criterion (iii): Byblos bears exceptional testimony to the beginnings of Phoenician civilization; Criterion (iv): In existence since the Bronze Age, Byblos provides one of the primary examples of urban organization in the Mediterranean world; and Criterion (vi): Byblos is directly and tangibly associated with the history of the diffusion of the Phoenician alphabet (on which humanity is still largely dependent today), with the inscriptions of Ahiram, Yehimilk, Elibaal and Shaphatbaal. The city is located midway between Beirut and Tripoli, which makes it easily accessible for tourists, visitors, and residents. Its location next to the Mediterranean Sea adds even more touristic value to the site. Figure 1 shows the road map of Byblos; however, the core part of the old city of Byblos is only open to pedestrians, specifically at night, to preserve its enchanting spirit.

2.1.2. Urban Development in Byblos

Byblos, originally a small coastal village, has grown over time to become a vibrant hub of residential, archaeological, and tourist activity. As the city's economic and tourist appeal has increased, its infrastructure has adapted to accommodate growth while preserving the value of its historical landmarks. Byblos is now a thriving destination for regional and international visitors, with residential areas expanding to the city's eastern region, while the western part is primarily used for agricultural and tourist purposes.

2.1.3. Demographical and Social Aspects in the City of Byblos

Byblos is renowned for its harmonious coexistence of multiple religions, particularly between Christians and Muslims. Boasting a population of roughly 40,000, Byblos is home to 28,486 Lebanese residents [26]. The remaining residents are comprised of Armenian, Syrian, and other nationalities. Social groups within Byblos are situated in social nodes located throughout the old city and a small area in the east. Workers can be found in commercial areas, while the younger demographic and tourists can be found frequenting the old port, souks, and restaurant-lined streets [24,26] (Figure 2).

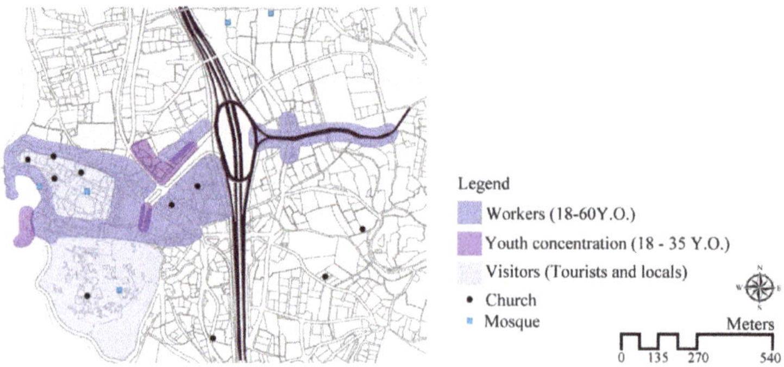

Figure 2. Map showing the social nodes in Byblos. Made by the author, 2024.

2.1.4. Byblos Landscape Assets

Byblos boasts a rich and varied landscape, encompassing a diverse coastline, river valleys, beaches, fertile orchards, and thriving agricultural areas. The city's natural heritage is a source of great pride for its residents [24]. Moreover, the municipality of Byblos reported Al Chamiye Beach as a pristine coastal reserve of remarkable ecological importance, captivated many scholars by its vibrant biodiversity. The abundance of life in Byblos extends from the shoreline to its rolling hills, where a vast expanse of woodland is nestled in the northeast of the city.

2.2. Methodology

The research comprised two stages: an initial macro-scale phase aimed at evaluating the canopy forest and diversity of the urban forest in Byblos, followed by a micro-scale phase focused on appraising the structural and aesthetic features of the urban forest, as shown in the methodology framework (Figure 3).

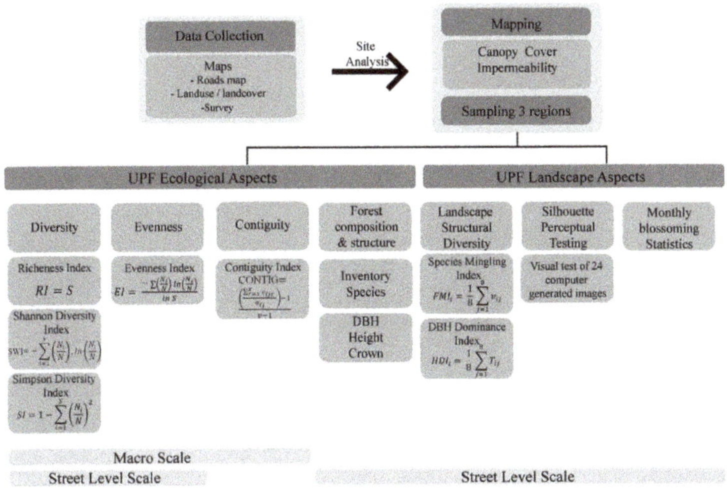

Figure 3. A flowchart showing the methodology applied in the study.

A random-sample point analysis conducted with i-Tree Canopy was employed at the macro-scale level to assess tree canopy cover in Byblos. This analysis, along with Land Use Land Cover (LULC) maps from the National Council for Scientific Research–Lebanon CNRS (2022) [27], were used to quantify fragmented and intact green spaces and calculate the urban area's permeability. Tree species richness, diversity, and connectivity metrics were computed via Fragstats software version 4.2, employing rasterized land-cover and land-use maps sourced from CNRS 2022.

On a street scale, to assess the tree composition and structure, twenty-four streets were selected for sampling, across the city center (nine streets), residentially (eight streets), and on vehicular streets (seven streets). The selection criteria were established on the parameters of usage, visitor frequency, and the expression of Byblos's identity. Streets were categorized based on their functional utility, prioritizing those exhibiting the highest visitor rates and symbolic significance to Byblos. The sampled streets are highlighted in Figure 4.

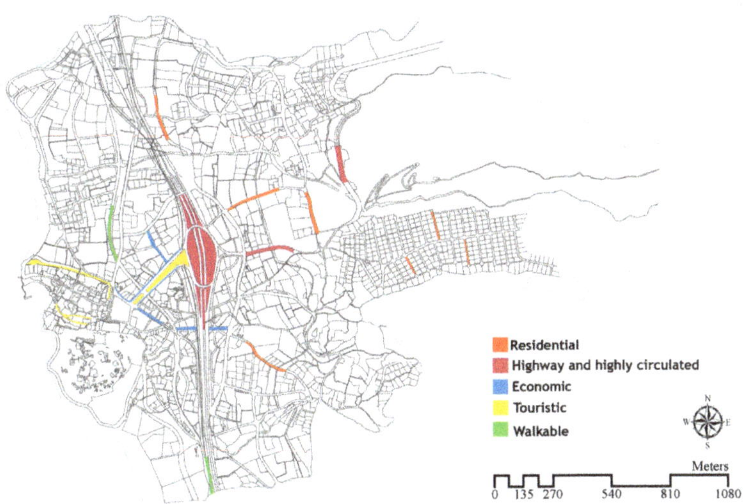

Figure 4. The 24 sampled streets in Byblos. Made by the author in 2024.

2.3. Forest Landscape Ecological Values

Ecological characteristics of the species were assessed according to their importance in providing better biodiversity and fostering urban resiliency.

The diversity index was assessed since species diversity is critical in promoting a healthy and resilient urban forest. Moreover, urban tree diversity gives urban forests the capacity to react to current and upcoming pressures and threats. Thus, the indexes of diversity are the richness index (*RI*), the Shannon–Wiener index (*SWI*), and the Simpson index (*SI*), as used in previous studies [28,29].

$$RI = S \tag{1}$$

$$SWI = -\sum_{i=1}^{S}\left(\frac{N_i}{N}\right)\cdot ln\left(\frac{N_i}{N}\right) \tag{2}$$

$$SI = 1 - \sum_{i=1}^{S}\left(\frac{N_i}{N}\right)^2 \tag{3}$$

where *RI* is the richness index, *S* is the number of species present, N_i is the number of trees of a specific species, and N is the total number of trees present.

The second index used is the evenness index, which provides valuable information about the stability of the ecosystem and the distribution of the species.

$$EI = \frac{-\Sigma\left(\frac{N_i}{N}\right)ln\left(\frac{N_i}{N}\right)}{lnS} \tag{4}$$

With

$$N = \sum_{i=1}^{S} N_i \tag{5}$$

where *EI* is the evenness index, *S* is the number of species present, N_i is the number of trees of a specific species, while N is the total number of trees present.

All the ecological indicators were also applied via Fragstats software version 4.2, to assess them all over Byblos.

2.4. Forest Composition and Structure

During the field visits, a tree inventory of the streets of Byblos was carried out. Tree measurements, such as DBH (diameter at breast height), height, and crown area, were gathered for trees that could be seen by visitors along the sampled streets. DBH was measured using a measuring tape and projecting the sampled points from the periphery of the canopy to detect its area, and the height was assessed by using the GLOBE Observer app. These measurements were sorted into four categories (as shown in Table 1) to gain insight into the types and distribution of trees. Evaluating the existing trees is an important aspect of assessing UPFs, so it is essential to understand the composition of the site and the condition of its trees. During the field visits, measures of DBH, height, and crown area were performed on trees that were visible to the visitors, whether they were directly on the sidewalks or a few meters away [30].

Table 1. Representing the specific class classifications for DBH, height, and crown area.

Class	DBH/cm	Height/m	Crown Area/m^2
C1	5–20	1–7	1–16
C2	20–40	7–11	16–64
C3	40–80	11–17	64–144
C4	>80	>17	>144

2.5. *Landscape Structural Diversity*

By assessing the structural diversity of trees on selected streets, we aimed to comprehend the complexity and diversity of the urban forest in Jbeil, as it is linked to emotional responses to the landscape. The indexes used are the form mingling index (FMI), which describes the diversity of the form of neighbor trees compared to the reference tree, and the height dominance index (HDI), which is useful for describing the dominance of the height of the reference tree compared to the eight closest neighboring trees. These indexes were calculated based on eight neighboring trees that were the closest to the reference tree, instead of four neighboring trees, due to the long streets and the abundance of trees [31,32].

The form mingling index:

$$FMI_i = \frac{1}{n}\sum_{j=1}^{n} v_{ij} \tag{6}$$

where *FMI* is the form mingling index, *i* is the reference tree, and *j* refers to the closest trees. When the neighbor *j* is not the same species as the reference tree *i*, then $v_j = 1$; otherwise, $v_j = 0$. The interpretation of the *FMI* values is as follows: no mixture (FMI = 0); low mixture (FMI = 0.25); medium mixture (FMI = 0.50); high mixture (FMI = 0.75); and complete mixture (FMI = 1.00).

The height dominance index:

$$HDI_i = \frac{1}{n}\sum_{j=1}^{n} T_{ij} \tag{7}$$

where *HDI* is the height dominance index, *i* is the reference tree, and *j* refers to the closest trees. When the neighbor *j* is different than the reference tree *i*, then $T_j = 1$; otherwise, $T_j = 0$. The interpretation of the U values is as follows: predominant (HDI = 0); subdominant (HDI = 0.25); medium (HDI = 0.50); disadvantaged (HDI = 0.75); and absolutely disadvantaged (HDI = 1.00). The trees on the streets were represented visually and grouped by a minimum of 5 trees and a maximum of 13. The tree in the middle is considered the reference tree, and the calculations of the form mingling index and height dominance index were applied. Figure 5 shows how this was represented.

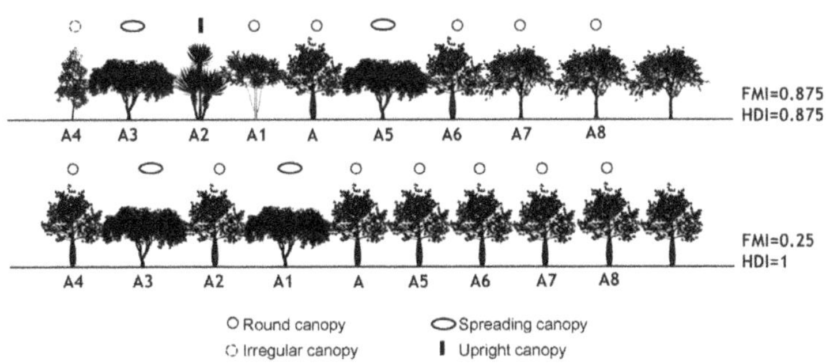

Figure 5. Representation of the FMI and HDI figures, where A is the reference tree and A1, A2, A3, A4, A5, A6, A7, and A8 are the neighbor trees.

2.6. *Silhouette and Perceptual Testing*

A survey combining verbal questioning and landscape profile photographs (26 photos) was conducted in Byblos to understand forest landscape preferences in both the core and peri-urban areas. One hundred people, including residents and visitors, participated in the survey. The survey adopted a random sampling methodology to ensure a representative sample of the Byblos population. This approach included residents of Byblos alongside visitors, encompassing a broad age range from 18 to 64 years and older. This strategy aimed to capture the perspectives of various generations within the Byblos community.

To ensure accessibility, those unfamiliar with technology were surveyed in person, while others completed the survey online. Moreover, 46% of the participants in this survey were residents of Byblos, while the rest were visitors, tourists, or workers from the neighboring villages. Regarding gender distribution, 52% of the participants identified as female, and the remaining 48% identified as male.

On the perceptual testing level, the landscape sceneries were evaluated through visual representation so we could determine the silhouette value of the landscape [32]. Landscape sceneries were shown to volunteers to assess their aesthetic quality. The landscape sceneries were representations of the sampled streets; the silhouette of the trees was used and classed in the same order as reality; and all sceneries were in black and white to encourage people to focus on the shapes of the trees, as tree colors are known to cause differences in perception [33]. After explaining the aesthetic function and the silhouette effect to the volunteers, they were asked to rank the silhouette effect using a 10-point scoring system. As for the silhouette value, Excel software version 2403 was used for statistical analysis to assess the scores given by the participants for each landscape scenery. Once the silhouette scores were calculated, a correlation test took place between the silhouette, FMI, HDI, and FMI + HDI.

3. Results

3.1. Canopy Cover

Byblos's canopy cover ranges from 28% to 30%, as estimated by the ArcGIS 10.8 application (Figure 6), with an i-Tree Canopy rating of 27.63 ± 1.58 SE. The northeast and east areas of the city have the highest amount of canopy cover. On the other hand, the residential area situated in the western part of the city only has 11% coverage. In terms of impermeability, approximately 71% of the area in Byblos is covered by both continuous and discontinuous urban fabric. The majority of the impermeable surface is located in the west, southwest, east, and southeast regions of the city.

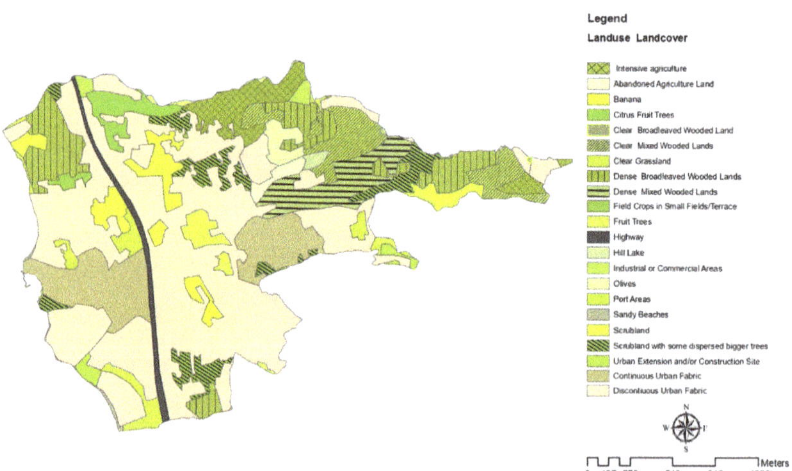

Figure 6. Map representing the different land-use and land-cover types in Byblos, retrieved from National Council for Scientific Research–Lebanon (CNRS) 2022, edited by the author, 2024.

3.2. Forest Landscape Ecological Values

The Shannon diversity index represented a value of SWI = 2.05, while the Simpson diversity index scored SI = 0.75 and the evenness index scored EI = 0.66. Among all surveyed areas, residential streets exhibited the highest diversity, as indicated by values of 5.75 for the richness index (RI), 1.35 for the Simpson diversity index (SWI), and 0.66 for the Shannon diversity index (SI). Additionally, residential streets attained the maximum evenness with an evenness index (EI) value of 0.8. Comparatively, the city center streets demonstrated slightly lower richness (RI = 5.56) than residential streets, and there were notable distinctions in the diversity indexes of Shannon (0.95) and Simpson (0.47). The evenness index for city center streets was recorded as EI = 0.56. Conversely, the thoroughfares with vehicular streets exhibited the lowest values across all four indexes (Table 2). The contiguity results indicated that, with the exception of the clear mixed wooded lands (0.4) and the clear broadleaved wooded land (0.28), all classes demonstrated high values.

Table 2. Represents the average of each ecological indicator in each type of street.

Area	No. of Species	RI	SWI	SI	EI
City center streets	30	5.56	1.02	0.48	0.56
Residential streets	30	5.75	1.35	0.66	0.80
Vehicular highly circulated streets	14	3.14	0.64	0.34	0.51
All over Byblos	56	-	2.05	0.75	0.66

3.3. *Tree Species and Composition*

The forest in Byblos showed a diverse array of trees, encompassing ornamental, evergreen, and fruit varieties. We found 56 tree species present in the surveyed streets of Byblos, and of those, 17 were native to the region, such as *Nerium oleander* and *Cupressus sempervirens*, while the remaining species were non-native, including *Ceiba speciosa*, *Jacaranda mimosifolia*, and *Delonix regia*, among others. The diversity of the tree species generally followed the 10–20–30 rule of Santamour (1990) [34], except for *Nerium oleander*, which exceeded the recommended 10% to 15% threshold. The diversity of trees in the city center streets was significantly higher than those in the streets with heavy car traffic and in residential areas (Figure 7). Within the city center, the number of trees per street varied from 18 to 76, typically encompassing a range of 4 to 16 distinct species. Streets experiencing heavy traffic contained 11 to 33 trees, commonly featuring only 2 species per street. In residential zones, the tree counts per street ranged from 5 to 21, highlighting a diversity of 4 to 9 species on each street (Figure 7).

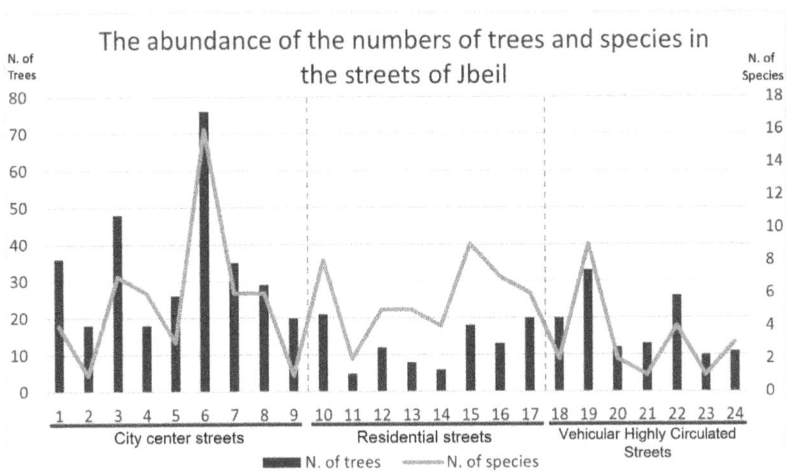

Figure 7. The abundance of tree species and tree numbers by streets.

3.4. Forest Composition and Structure

The city center streets exhibited one of the narrowest ranges of diameter at breast height (DBH) values, spanning from 34.9 cm to 99 cm, with heights ranging from 4.9 m to 8.4 m, and possessed the lowest crown area. In contrast, vehicular highly circulated streets displayed the widest range of DBH values, extending from 44.5 cm to 197.5 cm. These streets showcased both the shortest and tallest tree heights, ranging from 3.5 m to 13.9 m. Additionally, vehicular highly circulated streets had the highest average crown area of 65.7 m^2. As for the reversed J-shaped distribution, it was found in several indicators like the height (c1: 75, c2:22,c3: 3, c40) and the crown area (c1:60, c2:35, c3:3, c4:1) in the city center; the crown area (c1:75, c2:16, c3:10, c4:0) in the highly circulated streets; and the height (c1:49, c2:27, c3:18, c4:6) in the residential streets. In all streets, the first class (c1) had the highest value, and the rest of the classes had decreasing values, respectively.

One of the distributions fit the J-shaped type—the DBH (c1:6, c2:6, c3:28, c4:61). In the residential streets, it started with a low value in c1 and started increasing until it reached its highest value in the last class, c4. The last distribution type was the inversed J-shaped, like the height (c1:39, c2:37, c3:17, c4:7) in the vehicular highly circulated streets, and in the crown of the residential area (c1:36, c2:46, c3:17, c4:1), where they start at constant values and then begin decreasing.

Classification by Classes

Figure 8 shows the percentages of different classes of DBH, height, and crown. We detected several types of distribution: approximately normal distribution, J-shaped, reversed J-shaped, and inverted J-shaped. We found a normal distribution in the DBH in the city center streets (c1:6, c2:30, c3:33, c4:32) and the DBH in the vehicular highly circulated streets (c1:1, c2:18, c3:46, c4:35), where they started with a low value, reached the highest value in the middle and then decreased.

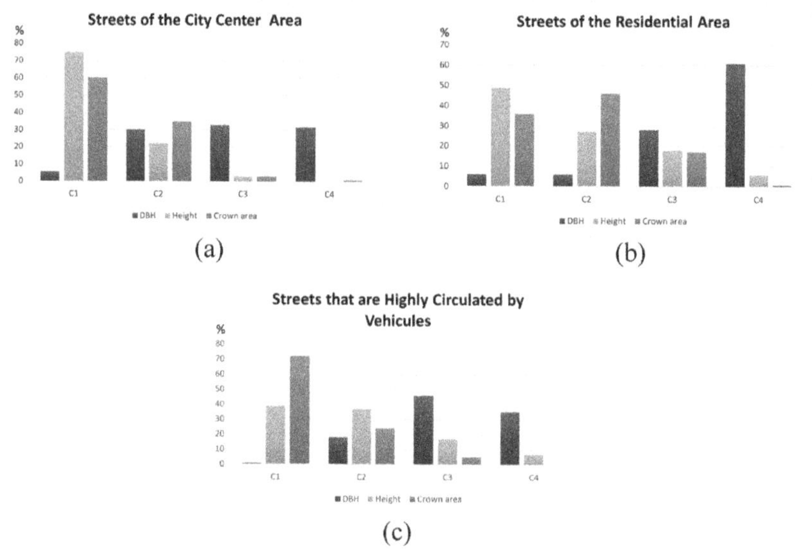

Figure 8. Charts showing the class distribution of DBH, height, and crown area in all three areas, (**a**) Streets of the city center area, (**b**) Streets of the residential area, and (**c**) Streets that are highly circulated by vehicles.

3.5. Landscape Assessment

3.5.1. Silhouette and Perceptual Testing

Our analysis revealed that Landscape 14, located within the residential streets, exhibited the highest levels of both forms of the mingling index and the height dominance index. The streets of the city center demonstrated lower scores in terms of structural diversity, with most of the sampled streets exhibiting low values for the height dominance index but higher values for the form mingling index. In highly circulated areas, several streets scored poorly on the form mingling index, with some even scoring 0.5 or lower on the height dominance index (Figure 9).

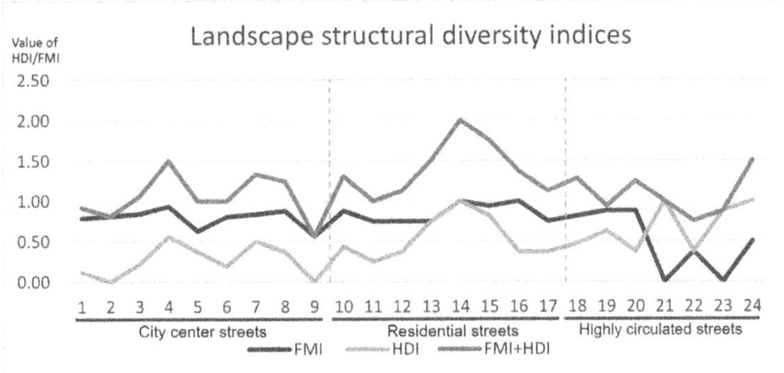

Figure 9. Chart representing the landscape structural diversity indices.

The perceptual test revealed landscape silhouette values ranging from 4.9 to 7.7 (as seen in Table 3). Interestingly, landscapes with the same species had different scores. For instance, Landscape No. 19 (Figure 10), with columnar-shaped trees of varying heights, scored the lowest (4.9). Meanwhile, Landscapes No. 9, 21, and 23, with the same species but with round or irregular crowns, scored higher at 6.4, 6.5, and 6.5, respectively. Landscape No. 2, with the same species but slightly different heights and a V-shaped crown, received the highest score of 7.7.

Table 3. Descriptive statistics of landscape images.

No. of the Street	Min.	Max.	Average Score	Stand. Dev.	No. of the Street	Min.	Max.	Average Score	Stand. Dev.
1	1	10	6.2	1.9	13	1	10	6.1	2
2	3	10	7.7	1.9	14	4	10	7.2	1.6
3	1	10	5.3	2.1	15	1	10	6.1	1.9
4	2	10	6.3	2.1	16	1	10	5.3	2.2
5	1	10	6.4	2.1	17	1	10	7	2.2
6	1	10	6.8	2.2	18	2	10	6.4	2.1
7	1	10	6.1	2	19	1	10	4.9	2.2
8	1	10	5.9	2.2	20	3	10	6.4	1.5
9	2	10	6.4	1.7	21	3	10	6.5	1.8
10	2	10	7.3	1.7	22	1	10	5.3	2.1
11	2	10	6.5	1.8	23	1	10	6.5	1.9
12	1	10	6.3	2	24	3	10	6.9	1.7

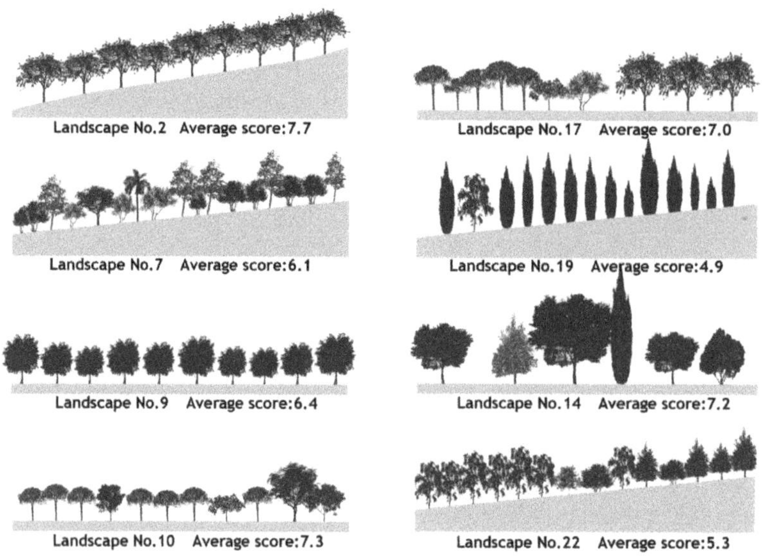

Figure 10. The landscape sceneries of the perceptual testing, with their average score.

Large trees of different species and crown shapes, like those found in Landscapes 11, 14, 20, and 24, scored between 6.4 and 7.2. On the other hand, smaller trees, like those in Landscapes 10 (7.3) and 12 (6.3), were also well-regarded by participants. Trees clustered within the same species, as seen in Landscapes 17 and 18, attained scores of 7 and 6.4, respectively. Conversely, Landscapes 7 and 8, showcasing a mixture of species and forms, received lower scores of 6.1 and 5.9. To prove a relation between the form of the trees, their height, and their silhouette scores, a Pearson correlation was applied. There was a positive correlation between the form mingling index and the silhouette perceptual scores, as well as between the height dominance index and the silhouette perceptual scores as shown in Table 4. The combination of HDI and FMI had a more pronounced impact on visual preference, showing a correlation of 0.37 with the silhouette perceptual test.

Table 4. Correlation coefficient between FMI, HDI, FMI + HDI, and the perceptual testing results.

Indicator	r	*p*-Value	R2
FMI	−0.06	0.25	0.24
HDI	0.03	0.42	0.17
FMI + HDI	0.14	0.07	0.37

3.5.2. Blossoming Monthly Statistics

Each area's monthly statistics are represented in Figure 11, indicating the abundance of colors in Byblos during the seasons, based on the species sampled in each street.

In residential streets, 30 distinct tree species were identified, primarily blooming in spring from mid-March to mid-June. In the city center, 30 tree species, mostly evergreen, bloom from April to August. Highly circulated streets feature only 14 tree species, mainly evergreen, with limited visibility of spring blossoming.

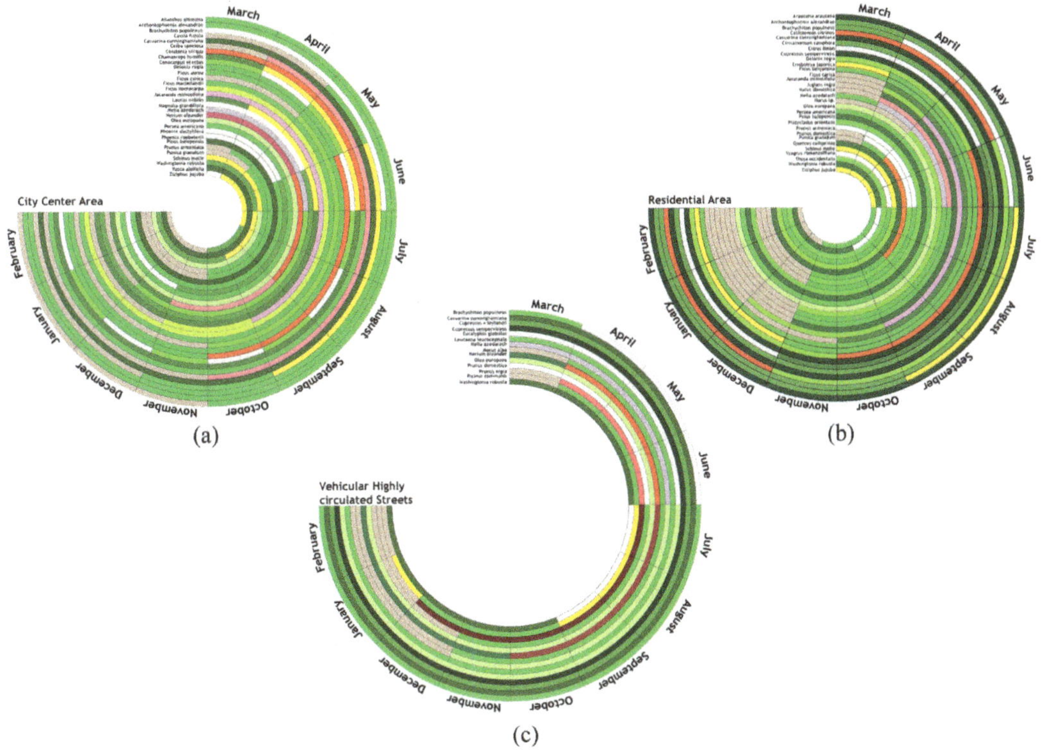

Figure 11. Blossoming calendar of each area based on the species present within, (**a**) City center area, (**b**) Residential area, (**c**) Vehicular highly circulated streets. Each ring in the figure represents the blossoming cycle of a species of trees assessed in the inventory. When it is evergreen, the ring is always green; light brown indicates deciduous trees, and the colors represent the color of tree flowers. Source: Made by the author, 2024.

4. Discussion

As urbanization accelerated in Byblos City, vegetation that once facilitated permeability was supplanted by impervious surfaces [35]. This change led to insufficient drainage and heat dissipation, ultimately contributing to an increase in urban flood disasters [35]. With climate change, extreme rainfall events are becoming more frequent [36], exacerbating the problem. Some incidents related to urban floods have been reported, including in November 2017 when water filled the walkways of the old city, and in January 2019 when stormwater caused damage to the water drainage channels, resulting in financial loss [24]. Byblos still has a large green area (28%); however, it cannot act as a sponge to retain stormwater since it is located in the northeast part of the city. Byblos boasts a significant green area (28%); however, its capacity to retain stormwater is limited due to its location in the northeast part of the city. Nevertheless, when specifically considering the canopy cover, the values align with the established benchmark of 20%, as determined by UPF guidelines [1]. In many cities, the typical green infrastructure ratio hovers around 16.8% [37]. Despite this accomplishment, only 11% of the green infrastructure is situated in the city center, underscoring the need to establish green corridors that connect these smaller green spaces with the larger expanses in the north. Such an approach not only facilitates the movement of flora and fauna but also fosters increased biodiversity within the city, particularly given the current low connectivity of wooded areas [38,39].

Byblos boasts a diverse selection of tree species. It has carefully curated a mix of both native and exotic trees, with native species abundance making up 47% of the plantation—a figure in line with UPF guidelines [1]. The city center, being a prominent tourist destination, strategically incorporates specific plants like *Ceiba speciosa* for ornamental purposes and culturally significant species like *Olea europaea* and *Nerium oleander*, which evoke the Roman era in Byblos [40]. This deliberate selection of trees to reflect Byblos' cultural heritage underscores the existence of a comprehensive management plan. The combination of both evergreen and deciduous species in Byblos City, both of which play a vital role in purifying the air and sequestering carbon, add visual interest to the landscape of Byblos [41,42]. Notably, the city center and residential area demonstrate nearly identical richness indices, each with thirty species of trees lining their streets. The richness indicator in this research presented a value ranging between 3.14 and 5.75, which shows good diversity compared to the richness index values of core areas in three villages in China (Guangzhou, Foshan, and Zhuhai), which demonstrated 3.27, 5.05, and 4.69, respectively [27]. Though Alvey [43] reported that, traditionally, urban forest areas have been regarded as locations of low biodiversity that are dominated by non-native species, evidence from this study, as well as from published information, suggests that urban and suburban areas can contain relatively high levels of biodiversity [44,45].

This suggests a well-balanced composition, leading to a stable and resilient urban forest that is not only richer in fauna but also provides residents with a broader array of benefits more effectively [46]. This diversity serves to minimize the risk of species-specific pests and diseases, making it a valuable asset [47]. However, thoroughfares with heavy vehicular traffic tend to exhibit lower species diversity and richness indices. This phenomenon stems from the prevalent practice of planting the same species along highways for visual consistency and ease of maintenance, resulting in a monotonous landscape [48]. Increasing diversity in vehicular roads is crucial, as a variety of plant species can offer a range of ecosystem benefits [48]. To achieve this, it is recommended to plant more tree species, particularly those that are native and can provide significant benefits such as pollination, erosion control, and soil stabilization. This approach can also create pathways that connect isolated habitats, promoting gene flow and improving population connectivity for plants, animals, and seeds [48]. According to studies conducted by Modlingerová et al. (2012) [49] and Bandara and Dissanayake (2021) [50], trees along roadways are expected to endure environmental pollution. Therefore, it is important to select appropriate types of trees that can reduce pollution and absorb noise.

The distribution of tree diameters (DBH), heights, and crown areas within the city center streets exhibits a normal distribution J shape, which is considered the most stable pattern for urban forests. The presence of a significant number of trees in medium-sized classes contributes to maintaining community stability, while the presence of both small- and large-sized classes ensures a balanced transition from younger to older trees [51]. Residential and vehicular highly circulated streets exhibit an abundance of medium and large diameters. The prevalence of *Olea europaea* and *Pinus halepensis* in the residential area can account for the high abundance of trees in the larger diameter at breast height (DBH) classes. The low percentage of trees falling into the small DBH classes suggests a lack of initiative to introduce new trees to these streets. However, it is important to note that one of the primary objectives of urban and peri-urban forests (UPFs) is to promote the physical and emotional well-being of residents [52]. Urban forests with elevated rates in large DBH classes can also support local livelihoods, improve community cohesiveness, and strengthen urban dwellers' connection to nature, as well as improve it.

The city center streets feature a moderate level of structural diversity, displaying various shapes and minimal variations in height. This diversity primarily stems from the types of trees planted rather than their growth patterns. On the contrary, residential streets exhibit high levels of structural diversity, largely influenced by the shapes of the trees and, to a lesser extent, their heights. Conversely, vehicular streets tend to have less structural diversity as they are often dominated by similarly sized trees, with variations in height being the main factor contributing to any observed diversity.

The significance of the largest blossoming season in the central area lies in its ability to attract people and visitors to the heart of the city, where most heritage sites and attractions are located. The colorful spaces and variety of species can have a positive impact on visitors' perceptions; therefore, they can generate feelings of enjoyment, self-esteem, and motivation, driving them to revisit the city [53]. The residential area has more deciduous trees than evergreen ones, creating a stronger connection between residents and nature. Warm colors in this area can generate feelings of comfort, power, and passion [54]. Residents can witness the natural cycle of the trees, experiencing the changing colors, falling leaves, and new growth in the spring. Highly circulated streets are mainly planted with evergreen species, providing drivers with pleasant scenery and a sense of security while also reducing driving-related stress [55].

The findings from the perceptual tests align with previous research on the impact of tree shapes on preference, which consistently demonstrates a preference for broad canopies, including spreading and globular canopies. This is likely due to the perception of spreading canopies providing protection [56]. Moreover, numerous studies have compared the shapes of trees and found that people prefer a spreading canopy over a round shape, while the conical shape is the least favored [56,57]. Our analysis indicates that the shape of the tree has a greater influence.

4.1. Recommendations

Based on the results gained from our multi-faceted approach, this study presents actionable recommendations to address key issues concerning Byblos' urban landscape. In particular, the following recommendations aim to enhance the city's green infrastructure, leading to improved environmental quality and resident well-being:

- The city of Byblos can enhance its canopy cover to a minimum of 30% by connecting the wooded lands in the north of the city to the smaller patches distributed throughout. The incorporation of man-made urban corridors or roads can increase the city's biodiversity and resilience by connecting green spaces within the city.
- To improve stormwater management, the city can implement new trees of medium size along the streets. These trees can aid in retaining water from drainage channels by transpiration and interception, reducing throughfall, and increasing infiltration.
- New implementation projects ought to be carried out in residential and highly circulated areas, with a selection of trees that fit the intended function. The trees planted should ideally reduce pollution and absorb noise while tolerating environmental difficulties. A mix of medium and small trees should be utilized to ensure long-term sustainable UPFs.
- A management plan that includes residential streets, not just tourist ones, should be applied to enhance and maintain tree health in Byblos. Residential streets have a significant impact on urban dwellers' health and well-being, and therefore, the 3–30–300 rule could be applied [58]. This rule suggests that every resident should be able to see three trees from their house, each neighborhood should have 30% canopy cover, and the nearest high-quality green public space should be 300 m away.
- Encouraging eco-roofs has numerous benefits. For one, it can offer green spaces for building residents while also being open to public use through collaboration between private owners and the municipality of Byblos. This initiative can significantly increase public green spaces in residential areas of the city. Eco-roofs can form a steppingstone for connectivity in Byblos while increasing its canopy cover. They can also reduce the flow of stormwater. Eco-roofs can be connected to green spaces or the base of trees where the water would be retained by the tree, and not added to the drainage system.

4.2. Study Limitations

The study acknowledges a limitation regarding the number of sampled streets. While a more extensive sample could have enhanced the study's external validity, constraints in resources and time dictated the final sample size. The findings should be interpreted within the context of this limitation, recognizing the potential impact on generalizability.

The present study primarily focused on tree silhouettes, regardless of their specific placement on different types of streets. However, it was inherently limited in terms of the depth of contextual information gathered. As a result, certain nuanced factors associated with street and building contexts influencing residents' preferences may not have been fully explored.

5. Conclusions

The outcome of this research has established foundational data on urban and peri-urban forests within Byblos City. This information can serve as a basis for developing a comprehensive tree species database for the region. The research has also highlighted the potential of urban and peri-urban forests in conserving biodiversity and offering crucial products and services for environmental management, economic empowerment, and social welfare within the community. An improvement in the connectivity between the rural part in the north and the rest of the city and an increase in biodiversity based on a specialized selection of trees is required. At the landscape level, the diverse scenery in Byblos, characterized by a variety of blooming colors and shapes, offers valuable insights for future studies focused on visually guiding trees based on their silhouette values and perception. This exploration could potentially lead to the establishment of guidelines for selecting and placing trees based on their visual characteristics. Additionally, future research could explore connectivity strategies aimed at enhancing the linkages between pocket green spaces, especially in urban areas where grey infrastructure dominates the landscape. These strategies could facilitate biodiversity restoration and contribute to the overall environmental quality of urban settings.

Author Contributions: Conceptualization, M.H. and V.D.; methodology, M.H. and V.D.; software, M.H.; formal analysis, M.H.; investigation, M.H.; data curation, M.H.; writing—original draft preparation, M.H., V.D. and A.R.; writing—review and editing, M.H., V.D., G.K. and A.R. All authors have read and agreed to the published version of the manuscript.

Funding: This research received no external funding.

Data Availability Statement: The data that support the findings of this study are available from the corresponding author upon reasonable request.

Acknowledgments: We thank the two anonymous reviewers for their constructive feedback, which helped us improve the quality of our paper.

Conflicts of Interest: The authors declare no conflicts of interest.

References

1. Salbitano, F.; Borelli, S.; Conigliaro, M.; Chen, Y.; FAO. *Guidelines on Urban and Peri-Urban Forestry*; FAO Forestry Paper No. 178; Food and Agriculture Organization of the United Nations: Rome, Italy, 2016; Available online: https://www.fao.org/3/i6210e/i6210e.pdf (accessed on 14 April 2023).
2. Georgi, J.; Zigkiris, S.; Ftika, Z.; Konstantinidou, E. Management and protection of peri-urban forests of three towns in Greece. In Proceedings of the Fourth International Conference on Remote Sensing and Geoinformation of the Environment (RSCy2016), Paphos, Cyprus, 4–8 April 2016; Volume 9688, p. 96881. [CrossRef]
3. Abd El Karim, A.; Alogayell, H.M.; Alkadi, I.I.; Youssef, I. Mapping of GIS-land use suitability in the rural–urban continuum between Ar Riyadh and Al Kharj Cities, KSA based on the integrating GIS Multi Criteria Decision Analysis and Analytic Hierarchy Process. *Environments* **2020**, *7*, 75. Available online: https://www.mdpi.com/840318 (accessed on 11 April 2023). [CrossRef]
4. Weissert, L.F.; Salmond, J.A.; Schwendenmann, L. A review of the current progress in quantifying the potential of urban forests to mitigate urban CO2 emissions. *Urban Clim.* **2014**, *8*, 100–125. [CrossRef]

5. Cueva, J.; Yakouchenkova, I.A.; Fröhlich, K.; Dermann, A.F.; Dermann, F.; Köhler, M.; Grossmann, J.; Meier, W.; Bauhus, J.; Schröder, D.; et al. Synergies and trade-offs in ecosystem services from urban and peri urban forests and their implication to sustainable city design and planning. *Sustain. Cities Soc.* **2022**, *82*, 103903. [CrossRef]
6. Russo, A.; Cirella, G.T. Urban Sustainability: Integrating Ecology in City Design and Planning. In *Sustainable Human—Nature Relations: Environmental Scholarship, Economic Evaluation, Urban Strategies*; Springer: Singapore, 2020; pp. 187–204. [CrossRef]
7. Grey, G.W.; Deneke, F.J. *Urban Forestry*; Wiley: New York, NY, USA, 1986; 299p.
8. Konijnendijk, C.C.; Sadio, S.; Randrup, T.B.; Schipperijn, J. Urban and peri-urban forestry in a development context—Strategy and implementation. *J. Arboric.* **2004**, *30*, 269–275. [CrossRef]
9. Escobedo, F.J.; Kroeger, T.; Wagner, J.E. Urban forests and pollution mitigation: Analyzing ecosystem services and disservices. *Environ. Pollut.* **2011**, *159*, 2078–2087. [CrossRef] [PubMed]
10. Roeland, S.; Moretti, M.; Amorim, J.H.; Branquinho, C.; Fares, S.; Morelli, F.; Niinemets, Ü.; Paoletti, E.; Pinho, P.; Sgrigna, G.; et al. Towards an integrative approach to evaluate the environmental ecosystem services provided by urban forest. *J. For. Res.* **2019**, *30*, 1981–1996. [CrossRef]
11. Russo, A.; Escobedo, F.J.; Zerbe, S. Quantifying the local-scale ecosystem services provided by urban treed streetscapes in Bolzano, Italy. *AIMS Environ. Sci.* **2016**, *3*, 58–76. [CrossRef]
12. Wolf, K.L.; Lam, S.T.; McKeen, J.K.; Richardson, G.R.A.; van den Bosch, M.; Bardekjian, A.C. Urban Trees and Human Health: A Scoping Review. *Int. J. Environ. Res. Public Health* **2020**, *17*, 4371. [CrossRef] [PubMed]
13. Jones, B.A. Planting urban trees to improve quality of life? The life satisfaction impacts of urban afforestation. *For. Policy Econ.* **2021**, *125*, 102408. [CrossRef]
14. Thompson, E.; Herian, M.; Rosenbaum, D. The Economic Footprint and Quality-of-Life Benefits of Urban Forestry in the United States. 2021. Available online: https://www.arborday.org/urban-forestry-economic/downloads/complete-report-findings.pdf (accessed on 25 March 2024).
15. Cirella, G.T.; Russo, A.; Benassi, F.; Czermański, E.; Goncharuk, A.G.; Oniszczuk-Jastrzabek, A. Energy Re-Shift for an Urbanizing World. *Energies* **2021**, *14*, 5516. [CrossRef]
16. McPherson, E.G.; Simpson, J.R. Potential energy savings in buildings by an urban tree planting programme in California. *Urban For. Urban Green.* **2003**, *2*, 73–86. [CrossRef]
17. Barona, C.O.; Eleuterio, A.A.; Vasquez, A.; Devisscher, T.; Baptista, M.D.; Dobbs, C.; Orozco-Aguilar, L.; Meléndez-Ackerman, E. Views of government and non-government actors on urban forest management and governance in ten Latin-American capital cities. *Land Use Policy* **2023**, *129*, 106635. [CrossRef]
18. Borghesi, S.; Ticci, E.; Climate change in the MENA region: Environmental risks, socioeconomic effects and policy challenges for the future. MED. 2019, pp. 289–292. Available online: https://www.iemed.org/publication/climate-change-in-the-mena-region-environmental-risks-socioeconomic-effects-and-policy-challenges-for-the-future/ (accessed on 10 May 2023).
19. Abumoghli, I.; Goncalves, A. Environmental Challenges in the MENA Region. Available online: https://wedocs.unep.org/20.500.11822/31645. (accessed on 10 May 2023).
20. Saudi Vision 2030. Green Riyadh. Available online: https://www.vision2030.gov.sa/en/projects/green-riyadh/ (accessed on 7 May 2023).
21. Ostoić, S.K.; Salbitano, F.; Borelli, S.; Verlič, A. Urban forest research in the Mediterranean: A systematic review. *Urban For. Urban Green.* **2018**, *31*, 185–196. [CrossRef]
22. National Urban Policies Program in Lebanon, Diagnosis Report | UN-Habitat. Available online: https://unhabitat.org/national-urban-policies-programme-in-lebanon-diagnosis-report (accessed on 10 May 2023).
23. UNDP. Climate-Proofing Lebanon's Development Plans. Beirut, Lebanon. 2021. Available online: https://climatechange.moe.gov.lb/viewfile.aspx?id=323 (accessed on 12 May 2023).
24. ARUP. Resilient Byblos Connecting with our Past, Creating Our Future. 2016. Available online: https://www.arup.com/perspectives/publications/research/section/resilient-byblos# (accessed on 19 May 2023).
25. Mark, J.J. Byblos. World History Encyclopedia. 2009. Available online: https://www.worldhistory.org/Byblos/ (accessed on 19 May 2023).
26. Najjar M.; Ortais C.; Piaalucha D.; Zerbib L. Byblos Toward an Inclusive City. 2017. Available online: https://resilientcitiesnetwork.org/byblos/ (accessed on 1 May 2023).
27. National Council for Scientific Research – Lebanon (CNRS). Land Use Land Cover Maps of Byblos Lebanon. 2022.
28. Lande, R. Statistics and partitioning of species diversity, and similarity among multiple communities. *Oikos* **1996**, *76*, 5–13. [CrossRef]
29. Gotelli, N.J.; Colwell, R.K. Estimating species richness. *Biol. Divers. Front. Measure. Assess.* **2011**, *12*, 39–54.
30. Zhao, Q.; Xu, D.; Qian, W.; Hu, R.; Chen, X.; Tang, H.; Zhang, C. Ecological and Landscape Perspectives on Urban Forest Planning and Construction: A Case Study in Guangdong-Hong Kong Macao Greater Bay Area of China. *Front. Sustain. Cities* **2020**, *2*, 44. [CrossRef]
31. Aguirre, O.; Hui, G.; Gadow, K.V.; Jime´nez, J. An analysis of spatial forest structure using neighbourhood-based variables. *Ecol. Manag.* **2003**, *183*, 137–145. [CrossRef]
32. Ozkan, U.Y.; Ozdemir, I. Assessment of landscape silhouette value in urban forests based on structural diversity indices. *Int. J. Environ. Sci. Technol.* **2015**, *12*, 3971–3980. [CrossRef]

33. Müderrisoğlu, H.; Eroğlu, E.; Özkan, Ş.; Ak, K. Visual perception of tree forms. *Build Env.* **2006**, *41*, 796–806. [CrossRef]
34. Santamour, F.S., Jr. Trees for urban planting: Diversity, uniformity, and common sense. In Proceedings of the 7th Conference Metropolitan Tree Improvement Alliance (METRIA), Lisle, IL, USA, 11–12 June 1990; Volume 7, pp. 57–65.
35. Walsh, C.J.; Fletcher, T.D.; Burns, M.J. Urban stormwater runoff: A new class of environmental flow problem. *PLoS ONE* **2012**, *7*, e45814. [CrossRef]
36. Guan, X.; Wang, J.; Xiao, F. Sponge city strategy and application of pavement materials in sponge city. *J. Clean. Prod.* **2021**, *303*, 127022. [CrossRef]
37. Forestry Commission. National Forest Inventory Report – Tree Cover outside Woodland in Great Britain. 2017. Available online: https://www.forestresearch.gov.uk/tools-and-resources/national-forest-inventory/trees-outside-woodland-tow/ (accessed on 26 August 2023).
38. Heidt, V.; Neef, M. Benefits of Urban Green Space for Improving Urban Climate. In *Ecology, Planning, and Management of Urban Forests*; Carreiro, M.M., Song, Y.C., Wu, J., Eds.; Springer: New York, NY, USA, 2008. [CrossRef]
39. BCN. Barcelona Green Infrastructure and Biodiversity Plan 2020. 2013. Available online: https://climate-adapt.eea.europa.eu/en/metadata/case-studies/barcelona-trees-tempering-the-mediterranean-city-climate/11302639.pdf (accessed on 7 April 2023).
40. Fărcaş, C.P.; Cristea, V.; Fărcaş, S.; Ursu, T.M.; Roman, A. The symbolism of garden and orchard plants and their representation in paintings (I). *Contrib. Bot.* **2015**, *50*, 189–200.
41. Gratani, L. Understanding the Benefits from Green Areas in Rome: The Role of Evergreen and Deciduous Species in Carbon Dioxide Sequestration Capability. *Am. J. Plant Sci.* **2020**, *11*, 1307–1318. [CrossRef]
42. Lu, C.; Kotze, D.J.; Setälä, H.M. Evergreen trees stimulate carbon accumulation in urban soils via high root production and slow litter decomposition. *Sci. Total Environ.* **2021**, *774*, 145129. [CrossRef] [PubMed]
43. Alvey, A.A. Promoting and preserving biodiversity in the urban forest. *Urban For. Urban Green.* **2006**, *5*, 195–201. [CrossRef]
44. Cornelis, J.; Hermy, M. Biodiversity relationships in urban and suburban parks in Flanders. *Landsc. Urban Plan.* **2004**, *69*, 385–401. [CrossRef]
45. Kühn, I.; Brandl, R.; Klotz, S. The flora of German cities is naturally species rich. *Evol. Ecol. Res.* **2004**, *6*, 749–764.
46. Cowett, F.D.; Bassuk, N. Street tree diversity in three northeastern US states. *Arboric. Urban For.* **2017**, *43*, 1–14. Available online: https://www.researchgate.net/publication/312231638_Street_Tree_Diversity_in_Three_Northeastern_US_States (accessed on 16 September 2023).
47. Setiawan, N.N.; Vanhellemont, M.; Baeten, L.; Dillen, M.; Verheyen, K. The effects of local neighborhood diversity on pest and disease damage of trees in a young experimental forest. *For. Ecol. Manag.* **2014**, *334*, 1–9. [CrossRef]
48. Lázaro-Lobo, A.; Ervin, G.N. A global examination on the differential impacts of roadsides on native vs. exotic and weedy plant species. *Glob. Ecol. Conserv.* **2019**, *17*, e00555. [CrossRef]
49. Modlingerová, V.; Száková, J.; Sysalová, J.; Tlustoš, P. The effect of intensive traffic on soil and vegetation risk element contents as affected by the distance from a highway. *Plant Soil Environ.* **2012**, *58*, 379–384. [CrossRef]
50. Bandara, W.A.R.T.W.; Dissanayake, C.T.M. Most tolerant roadside tree species for urban settings in humid tropics based on Air Pollution Tolerance Index. *Urban Clim.* **2021**, *37*, 100848. [CrossRef]
51. Yang, X.; Hong, W.; Wu, C. Distribution characteristic of DBH and tree height for gap edge trees of Acacia confusa in mid-subtropical zone. *J. Southwest For. Univ.* **2010**, *30*, 25–28. [CrossRef]
52. Braubach, M.; Egorov, A.; Mudu, P.; Wolf, T.; Ward Thompson, C.; Martuzzi, M. Effects of urban green space on environmental health, equity and resilience. In *Nature-Based Solutions to Climate Change Adaptation in Urban Areas: Linkages between Science, Policy and Practice*; SpringerOpen: Cham, Switzerland, 2017; pp. 187–205.
53. Akers, A.; Barton, J.; Cossey, R.; Gainsford, P.; Griffin, M.; Micklewright, D. Visual color perception in green exercise: Positive effects on mood and perceived exertion. *Environ. Sci. Technol.* **2012**, *46*, 8661–8666. [CrossRef] [PubMed]
54. Kurt, S.; Osueke, K.K. The effects of color on the moods of college students. *Sage Open* **2014**, *4*, 2158244014525423. [CrossRef]
55. Elsadek, M.; Liu, B.; Lian, Z.; Xie, J. The influence of urban roadside trees and their physical environment on stress relief measures: A field experiment in Shanghai. *Urban For. Urban Green.* **2019**, *42*, 51–60. [CrossRef]
56. Lohr, V.I.; Pearson-Mims, C.H. Responses to scenes with spreading, rounded, and conical tree forms. *Environ. Behav.* **2006**, *38*, 667–688. [CrossRef]
57. Balling, J.D.; Falk, J.H. Development of visual preferences for natural landscapes. *Environ. Behav.* **1982**, *14*, 5–28. [CrossRef]
58. Konijnendijk, C. Promoting Health and Wellbeing through Urban Forests–Introducing the 3-30-300 Rule. 2021. Available online: https://iucnurbanalliance.org/promoting-health-andwellbeing-through-urban-forests-introducing-the-3-30-300-rule (accessed on 12 October 2023).

land

MDPI

Article

Integrated Eye-Tracking Response Surface Analysis to Optimize the Design of Garden Landscapes

Xinman Wang [1], Baoqi Che [1], Qi Lou [2] and Rong Zhu [1,*]

[1] School of Design, Jiangnan University, Wuxi 214122, China; wangxinman@jndx.wecom.work (X.W.); 6210306003@stu.jiangnan.edu.cn (B.C.)
[2] College of Energy Engineering, Zhejiang University, Hangzhou 310058, China; louqi1218@zju.edu.cn
* Correspondence: 7210306007@stu.jiangnan.edu.cn

Abstract: Gardens not only provide people with a place for leisure and relaxation, they also contribute to improving urban ecological environments and promoting social interactions and cohesion. Additionally, from a psychological perspective, gardens play a role in alleviating stress, enhancing happiness, and improving the quality of life. Current research on gardens has primarily employed methods such as questionnaire surveys, environmental psychology analyses, and eye-tracking analyses; however, comprehensive studies on the relationships between multiple factors and levels in garden designs are lacking. Here, we propose a response surface analysis approach based on eye-tracking technology for the design and optimization of gardens. Firstly, the impacts of different garden elements on visitors' psychology and fixation counts were analyzed using environmental psychology and eye-tracking analyses. Subsequently, the optimal range of each garden feature was determined through single-factor experiments, followed by response surface analysis to obtain the optimal value for each element. The results revealed that changes in garden elements such as the greenery ratio, number of buildings, and water saturation significantly affected visitors' psychology. The greenery ratio had a greater impact than the number of buildings, which in turn had a greater impact than water saturation. This study is the first to analyze the relationships between multiple garden elements. A strong relationship was found between the greenery ratio and the number of buildings, as well as between the number of buildings and water saturation, while the relationship between the greenery ratio and water saturation was weaker. This approach can not only optimize garden designs but can also be widely applied in fields such as urban planning and public space transformation to enhance visitors' comfort and satisfaction with the environment and promote sustainable urban development.

Keywords: garden landscape; environmental psychology; eye tracking; response surface analysis; parameter optimization

1. Introduction

Environmental psychology is a discipline that investigates the perception, cognition, and emotional responses of individuals towards environmental landscapes [1]. Its primary focus is on studying the interplay between individuals and their surroundings, as well as examining the impact of the environment on an individual's psychology and behavior [2]. On one hand, the relationship between humans and the environment is mainly influenced by culture, which promotes an understanding of human–environment interactions (i.e., cultural environmental psychology) [3]. On the other hand, researchers have explored the impact of the environment on an individual's psychology, specifically by examining how the environment affects environmental sensitivity [4], place satisfaction [3], and the attachment effect [5]. In the field of landscape architecture, the study of environmental psychology primarily investigates the psychological responses of individuals to various garden elements, including their attention, preferences, and feelings of comfort [6]. For instance, studies have focused on people's comfort levels in green spaces [7,8], the extent of their preferences for architectural aesthetics [9,10], and their levels of satisfaction with water features [11]. Psychological research on these factors facilitates the optimization of garden design and the creation of exceptional aesthetic experiences for individuals while also promoting relaxation and recreational activities, with the ultimate goal of creating environments that are beneficial to people [12–14].

Currently, research methods in environmental psychology primarily include self-report questionnaires, verbal interview protocols such as think-aloud procedures [15,16], electroencephalography (EEG) [17], and eye-tracking analyses [18]. In recent years, eye-tracking technology has been extensively applied in various fields, including advertising, user interface design, and human–computer interactions [19–21]. Eye-tracking analysis is a highly accurate tool that analyzes visual attention and cognitive processes by recording eye movements. It provides precise quantitative data, capturing users' gaze points, gaze duration, and eye movement trajectories while observing landscapes [22]. Due to its real-time and objective nature, it is an ideal tool for studying landscape design and user experiences, with significant application value in these domains [18,23]. Applying eye-tracking technology in landscape research allows for attention and interest measurements to be recorded for different landscape elements, thus reducing biases during data analysis [18]. For example, eye tracking has been used to study the influence of individual landscape features, such as the proportion, shape, or height of green spaces, on visitors' visual preferences [24,25]. In another case, visual heatmaps were generated to show the visual completeness between two elements (buildings or trees) or multiple landscape features (buildings, rivers, or greenery) using eye tracking from a micro-to-macro perspective [26,27]. Given the unique advantages of eye-tracking technology in landscape design research, we chose to apply this technique in our study to provide an in-depth analysis of users' visual focal points and emotional responses to different landscape elements [28]. Through the analysis of eye-tracking data, we can uncover the specific effects of varying greenery ratios, architectural layouts, and water saturation on users' visual attention and emotional experiences. This not only contributes to optimizing landscape design and enhancing the quality of user experiences, it also provides landscape designers with scientific evidence and supporting data, facilitating the development of landscape design theory and practice.

Response surface methodology (RSM), also known as the response surface design method, utilizes a multivariate quadratic regression equation to model the functional relationship between factors and response values. In 1951, Box and Wilson first introduced the Central Composite Design (CCD) [29], which has become one of the most commonly used designs for quadratic models, alongside the Box–Behnken design [30]. By combining multiple responses, the CCD enables the study of their behaviors and has led to the development of various optimization and desirability routines [31]. Through analysis of the regression equation, the RSM seeks to identify optimal process parameters and address the challenges associated with multivariate problems [32]. The response surface analysis method enables the examination of the regression relationship between experimental indicators (dependent variables) and multiple experimental factors (independent variables), which can take the form of either a curve or a surface. Currently, response surface analyses are primarily applied in industrial structural design and parameter optimization. For instance, design optimization has been performed for turbine drill bit structures by using the response surface method to analyze the relationships between various parameters in order to enhance the drilling efficiency of the bit [33]. Response surface analysis typically involves conducting single-factor experiments and designing experiments for the response surface through which optimal ranges of multiple factors, with each containing the optimal value, can be determined and a regression equation can be derived [34]. Due to its high efficiency, the response surface methodology has witnessed rapid development in the fields of environmental protection and urban planning [35]. For instance, in relation to the components of fermentation of vegetable residues mixed with a ratio of moisture (M) to solid content (S), initial pH value, and organic loading (OL), the fertilizer yield can be determined by using a regression analysis to determine the optimal fertilizer yield formula [36]. Research on landscape design has indicated that different landscape elements in courtyards and peripheral areas have an impact on the overhead layer, which can be determined using response surface analysis (RSA) [37]. Moreover, the proportions and spatial layouts of park elements such as lawns, trees, water bodies, hard surfaces, and buildings also influence urban parks [38]. However, the application of RSA in landscape design is still in the early stages of development. Furthermore, studying gardens as holistic entities rather than focusing on specific elements or features is crucial in order to avoid biased research outcomes. A comprehensive and objective assessment of research findings considers subjective perception and the interrelationships among different garden elements, as well as the overall compositional effects on garden design. Consequently, there is a lack of research methods that incorporate multiple factors and levels. This study aimed to employ eye-tracking technology and RSA to quantify and evaluate the comprehensive impacts of different garden elements (such as the greenery ratio, architectural style, water saturation, and stone design) on people's visual attention and emotional experiences. Ultimately, this research provides scientific evidence for optimizing landscape designs.

Here, we aimed to optimize the design of gardens by leveraging the advantages of the response surface methodology and a multifactorial analysis. Using tourists' focal points as the dependent variable, and the vegetation, architecture, and water as the independent variables, we employed eye-tracking technology to determine the optimal values of independent variables that resulted in the optimal value of the dependent variable. This approach allowed us to achieve the goal of optimizing the garden design while significantly reducing the design costs [39]. Therefore, we employed response surface analysis based on eye-tracking data. This method allowed for analysis of the relationship between multiple factors and multiple levels using statistical modeling, resulting in the determination of the optimal values for each element. In contrast to a single-factor analysis, response surface analysis takes into consideration the relationship between multiple garden elements, making the research results more comprehensive and accurate. The integration of eye tracking and response surface analysis presents significant advantages in the field of landscape design. Firstly, it can help designers understand the extent of the impact that different elements have on people's attention, preferences, and comfort, enabling targeted adjustments during the design process. Secondly, by considering the relationship between different elements, this method can facilitate overall coherence and harmony in landscape design, enhancing people's experience. Additionally, eye tracking and response surface analysis can be widely applied in landscape design, urban planning, and public space transformations. Through an in-depth examination of people's behavior and perceptions in different environments, this method can provide scientific evidence for creating more attractive and comfortable surroundings, thereby improving quality of life for urban residents.

2. Materials and Methods

2.1. Study Area

This study focused on six selected gardens that served as key representatives of the Jiangnan style (Surging Wave Pavilion, Lion Grove Garden, Humble Administrator's Garden, Lingering Garden, Jichang Garden, and Master-of-Nets Garden; Figure 1). The criteria and rationale behind the selection of these six gardens were as follows [40–42]:

(1) Representativeness: These gardens hold significant historical and cultural value in the Jiangnan region, representing the typical style and design concepts of Jiangnan gardens. Specifically, they showcase the essence of Jiangnan gardens in terms of their layout, architecture, and plant arrangements.

(2) Geographical coverage: The chosen gardens are distributed across different cities in the Jiangnan region, including Suzhou and Wuxi, reflecting their diversity and regional characteristics.

(3) garden elements: The presence of pavilions, towers, small bridges, flowing water, and rock formations within these gardens fully embodies the artistic features and aesthetic pursuits of the Jiangnan garden design.

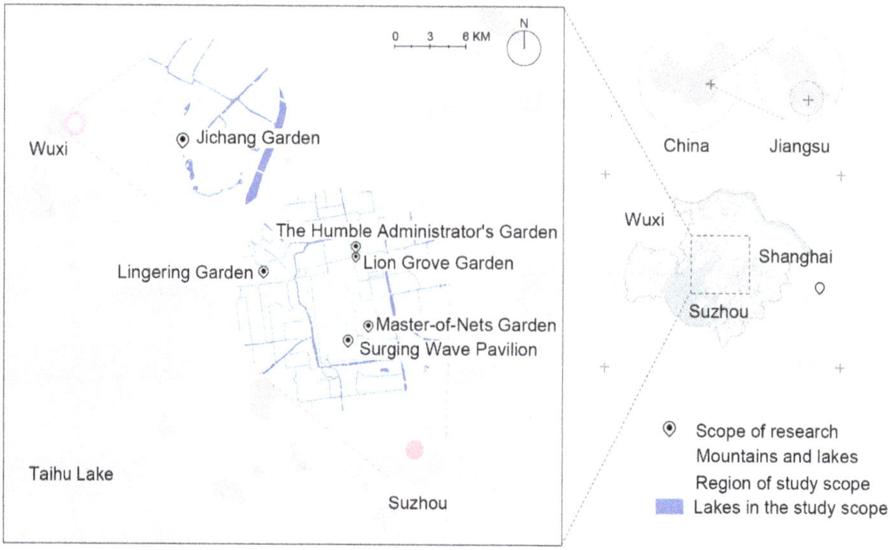

Figure 1. Location of the study area.

Through literature review and field investigations, we examined the characteristics of these garden elements across the six chosen gardens (Table S1).

2.2. Photo Acquisition and Processing

The use of photographic images as substitutes for real landscape scenes has been shown to be reliable and has been widely adopted in landscape perception research [43–47]. Therefore, it is feasible to conduct experiments using photographs as stimuli. Three types of garden element images representing different landscape types—greenery, buildings, and water features—were used as stimuli. These elements were selected because they are common in Jiangnan gardens. Representative images were chosen for each type of landscape based on the similarity of the landscape structures in the images, the complexity of the garden settings, and the feasibility of modifying content, eliminating other covariates, and controlling landscape complexity by removing or adding certain landscape components. All the photographs were taken under similar weather and seasonal conditions, including sunny weather and lower traffic flow. Considerations included mounting the camera on a tripod at a fixed height and capturing photos at a resolution of 2970 × 1980 pixels. A focal length of 50 mm was maintained to ensure a consistent field of view (+31 × 21°). All images were selected from a photo library, with a focus on three key garden elements: the greenery ratio, the number of buildings, and water saturation. Collage techniques were used to manipulate the scenes and create well-integrated images by adding, removing, or synthesizing landscape elements [24,48–51]. All added landscape elements referenced actual garden photos with similar landscape structures to enhance their realism [42]. This resulted in 18 images (Figure S1), with 6 images per setting type, used to compare the differences in preference ratings and eye movement metrics across different levels of landscape complexity in each environment.

2.3. Subject

We recruited participants from diverse age groups and genders by sending requests. The criteria for eligibility included normal vision and color perception, a willingness to participate, and the absence of any assistance. The exclusion criteria included excessive eye blinking, a prolonged dwelling time, and erratic eye-tracking trajectories. The final sample consisted of 94 individuals (44 males and 50 females) between the ages of 18 and 65. Before their enrollment, the participants were provided with concise instructions about the testing procedures, while the specific research objectives remained undisclosed. Additionally, the participants were instructed to avoid using eye makeup and mascara to prevent potential interference with accurate pupil tracking by the eye-tracking glasses.

2.4. Eye Tracking and Questionnaire Data Collection

Research has shown that attractive features lead to longer gaze durations and higher fixation rates [52,53] (Table S2). Therefore, it was imperative to examine the differences in the fixation counts and dwelling times when perceiving variations in the complexity of different elements to gain a better understanding of individual differences within the general population in order to inform the future development of gardens [54,55].

Eye-tracking data were analyzed using the ETG2 Wireless Analysis Pro (18009771) software (SMI, Germany). This software utilizes infrared eye-tracking technology [56], enabling the accurate recording of both reflective signals and pupil positions. As a result, all the fixations (fixation count) and saccades (scan path) were recorded [57]. In addition, the SMI analysis software BeGaze 3.7, which facilitated the export of the eye-tracking metrics (ETMs) into well-structured Excel files, was employed.

2.5. Procedure

To analyze the visual preferences and differences in various elements in Jiangnan gardens, we conducted a specific scene-based visual and psychological cognition experiment. The participants experienced the same location continuously for five minutes at a time, and their experiences were recorded and analyzed through a questionnaire survey (Table S3). Using qualitative methods, our aim was to reveal the impact of garden elements on psychological responses and validate the relationship between garden elements and psychological reactions.

The eye-tracking landscape optimization method was used to carry out the experiments and procedures (Figure 2). The eye-tracking experiments were performed using 18 garden photographs and the participants were allowed to freely view and control the observation duration. To address potential biases introduced by a fixed order, the presentation sequence of the photographs was randomized, creating a relaxed experimental environment [50,58]. The space was carefully controlled to ensure the strict regulation of lighting and ambient noise. Prior to testing, all the participants received identical instructions (training and experimental operation guidelines), and three calibration points were utilized to ensure accurate eye-tracking measurements across the entire screen. During the experiment, the participants observed three sets of different garden elements that were recorded by an independent eye-tracking device in conjunction with a computer. After the experiment, the participants were asked to complete questionnaires (Table S4) and received corresponding rewards.

After completing the experiment, the eye-tracking data were analyzed using the BeGaze 3.7 software package (SMI). The raw data obtained from the eye tracker were transformed to derive a meaningful measure, the fixation count. Subsequently, response surface analysis was performed using the Design-Expert 8.0.5b software (Design-Expert, Minneapolis, MN, USA). The independent variables were the greenery ratio, number of buildings, and water saturation. The response surface analysis employed a Box–Behnken central composite experimental design with three factors and three levels. To validate the reliability of the experiment, another eye-tracking experiment was conducted, ultimately leading to the obtained results.

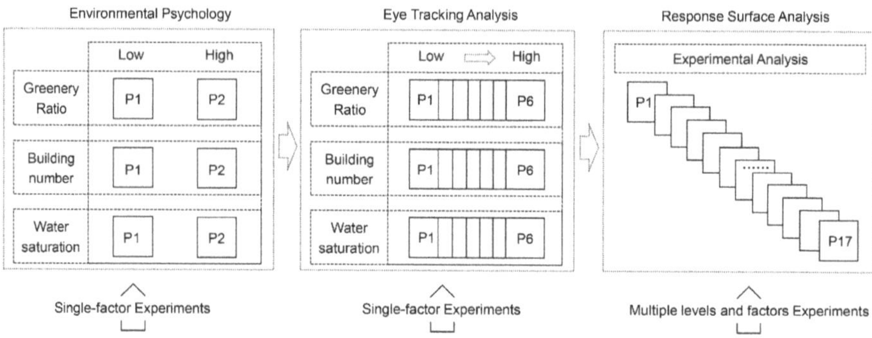

Figure 2. The process of the eye-tracking landscape optimization method.

2.6. Data Analysis

Statistical analysis was conducted using SPSS to determine whether there are significant differences in preference ratings between different garden elements and visitors. A bivariate correlation analysis was performed to examine the relationships between three factors—greenery ratio, number of buildings, and water saturation—and the fixed count data. This analysis quantified the impact of these variables on different indicators and assessed the normality of data distribution. The factors included in this analysis were the" number of buildings" (rated on a six-point scale), "water saturation" (rated on a six-point scale), and "greenery ratio" (rated on a six-point scale). Origin and heatmap analyses were employed to investigate differences in visual behavior and cognitive assessment when observing the three park elements based on an assumption of normal data distribution (Figure S2). Furthermore, response surface methodology was used to evaluate the overall trend and relationships between variables in the dataset, as well as the relationships between park elements. Finally, quadratic regression and analysis of variance were applied to test the interactions between the three factors.

3. Results

3.1. Research on Environmental Psychology of Garden Elements

3.1.1. Impact of the Greenery Ratio on Visitors' Psychology

Landscaping elements primarily comprise the greenery ratio, number of buildings, and water saturation. In order to examine the impact of these elements on visitors' environmental psychology using field research with three-dimensional scenes (Figure 3a,b), we compared the psychological experiences of visitors under scenarios of low and high levels of vegetation. The findings revealed a significant influence of vegetation on visitors' psychology and behavior (Table S5).

When the greenery ratio in a garden was relatively low, the absence of greenery created a monotonous and oppressive atmosphere that failed to capture the interest and attention of visitors. In contrast, scenes with a higher greenery ratio were perceived as more beautiful and pleasant. An increase in the greenery ratio added vitality to the environment, thereby enhancing visitors' level of enjoyment. However, an excessive density of greenery often resulted in feelings of fear, similar to findings from studies on people's fear of densely vegetated areas [59]. The greenery ratio plays a significant role in environmental psychology, and an appropriate increase in the greenery ratio is often associated with positive emotions [24,60]. This suggests that visitors' perception, cognition, and emotional responses to their environment are directly influenced by the surrounding landscape (greenery), as observed in studies on environmental psychology.

(a) (b)

Figure 3. Landscape renderings with different greenery ratios. (**a**) Low greenery ratio; (**b**) high greenery ratio.

3.1.2. Impact of the Number of Buildings on Visitors' Psychology

By conducting field research using three-dimensional scenes (Figure 4a,b) and comparing the psychological experiences of visitors under scenarios with fewer and more buildings, it was observed that the number of buildings had a profound impact on visitors' environmental psychology (Table S5).

(a) (b)

Figure 4. Landscape renderings with different numbers of buildings. (**a**) Low number of buildings; (**b**) high number of buildings.

Scenarios with fewer buildings, which were characterized by a low level of architectural detail and a lack of complexity in the building facades, were considered compact and devoid of variation, inducing a sense of urgency. In contrast, scenes with an excessive number of buildings led to feelings of anxiety and suppression. The presence of numerous buildings contributes to a crowded and chaotic environment, evoking a sense of oppression and unease [61,62]. Psychological studies have shown that a single architectural scene can be considered relatively dull and lonely; without the accompaniment of other buildings, such a scene lacks diversity and vitality, imparting a feeling of isolation and monotony. Inversely, scenes with a multitude of diverse buildings can be seen as oppressive and noisy, failing to capture the interest and favor of visitors [63]. In summary, the number of buildings plays a significant role in environmental psychology. A suitable increase in the number of buildings can enhance subjective comfort and sense of security [64]. Thus, it is necessary to design the interval between buildings reasonably.

3.1.3. Impact of Water Saturation on Visitors' Psychology

In the field of color theory, saturation, also known as purity, refers to the intensity or brightness of a color. In the hue–saturation–value (HSV) color model, saturation is one of three attributes of color, alongside hue and value. In this model, saturation ranges from 0 to 100%. As saturation decreases, the color becomes duller, eventually losing its hue and becoming achromatic, with a saturation value of 0. In our analysis of numerous photographs showcasing garden water features, the highest saturation level frequency identified was 50%, which we established as the standard (Figure S3). Experimentally, we observed that when the saturation of this photograph was adjusted to 100%, the colors approached full saturation. Therefore, the definition of 50% saturation aligns with our requirements. Subsequently, based on the regular patterns observed in the single-factor experiments, we assigned saturation levels of 20%, 30%, 40%, 50%, 60%, and 70% to the water feature photographs.

Through field research using three-dimensional scenes, the impact of water saturation on the environmental psychology of visitors was compared (Figure 5a,b). It was found that changes in water saturation elicited different psychological experiences and behavioral responses (Table S5). Low water saturation (dark-colored water) evoked a sense of oppression. Black is typically perceived as a negative and heavy color, and thus, water with a dark hue may induce feelings of sadness or repression. This color may trigger negative emotional responses, such as anxiety or depression, as it may evoke associations with pollution or a dirty and disorderly environment [65]. In contrast, high water saturation (yellow-colored water) created a feeling of nausea [66]. Yellow is often associated with dirt or impurities; therefore, water with a yellow hue may cause discomfort and aversion among visitors [67]. This color can evoke associations with pollution or the presence of harmful substances and may even elicit feelings of nausea or unease. Water saturation significantly influences the environmental psychology of individuals; thus, in landscape design, it is crucial to consider the impact of water saturation on visitors' emotions and mental states in order to create a more comfortable and pleasant environment.

(a)

(b)

Figure 5. Landscape renderings with different levels of water saturation. (**a**) Low water saturation; (**b**) high water saturation.

Based on the analysis of the questionnaire, we validated and discussed the relationship between garden elements and psychological responses. The experimental data clearly demonstrated that varying vegetation proportions, building quantities, and water saturation levels had differential impacts on the visual attention and psychological perceptions of the participants. These findings are in line with existing theories in landscape psychology [68–71].

3.2. *Eye-Tracking Analysis of Elements of Garden Landscapes*

3.2.1. Eye-Tracking Analysis of Greenery Ratio

The influence of the greenery ratio on attention distribution and visual preferences was assessed using eye-tracking analysis. The eye-tracking device recorded the visitors' eye movements, and heatmaps were generated to illustrate the influence of different greenery ratios on eye behavior. It was observed that in environments with a low greenery ratio (Figure 6a), the heatmaps exhibited scattered and sparse features, indicating that the visitors' visual attention was not concentrated on specific areas. The visitors tended to fixate more frequently on the empty sky or buildings, as the lack of greenery failed to capture their attention. Conversely, in scenes with high greenery ratios (Figure 6b), the clear concentration of the fixation counts in areas with sparse greenery was observed, forming distinct hotspots. The visitors tended to focus on areas not covered by vegetation, as these areas presented simpler features that attracted their visual attention, accompanied by a sense of security and comfort [72,73]. However, when the visitors concentrated on highly vegetated areas, a feeling of fear emerged, driving them to seek escape.

(a) (b)

Figure 6. Heatmaps of different greenery ratios. (**a**) Low greenery ratio; (**b**) high greenery ratio.

Therefore, an appropriate greenery ratio can guide visitors' eye tracking and enhance their overall cognition and preferences toward the landscape. Excessive or insufficient greenery may lead to visual fatigue or monotony, thereby affecting visitors' perception and emotional experiences in response to their environment.

3.2.2. Eye-Tracking Analysis of The Number of Buildings

Using eye-tracking technology, this study investigated the influence of the number of buildings on visitors' eye movement behavior, as well as their attention distribution and visual preferences. The results showed that the number of buildings significantly influenced participants' eye movement patterns. On one hand, in scenes with fewer buildings (Figure 7a), the heatmaps exhibited concentrated features, indicating that the visitors' visual attention was predominantly focused on specific areas. The visitors tended to frequently fixate on the few available buildings, but these individual buildings struggled to sustain their attention for long periods. Consequently, while the fixation counts were concentrated, they also displayed fast scanning or transient dwelling behavior, reflecting a state of uneasiness or tension. On the other hand, in scenes with an excessive number of buildings (Figure 7b), the heatmaps displayed a more concentrated eye movement trajectory around the areas where the buildings were located, forming distinct hotspots. The visitors tended to prioritize the structures and details of each individual building in these areas because they presented richer and more visually appealing features.

(a) (b)

Figure 7. Heatmaps of different numbers of buildings. (**a**) Low number of buildings; (**b**) high number of buildings.

However, when the visitors concentrated on areas with too many buildings, they might have experienced a sense of suppression or anxiety [61,62]. Therefore, an appropriate number of buildings is crucial for diverting visitors' attention distribution and guiding them toward a more pleasant emotional and psychological experience.

3.2.3. Eye-Tracking Analysis of Water Saturation

We conducted an eye-tracking study to examine the effects of water saturation on attention distribution and visual preferences. The results of the eye-tracking study demonstrated that water saturation significantly influenced the visitors' eye-tracking behavior. In scenes with low water saturation (presenting dark hues) (Figure 8a), the heatmaps exhibited concentrated fixation counts on the water's surface, forming distinct hotspots. Conversely, in scenes with high water saturation (presenting yellow hues) (Figure 8b), the heatmaps showed relatively dispersed fixation counts without clear concentration on the water's surface. Further analysis revealed that, on dark water surfaces, the participants' fixation counts were primarily centered on or above the water's surface, possibly due to their discomfort or unease regarding the color of the water. Therefore, they attempted to avoid direct fixation on the water's surface. On the yellow water surface, the visitors' fixation counts tended to cluster more around the edges or underwater areas, potentially due to feelings of nausea or aversion towards the color of the water; thus, they avoided direct fixation on the water's surface. Additionally, on the yellow water surface, the visitors' fixation counts were more likely to demonstrate rapid scanning or transient dwelling behavior, as the visitors tried to quickly avert their eyes away from the uncomfortable visual stimuli as they were unwilling to remain fixated on the water's surface for an extended period.

Collectively, the saturation level of water significantly influenced visitors' eye movement behavior by greatly impacting the distribution of fixations. Optimal water saturation can effectively redirect visitors' visual attention distribution, thereby guiding them toward a more enhanced emotional and psychological experience.

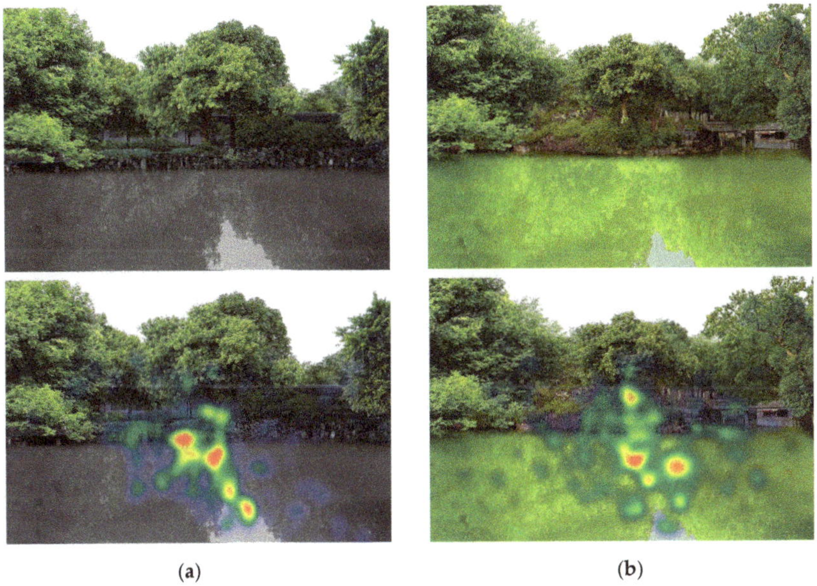

Figure 8. Heatmaps of different levels of water saturation. (**a**) Low water saturation; (**b**) high water saturation.

3.3. The Relationship between Environmental Psychology and Eye Tracking

Through the analysis of environmental psychology and eye tracking on the perception and psychology of visitors towards garden elements, it was found that the eye-tracking analysis not only encompassed the content of the environmental psychology analysis but also provided more intuitive and objective support through eye movement data. For example, in the eye movement recorded in response to two-dimensional images, we clearly observed the eye movement behavior of visitors when observing scenes with low or high degrees of vegetation. These results reflected the variations in visitors' psychology in response to garden elements. For instance, in the study on the number of buildings, when visitors concentrated on a specific area with a single building, it induced a sense of depression or anxiety, which aligned with the results of the psychological analysis. Similarly, the psychological results of the saturation level of water were also highly consistent between the two studies.

Therefore, eye-tracking analysis can be used as a substitute for environmental psychology when conducting landscape analysis within the realms of three-dimensional analysis and psychological analysis. Not only does this allow for the study of visitors' psychological changes in a sensory manner, but it also provides rational and intuitive eye-tracking research data for two-dimensional images. This enables a more accurate analysis and a better understanding of visitors' perceptions and responses to garden elements, thus providing a scientific basis for the design and optimization of such elements.

3.4. Single-Factor Experiments

Using an eye-tracking device, our research examined the impact of different elements of landscaping on the psychology of visitors. This study showed that the greenery ratio, number of buildings, and water saturation significantly affected visitors' visual fixation and psychological perceptions. However, a simple analysis using a real-world situation with an eye-tracking device cannot fully capture the influence of individual elements on visitors' psychology. Due to the inability to accurately measure the variations in each element in the actual three-dimensional space, the research results of a single element are heavily influenced by other landscaping elements, making it difficult to precisely determine the range of each element that satisfies visitors. To better investigate the impact of individual landscaping elements on visitors' psychology and effectively minimize the interference of other elements on visitors' preferences, we adopted single-factor experiments to determine the optimal range of landscaping elements.

3.4.1. The Impact of the Greenery Ratio on Visitors

As shown in Figure 9a, a single-factor experiment revealed that the visual score (fixation counts) for tourists exhibited an increasing–decreasing trend with an increase in the greenery ratio (Table S6). When the greenery ratio reached 60%, the visual fixation count of visitors reached its highest value, indicating that appropriately increasing the greenery ratio to replace other elements in the garden can make the space more natural and vibrant, thus increasing visitors' satisfaction [74,75]. This finding is consistent with the results of the eye-tracking analysis, which suggests that a certain level of greenery significantly enhances visitors' sense of security and comfort (Figure 9b). When the greenery ratio exceeded 70%, the visual fixation counts of visitors sharply declined, in line with the results of the eye-tracking analysis. This is because a large area of green space interferes with visitors' visual scanning and leads to feelings of fear and disgust. At the visual level, it has been observed that the hierarchy and focal points within a landscape guide the attention and gaze of visitors through the use of elements such as shape and color. A 60% ratio of greenery can potentially create the most optimal visual contrast and sense of hierarchy, thereby enhancing the attractiveness and interest of the landscape. Therefore, appropriately controlling the greenery ratio can optimize visitors' experience in the garden, with an optimal greenery ratio of around "60%".

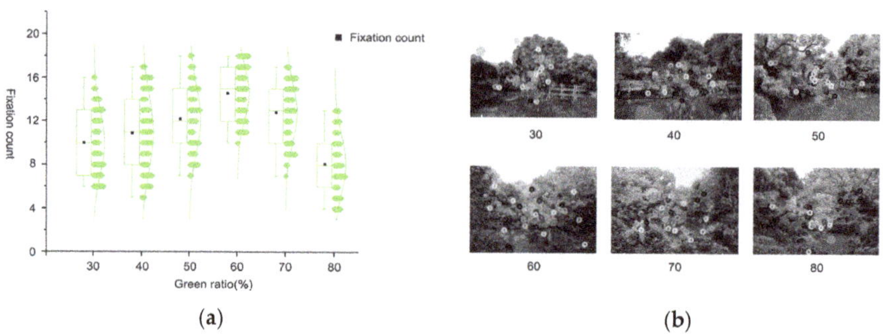

Figure 9. Relationship between greenery ratio and visitor fixation counts. (**a**) Fixation counts; (**b**) scan path.

3.4.2. The Impact of the Numbers of Buildings on Visitors

As depicted in Figure 10a, a single-factor experiment revealed a trend of increasing and then decreasing visual fixation counts with an increase in the number of buildings (Table S7). The optimal state of visitors' visual fixation counts was observed when the number of buildings was three. In the case of a lower number of buildings, buildings served as decorative elements between the greenery and water saturation, providing immediate architectural information to visitors [70,76]. A moderate increase in the number of buildings enhanced the diversity of architectural styles in visitors' perspectives, thereby attracting their visual attention and improving their garden experience (Figure 10b). Nonetheless, an excessive number of buildings may blur the boundaries between garden elements (the greenery ratio, number of buildings, and water saturation), causing a cognitive burden and triggering feelings of depression or anxiety among visitors [77,78]. Therefore, an appropriate number of buildings can provide a positive aesthetic experience, with the optimal number being around "three".

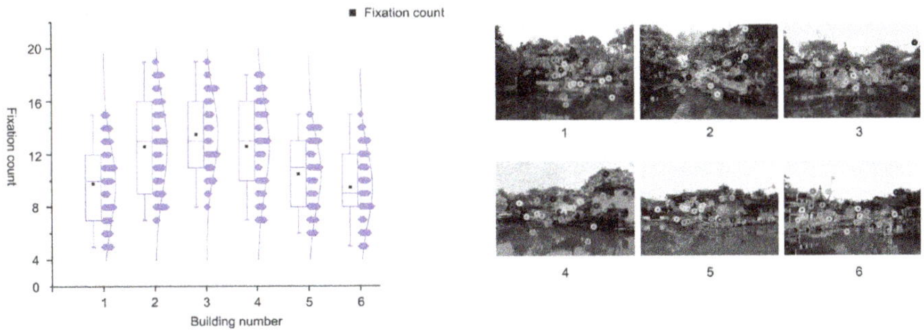

(a) (b)

Figure 10. Relationship between the number of buildings and visitor fixation counts. (**a**) Fixation counts; (**b**) scan path.

3.4.3. The impact of water saturation on visitors

As shown in Figure 11a, the results of the single-factor experiment revealed a trend whereby the visitors' fixation counts initially increased and then decreased as the saturation level of water increased (Table S8). The highest level of visual attention from visitors was observed when the water saturation level approached zero. Water saturation is considered a crucial factor in enhancing the aesthetic appeal of the landscape as it significantly contributes to the overall visual experience [71]. Assuming that the greenery ratio and the number of buildings remain constant, a water saturation level of zero (green water) enhanced the spatial atmosphere of the landscape. Gradually increasing the saturation level of water better integrated it into the garden environment, thereby improving visitors' sense of comfort (Figure 11b). However, excessive water saturation (yellow water) led to a gradual decrease in visitors' fixation counts, potentially due to the intense visual stimulation caused by the overly bright yellow water surface, which discourages a prolonged gaze. This phenomenon demonstrates the mechanism of a color–emotion response. Color is a significant component in landscape design, as different colors elicit distinct emotional reactions. For instance, green evokes feelings of tranquility and relaxation, while red can elicit excitement and arousal. Consequently, the color of water can also influence an individual's psychological and physiological state, leading to stress reduction or enhancing happiness [79]. Therefore, an appropriate water color provides a positive visual experience, with the optimal saturation level at approximately "60".

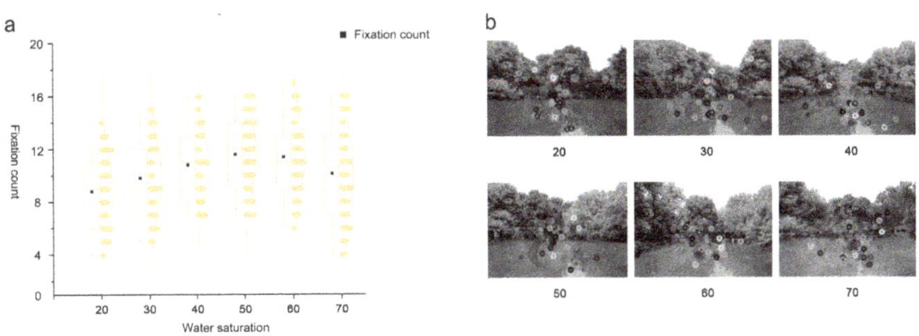

(a) (b)

Figure 11. Relationship between water saturation and visitor fixation counts. (**a**) Fixation counts; (**b**) scan path.

3.4.4. Validation of the Results of Single-Factor Experiments

We re-evaluated novel approaches for collecting and analyzing eye-tracking data to explore the changes in participants' attention in comparison to real observations. This was achieved by optimizing real-life scenarios and varying the placement of elements in multiple scenes while conducting multiple experiments and reproducing the scenes. For instance, we investigated the influence of pavilion positioning on eye-tracking heatmaps of visitors and found that the pavilion's placement slightly impacted the distribution of participants' visual attention, although the overall effect was minor (Table S9). Hence, the distribution of participants' attention in the eye-tracking heatmaps was convincing compared to actual circumstances.

Additionally, we conducted a questionnaire study to further validate the reliability of the findings. This research revealed that the results of the single-factor experiment aligned with the trends observed in the questionnaire survey. Therefore, the conclusions drawn from the single-factor experiment regarding the garden elements can serve as experimental parameters for subsequent response surface experiments (Table S10).

3.4.5. Mechanisms and Principles

The potential health benefits of exposure to novel natural landscapes may be better understood via the lens of the supportive environment theory (SET). Landscapes that are simple to comprehend and maintain are referred to as supportive environments. According to the SET, people require these kinds of surroundings—garden components—in order to preserve both physical and mental health [80].

Like all natural landscapes, natural landscapes devoid of plants, rocks (buildings), and water have various limitations and potential hazards despite to their many potential health advantages [81]. Thus, in order to effectively plan and create natural landscapes, we need to investigate the relationships between garden elements in more detail.

3.5. *Response Surface Analysis of Garden Elements*

3.5.1. Response Surface Optimization Simulation

After establishing the optimal range for each element of a garden, further investigation was conducted using response surface analysis to explore the interrelationships among these elements to determine the optimal value for each garden element. The specific research process is as follows: The experimental factors and levels are shown in Table 1, while Table 2, Tables S11 and S12 contain the design plans and analysis of the results.

Table 1. Factors and levels of response surface experience.

Number	Greenery Ratio (%)	B Number of Buildings	C Water Saturation (%)
−1	50	2	40
0	60	3	50
1	70	4	60

Multiple quadratic regression and analysis of variance were applied to the data presented in Table 2. These analyses yielded a quadratic regression equation (Equation (1)):

$$Y = 11.66 - 2.22A + 1.06B - 0.84C - 0.73AB - 0.27AC - 0.05BC + 1.79A^2 - 2.52B^2 + 3.37C^2 \quad (1)$$

To analyze the effectiveness of the quadratic regression equation, further variance analysis was conducted on the regression model. Table 3 shows that the garden model was highly significant ($p < 0.01$). The lack-of-fit terms (*AB*, *AC*, and *BC*) for all three models were not significant ($p > 0.05$), indicating that the influence of non-experimental factors on the results was minimal and the models were reasonable and acceptable. The coefficient of determination (R^2) for the particle size distribution was 0.9984, denoting a strong linear relationship between the factors and visual attractiveness to visitors. The adjusted coefficient of determination (R^2 = 0.9953) indicated that the model explained 99.53% of the variation in the response variable. The experimental coefficient of variation (CV) was 0.56% (<10%), and the signal-to-noise ratio was 81.827 (>4). In conclusion, this model can be used for predicting and simulating visitors' psychological preferences toward garden elements. Furthermore, it was found that variables A, B, and C had extremely significant effects on Y ($p < 0.01$). The interaction terms AB, AC, and BC did not significantly impact Y ($p > 0.05$), while the quadratic terms A^2, B^2, and C^2 had highly significant effects on Y ($p < 0.01$). Further analysis based on the F-value revealed that the order of influence in the garden was $A > B > C$, suggesting that vegetation coverage had the greatest impact, followed by the number of buildings and then water saturation.

Table 2. Design and results of response surface experiments.

Number	A Greenery Ratio (%)	B Number of Buildings	C Water Saturation (%)	Combination (Fixation Count)
1	−1 (50%)	−1 (2)	0 (50%)	14
2	1 (70%)	−1 (2)	0 (50%)	13.3
3	−1 (50%)	1 (4)	0 (50%)	13.5
4	1 (70%)	1 (4)	0 (50%)	12.8
5	−1 (50%)	0 (3)	−1 (40%)	15.1
6	1 (70%)	0 (3)	−1 (40%)	14.5
7	−1 (50%)	0 (3)	1 (60%)	14.8
8	1 (70%)	0 (3)	1 (60%)	14.3
9	0 (60%)	−1 (2)	−1 (40%)	13.2
10	0 (60%)	1 (4)	−1 (40%)	12.8
11	0 (60%)	−1 (2)	1 (60%)	13
12	0 (60%)	1 (4)	1 (60%)	12.6
13	0 (60%)	0 (3)	0 (50%)	17.9
14	0 (60%)	0 (3)	0 (50%)	17.8
15	0 (60%)	0 (3)	0 (50%)	17.7
16	0 (60%)	0 (3)	0 (50%)	17.9
17	0 (60%)	0 (3)	0 (50%)	17.9

Table 3. Analysis of variance of the response surface model.

Sources of Variance	Sum of Squares	Degrees of Freedom	Mean Square	*F*-Value	*p*-Value
Model	69.71	9	7.75	1095.30	<0.0001 **
A	0.78	1	0.78	110.48	<0.0001 **
B	0.41	1	0.41	57.27	<0.0001 **
C	0.10	1	0.10	14.32	0.0069
AB	0.000	1	0.000	0.000	1.0000
AC	2.500×10^{-3}	1	2.500×10^{-3}	0.35	0.5708
BC	0.000	1	0.000	0.000	1.0000
A^2	7.48	1	7.48	1057.22	<0.0001 **
B^2	40.66	1	40.66	5749.78	<0.0001 **
C^2	14.14	1	14.14	1999.48	<0.0001 **
Residuals	0.049	7	7.071×10^{-3}		
Lack-of-fit terms	0.017	3	5.833×10^{-3}	0.73	0.5860
Pure error	0.032	4	8.000×10^{-3}		
Total variation	69.76	16			

The '**' signifies highly significant differences ($p < 0.01$).

3.5.2. Analysis of the Relationships between Garden Elements

Moreover, to enhance the design of garden elements, a further analysis was conducted to investigate the synergistic effects between different elements in the courtyard.

1. The relationship between the greenery ratio and the number of buildings

The relationship between the greenery ratio and the number of buildings exhibited a steep trend in the response surface model, indicating significant impacts of the greenery ratio and the number of buildings on visitors' psychological preferences. Additionally, there existed a certain relationship between them (Figure 12a,b).

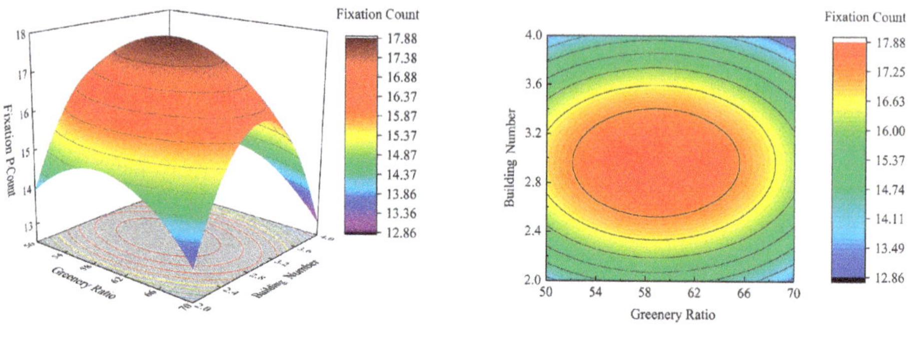

(a) (b)

Figure 12. Response surface analysis of the relationship between the greenery ratio and the number of buildings. (**a**) The response surface; (**b**) contour plots of interaction effects of the greenery ratio and number of buildings on visual appeal to visitors.

Research has revealed a coupled relationship between the greenery ratio and the number of buildings, where a higher greenery ratio generally implies greater vegetation coverage, consequently imposing potential constraints on the quantity and layout of buildings. In environments with a high greenery ratio, designers may be inclined to select a smaller number of buildings with distinctive features to avoid excessive interference with the visual effect of vegetation and maintain a natural sense of the green environment. Simultaneously, the density and layout of buildings can influence people's perception of the greenery ratio. Excessive numbers of buildings may reduce the visibility and perception of green spaces, making the greenery ratio appear lower. Conversely, a moderate layout of buildings can highlight green spaces and enhance the perception of the greenery ratio. Additionally, a building's design and height also affect the perception of the surrounding greenery. For instance, tall buildings may obstruct sunlight and hinder plant growth, thereby impacting the actual effect of the greenery ratio [82].

2. The relationship between the greenery ratio and water saturation.

The relationship between the greenery ratio and water saturation exhibited an elliptical trend in the response surface model plot, indicating significant influences of both factors on visitors' psychological preferences. However, the relationship was relatively weak (Figure 13a,b).

By further analyzing the correlation between the greenery ratio and water saturation, it was found that the impacts of the greenery ratio and water saturation on environmental landscapes are relatively independent. In the absence of alterations to water and vegetation colors, this implies that increasing or decreasing the greenery ratio does not significantly affect the water saturation level and vice versa. Therefore, in this study, which focused on investigating plant color without any alterations, the greenery ratio and water saturation in the environment can be adjusted independently without considering their relationship [83].

3. Relationship between the number of buildings and water saturation

The relationship between the number of buildings and water saturation exhibited a steep trend in the contour plot, indicating a significant impact of both factors on visitors' psychological preferences. Moreover, there existed a clear mutual influence between the number of buildings and water saturation (Figure 14a,b).

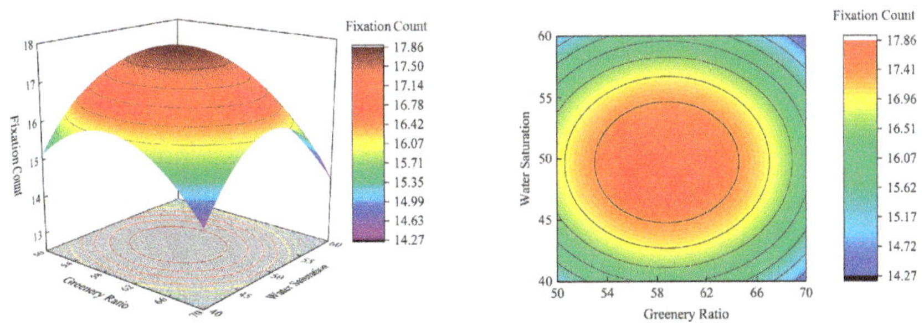

(a) (b)

Figure 13. Response surface analysis of the relationship between the greenery ratio and water saturation. (**a**) Response surface; (**b**) contour plots for interaction effects of greenery ratio and water saturation on visual appeal to visitors.

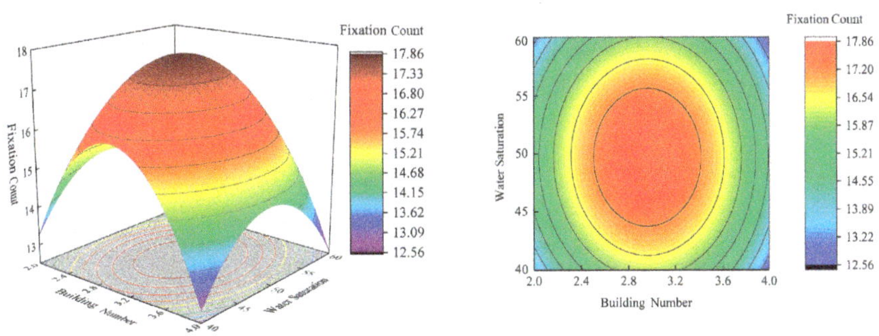

(a) (b)

Figure 14. Response surface analysis of the relationship between the number of buildings and water saturation. (**a**) Response surface; (**b**) contour plots of interaction effects of greenery ratio and number of buildings on visual appeal to visitors.

Analysis of the correlation between the greenery ratio and the number of buildings showed that an increase or decrease in the number of buildings may change the overall sense of balance of the environmental landscape. For example, too many buildings may suppress the visual effect of water saturation and make it appear less prominent, while a moderate number of buildings may better emphasize water saturation and enhance the visual appeal of the landscape.

Additionally, changes in the number of buildings may influence a visitor's perception of space. A lower number of buildings can create a more open and expansive spatial experience, with water saturation being more prominent within it. Inversely, a higher number of buildings can lead to a crowded and congested perception of space, diminishing the significance of water saturation. Hence, during the planning and design process, holistic consideration should be given to the layout, height, and form of buildings, as well as the positioning, size, and characteristics of water saturations, to achieve an optimal combination of the number of buildings and water saturation [79].

3.6. Optimizing the Design of Garden Elements

Thus, by analyzing the regression model for gardens, we obtained the following optimal conditions for garden elements: a greenery ratio of "58.82%", number of buildings of "3", and water saturation level of "47.2%". Under these conditions, the predicted visual attractiveness of the buildings was "17.86". We conducted three independent parallel experiments to validate the effectiveness and reliability of the model (Table S13). The experimental results indicated that the garden's visual attractiveness to tourists was rated as "18". The deviation of the experimental results from the predicted value was only "0.22%", suggesting that the model demonstrated a strong predictive capacity to capture the psychological preferences of garden tourists.

4. Discussion

4.1. The Integration of Landscape Psychology and Environmental Psychology

The field of environmental psychology involves extensively examining individuals' perception, cognition, and emotional responses to environmental landscapes, with a primary focus on the interplay between humans and their surrounding environment [84,85]. In the realm of horticulture, landscape psychology investigates individuals' psychological reactions, such as their attention, preferences, and comfort, towards various garden elements [86–88]. Adopting this perspective allows for a deeper understanding of the relationship between individuals and landscape features and enhances our comprehension of the dynamic relationship between humans and the environment. Our research findings support this theory, demonstrating the significant influence of garden elements such as the vegetation ratio, architectural layout, and water color on individuals' psychological and emotional responses.

4.2. The Impact of Different Garden Elements on Visitors' Psychology

The greenery ratio, number of buildings, and water saturation in garden landscapes have multifaceted effects on visitors' psychological experiences. Firstly, a high greenery ratio generally creates a more pleasant and vibrant environment, but excessive greenery may also induce feelings of oppression and unease. Therefore, striking a balance between visual pleasure and spatial comfort is necessary when designing green areas [24,60]. Secondly, the number of buildings also significantly influences visitors' psychology. Fewer buildings may result in a monotonous environment, while an excess of buildings can lead to visual confusion and a sense of oppressiveness. Consequently, architectural designs need to consider the diversity and openness of spaces to avoid a singular or overcrowded layout [63,64]. Lastly, water saturation has a noticeable impact on emotions. Low saturation of dark-colored water often triggers feelings of depression, whereas high saturation of yellow water may induce aversion and discomfort. This highlights the importance of selecting appropriate colors and transparency levels when designing water features [65]. Collectively, landscape design needs to take these factors into account in order to create an environment that is both aesthetically pleasing and meets the psychological needs of visitors.

4.3. The Impact of Different Garden Elements on Visitors' Visual Preferences

The greenery ratio, number of buildings, and water saturation in landscapes have complex and far-reaching effects on visitors' visual preferences. First of all, a greenery ratio of about 60% can most effectively attract visitors' attention and increase their visual interest and pleasure; however, too high a greenery ratio can lead to visual fatigue and a sense of oppression, indicating the need to balance the relationship between visual comfort and natural elements in design [72,73]. Furthermore, an appropriate number of buildings not only adds visual focus and diversity to the landscape but also enhances the overall aesthetic experience of visitors; however, too many buildings may cause visual confusion and depression, suggesting that the number of buildings needs to be carefully controlled during design to avoid visual overload [61,62]. Finally, moderate control of water saturation can effectively enhance visual experience and environmental aesthetics, while excessive saturation may cause discomfort and aversion due to strong visual stimulation. The balance and coordination between these elements are crucial, indicating that landscape design needs to comprehensively consider the visual and psychological needs of visitors in order to create both aesthetic and comfortable garden environments.

4.4. The Advantages of Integrated Eye-Tracking Response Surface Analysis

The integrated eye-tracking response surface analysis demonstrates significant advantages in landscape design. Primarily, this approach investigates the complex relationships between different topographic garden elements and visitor emotional preferences through multi-factor and multi-level analysis, effectively guiding topography design and planning. Moreover, by evaluating visitor preferences for garden elements and considering their coupling effects, optimal ranges for multiple elements are established, enhancing the comprehensiveness and scientific basis of the design. In addition, the combination of comprehensive analysis methods, eye-tracking technology, and psychological theories enables the analysis of visitor behaviors and preferences using big data, providing a scientific basis for landscape design that aligns with visitors' visual preferences and psychological needs. Importantly, this approach accurately reveals the interaction and influence mechanisms between different elements. Compared to traditional single-factor qualitative analysis, it enhances the accuracy of design decisions, reduces design costs, and facilitates more effective and economic landscape design optimization.

4.5. Directions for Future Research

Although our study preliminarily demonstrated the potential application of eye tracking and response surface analysis in landscape design, there are still some limitations that require further investigation. For example, the collection and analysis of eye-tracking data may have been influenced by external factors such as changes in lighting conditions and fluctuations in pedestrian flow. Future research could explore the impacts of these factors and consider other variables that may affect eye-tracking data. In addition, we plan to expand the scope of our study to encompass a wider range of garden elements and environmental conditions, allowing for further exploration of applications in landscape ecology, urban planning, biodiversity conservation, and health promotion.

5. Conclusions

Herein we propose an eye-tracking-based response surface analysis method for the design and optimization of gardens with the aim of enhancing visitors' comfort and satisfaction. Firstly, we performed a study of different garden elements using landscape psychology and identified that variations in certain factors, such as the greenery ratio, number of buildings, and water saturation, have an impact on visitor psychology. We also found that eye-tracking analysis was able to accurately capture visitors' psychological activities and provide intuitive eye-tracking data. Subsequently, through single-factor experiments and heatmap analysis, we established the optimal ranges for each garden element. The optimal values for the greenery ratio, number of buildings, and water saturation were around "60%" (greenery ratio), "3" (number of buildings), and "50%" (water saturation), respectively. Finally, we conducted the first-ever response surface analysis to investigate the relationships between different garden elements. The results revealed a significant relationship between the greenery ratio and the number of buildings, as well as between the number of buildings and water saturation, while the relationship between the greenery ratio and water saturation was relatively low. Among these elements, the greenery ratio was the primary factor influencing visitors' garden experiences, followed by the number of buildings and the level of water saturation, which aligned with actual visitor experiences. The optimized values of greenery ratio, number of buildings, and water saturation were determined, providing guidance for the optimization of landscape design.

The eye-tracking response surface analysis optimization method integrates visual perception analysis, multi-factor multi-level design, and nonlinear data processing, offering broad prospects for diverse landscape designs. This method provides a novel approach to multi-factor, multi-level landscape design that incorporates visual, sensory, and psychological analyses. It is not only applicable to garden design optimization but also has potential applications in landscape design, urban planning, and public space renovation. While this study demonstrates the significant potential of combining eye tracking and response surface analysis in garden design, there are still some limitations. For instance, this study selected only greenery ratio, number of buildings, and water saturation as the main variables and did not consider other potential influencing factors. Future research will further explore the impact of these factors on eye-tracking data and expand the scope of the study to include more landscape elements and environmental conditions.

Supplementary Materials: The following supporting information can be downloaded at: https://www.mdpi.com/article/10.3390/land13071045/s1, Figure S1: Research samples for eye tracking; Figure S2: Heat map for different elements; Figure S3: Water saturation research and analysis; Table S1: The types of Jiangnan gardens and the corresponding garden elements; Table S2: Indicators for eye tracking; Table S3: Questionnaire on Psychological Perception of Jiangnan Garden; Table S4: Different garden elements questionnaire; Table S5: Questionnaire on Psychological Perception of Jiangnan Garden (data analysis); Table S6: Eye tracking for greenery ratio (average data); Table S7: Eye tracking for building number (average data); Table S8: Eye tracking for water saturation (average data); Table S9: The non-typical focal center testing (e.g. 6 buildings); Table S10: Different garden elements questionnaire (data analysis); Table S11: Eye tracking for response surface experiment(average data); Table S12: Samples for response surface experiment; Table S13: Three independent parallel experiments of the model.

Author Contributions: Conceptualization, X.W. and Q.L.; methodology, X.W.; software, X.W.; validation, X.W., B.C. and Q.L.; formal analysis, X.W.; investigation, X.W.; resources, X.W.; data curation, X.W.; writing—original draft preparation, X.W. and Q.L.; writing—review and editing, X.W.; visualization, B.C.; supervision, R.Z.; project administration, R.Z.; funding acquisition, X.W. and B.C. All authors have read and agreed to the published version of the manuscript.

Funding: Project of the Degree and Graduate Education Development Center of the Ministry of Education (ZT-221029507); Research and practical project on graduate education and teaching reform at Jiangnan University (YJSJGZD22_006).

Data Availability Statement: The data presented in this study are available on request from the corresponding author due to privacy and property rights issues.

Acknowledgments: We thank all participants from Jiangnan Univerisity and Zhejiang University for taking part in our study, as well as Wei Xie, Dazhuan Wu, Xiuyu Wang, and Jizhou Chen for their help as instructors or during the field experiments.

Conflicts of Interest: The authors declare no conflicts of interest.

References

1. Gifford, R. Environmental psychology matters. *Annu. Rev. Psychol.* **2014**, *65*, 541–579. [CrossRef] [PubMed]
2. Ramkissoon, H.; Smith, L.; Weiler, B. Testing the dimensionality of place attachment and its relationships with place satisfaction and pro-environmental behaviours: A structural equation modelling approach. *Tour. Manag.* **2013**, *36*, 552–566. [CrossRef]
3. Tam, K.; Milfont, T.L. Towards cross-cultural environmental psychology: A state-of-the-art review and recommendations. *J. Environ. Psychol.* **2020**, *71*, 101474. [CrossRef]
4. Pluess, M. Individual differences in environmental sensitivity. *Child Dev. Perspect.* **2015**, *9*, 138–143. [CrossRef]
5. Cheng, T.M.; Wu, H.C. How do environmental knowledge, environmental sensitivity, and place attachment affect environmentally responsible behavior? An integrated approach for sustainable island tourism. *J. Sustain. Tour.* **2015**, *23*, 557–576. [CrossRef]
6. Gillson, L.; Hoffman, M.; Gell, P.; Ekblom, A.; Bond, W.J. Trees, carbon, and the psychology of landscapes. *Trends Ecol. Evol.* **2024**, *39*, 359–367. [CrossRef] [PubMed]
7. Klemm, W.; Heusinkveld, B.; Lenzholzer, S.; Van Hove, B. Street greenery and its physical and psychological impact on thermal comfort. *Landsc. Urban Plan.* **2015**, *138*, 87–98. [CrossRef]
8. Dzhambov, A.M.; Lercher, P.; Browning, M.H.E.M.; Stoyanov, D.; Petrova, N.; Novakov, S.; Dimitrova, D.D. Does greenery experienced indoors and outdoors provide an escape and support mental health during the COVID-19 quarantine? *Environ. Res.* **2021**, *196*, 110420. [CrossRef]
9. Kuliga, S.; Thrash, T.; Dalton, R.; Höelscher, C. Virtual reality as an empirical research tool—Exploring user experience in a real building and a corresponding virtual model. *Comput. Environ. Urban Syst.* **2015**, *54*, 363–375. [CrossRef]
10. Liu, Y.; Hong, Z.; Zhu, J.; Yan, J.; Qi, J.; Liu, P. Promoting green residential buildings: Residents' environmental attitude, subjective knowledge, and social trust matter. *Energy Policy* **2018**, *112*, 152–161. [CrossRef]
11. Ha, J.; Kim, H.J.; With, K.A. Urban green space alone is not enough: A landscape analysis linking the spatial distribution of urban green space to mental health in the city of chicago. Landsc. *Urban Plan.* **2022**, *218*, 104309. [CrossRef]

12. Li, X.; Zhang, X.; Jia, T. Humanization of nature: Testing the influences of urban park characteristics and psychological factors on collegers' perceived restoration. *Urban For. Urban Green.* **2023**, *79*, 127806. [CrossRef]
13. Ojala, A.; Korpela, K.; Tyrväinen, L.; Tiittanen, P.; Lanki, T. Restorative effects of urban green environments and the role of urban-nature orientedness and noise sensitivity: A field experiment. *Health Place* **2019**, *55*, 59–70. [CrossRef] [PubMed]
14. Tyrväinen, L.; Ojala, A.; Korpela, K.; Lanki, T.; Tsunetsugu, Y.; Kagawa, T. The influence of urban green environments on stress relief measures: A field experiment. *J. Environ. Psychol.* **2014**, *38*, 1–9. [CrossRef]
15. Li, S.; Walters, G.; Packer, J.; Scott, N. A comparative analysis of self-report and psychophysiological measures of emotion in the context of tourism advertising. *J. Travel Res.* **2018**, *57*, 1078–1092. [CrossRef]
16. Scott, N.; Green, C.; Fairley, S. Investigation of the use of eye tracking to examine tourism advertising effectiveness. *Curr. Issues Tour.* **2016**, *19*, 634–642. [CrossRef]
17. Aspinall, P.; Mavros, P.; Coyne, R.; Roe, J. The urban brain: Analysing outdoor physical activity with mobile EEG. *Br. J. Sports Med.* **2015**, *49*, 272–276. [CrossRef] [PubMed]
18. Scott, N.; Zhang, R.; Le, D.; Moyle, B. A review of eye-tracking research in tourism. *Curr. Issues Tour.* **2019**, *22*, 1244–1261. [CrossRef]
19. Savin, G.; Fleșeriu, C.; Batrancea, L. Eye tracking and tourism research: A systematic literature review. *J. Vacat. Mark.* **2022**, *28*, 285–302. [CrossRef]
20. Alemdag, E.; Cagiltay, K. A systematic review of eye tracking research on multimedia learning. *Comput. Educ.* **2018**, *125*, 413–428. [CrossRef]
21. Novák, J.Š.; Masner, J.; Benda, P.; Šimek, P.; Merunka, V. Eye Tracking, Usability, and User Experience: A Systematic Review. *Int. J. Hum-Comput. Int.* **2023**, *39*, 1–17. [CrossRef]
22. Tao, Z. Research on the Degree of Coupling between the Urban Public Infrastructure System and the Urban Economic, Social, and Environmental System: A Case Study in Beijing, China. *Math. Probl. Eng.* **2019**, *9*, 8206902. [CrossRef]
23. Skaramagkas, V.; Giannakakis, G.; Ktistakis, E.; Manousos, D.; Karatzanis, I.; Tachos, N.; Tripoliti, E.; Marias, K.; Fotiadis, D.I.; Tsiknakis, M. Review of eye tracking metrics involved in emotional and cognitive processes. *IEEE Rev. Biomed. Eng.* **2023**, *16*, 260–277. [CrossRef] [PubMed]
24. Li, J.; Zhang, Z.; Jing, F.; Gao, J.; Ma, J.; Shao, G.; Noel, S. An evaluation of urban green space in Shanghai, China, using eye tracking. *Urban For. Urban Green.* **2020**, *56*, 126903. [CrossRef]
25. Lin, W.; Zeng, C.; Bao, Z.; Jin, H. The therapeutic look up: Stress reduction and attention restoration vary according to the sky-leaf-trunk (SLT) ratio in canopy landscapes. *Landsc. Urban Plan.* **2023**, *234*, 104730. [CrossRef]
26. Chen, B.; Gong, C.; Li, S. Looking at buildings or trees? association of human nature relatedness with eye movements in outdoor space. *J. Environ. Psychol.* **2022**, *80*, 101756. [CrossRef]
27. Liu, L.; Qu, H.; Ma, Y.; Wang, K.; Qu, H. Restorative benefits of urban green space: Physiological, psychological restoration and eye movement analysis. *J. Environ. Manag.* **2022**, *301*, 113930. [CrossRef] [PubMed]
28. Liu, F.; Kang, J.; Wu, Y.; Yang, D.; Meng, Q. What do we visually focus on in a World Heritage Site? A case study in the Historic Centre of Prague. *Hum. Soc. Sci. Commun.* **2022**, *9*, 400. [CrossRef]
29. Box, G.E.P.; Wilson, K.B. On the Experimental Attainment of Optimum Conditions. *J. R. Stat. Soc. B* **1951**, *13*, 1–38. [CrossRef]
30. Meloun, M.; Militký, J.; Hill, M.; Brereton, R.G. Crucial problems in regression modelling and their solutions. *Analyst* **2002**, *127*, 433–450. [CrossRef]
31. Mäkelä, M. Experimental design and response surface methodology in energy applications: A tutorial review. *Energ. Convers. Manag.* **2017**, *151*, 630–640. [CrossRef]
32. Ockuly, R.A.; Weese, M.L.; Smucker, B.J.; Edwards, D.J.; Chang, L. Response surface experiments: A meta-analysis. *Chemometr. Intell. Lab.* **2017**, *164*, 64–75. [CrossRef]
33. Zhang, X.; Yu, S.; Gong, Y.; Li, Y. Optimization design for turbodrill blades based on response surface method. *Adv. Mech. Eng.* **2016**, *8*, 1. [CrossRef]
34. Gao, C.X. Theory and applications of swarm intelligence. *Neural Comput. Appl.* **2012**, *21*, 205–206.
35. Alam, M.D.; Hawas, Y.E. Multiobjective integrated signal-control system calibration in urban areas: Application of response surface methodology. *Transp. Res. Interdiscip. Perspect.* **2019**, *1*, 100011. [CrossRef]
36. Meng, Y.; Li, Y.; Han, R.; Du, Z. Optimization of the process conditions for methane yield from co-digestion of mixed vegetable residues and pig manure using response surface methodology. *Waste Biomass Valoriz.* **2024**, *15*, 4117–4130. [CrossRef]
37. Ji, K.F. Research on Landscape Design Strategy of Overhead Space Surroundings in Lingnan University Based on Thermal Comfort in Summer. Master's Thesis, Guangzhou University, Guangzhou, China, 2023.
38. Sun, S.B. Research on the Design Strategy of Urban Parks Based on Hot Climatic Adaptability. Ph.D. Thesis, Beijing Forestry University, Beijing, China, 2018.
39. Bucher, C. Metamodels of optimal quality for stochastic structural optimization Christian Bucher. *Probabilistic Eng. Mech.* **2018**, *54*, 131–137. [CrossRef]
40. Guo, M. Recognition and reconstruction of the historical development of chinese garden art system. *J. Jiangnan Univ.* **2015**, *14*, 110–114. (In Chinese)
41. Wu, W.; Zhou, K.; Li, T.; Dai, X. Spatial configuration analysis of a traditional garden in yangzhou city: A comparative case study of three typical gardens. *J. Asian Archit. Build.* **2024**, *23*, 1–12. [CrossRef]

42. Akpınar, A.; Barbosa-Leiker, C.; Brooks, K.R. Does green space matter? Exploring relationships between green space type and health indicators. *Urban For. Urban Green.* **2016**, *20*, 407–418. [CrossRef]
43. Nausser, J.I. Framing the landscape in photographic simulation. *J. Environ. Manag.* **1982**, *17*, 1–16.
44. Palmer, J.F.; Hoffman, R.E. Rating reliability and representation validity in scenic landscape assessments. *Landsc. Urban Plan.* **2001**, *54*, 149–161. [CrossRef]
45. Dupont, L.; Antrop, M.; Van Eetvelde, V. Eye-tracking analysis in Landscape Perception Research: Influence of photograph properties and landscape characteristics. *Landsc. Res.* **2013**, *39*, 417–432. [CrossRef]
46. Suppakittpaisarn, P.; Jiang, B.; Slavenas, M.; Sullivan, W.C. Does density of green infrastructure predict preference? *Urban For. Urban Green.* **2019**, *40*, 236–244. [CrossRef]
47. Wang, X.; Rodiek, S.; Wu, C.; Chen, Y.; Li, Y. Stress recovery and restorative effects of viewing different urban park scenes in Shanghai, China. *Urban For. Urban Green.* **2016**, *15*, 112–122. [CrossRef]
48. Waldheim, C.; Hansen, A.; Ackerman, J.S.; Corner, J.; Brunier, Y.; Kennard, P. *Composite Landscapes: Photomontage and Landscape Architecture*; Hatje Cantz: Ostfildern, Germany, 2014.
49. Shi, Y.; Zhang, J.; Shen, X.; Chen, L.; Xu, Y.; Fu, R.; Su, Y.; Xia, Y. Designing Perennial Landscapes: Plant Form and Species Richness Influence the Gaze Perception Associated with Aesthetic Preference. *Land* **2022**, *11*, 1860. [CrossRef]
50. Luo, H.; Deng, L.; Song, C.; Jiang, S.; Huang, Y.; Wang, W.; Liu, X.; Li, S.; Guo, B.; Peng, L.; et al. Which characteristics and integrations between characteristics in blue–green spaces influence the nature experience? *J. Environ. Plan. Manag.* **2022**, *66*, 1253–1279. [CrossRef]
51. Shafer, E.L.; Hamilton, J.; Schmidt, E.A. Natural Landscape Preferences: A Predictive model. *J. Leis. Res.* **1969**, *1*, 1–19. [CrossRef]
52. Maner, J.K.; Kenrick, D.T.; Becker, D.V.; Delton, A.W.; Hofer, B.; Wilbur, C.J.; Neuberg, S.L. Sexually selective cognition: Beauty captures the mind of the beholder. *J. Personal. Soc. Psychol.* **2003**, *85*, 1107–1120. [CrossRef]
53. Sun, M.; Herrup, K.; Shi, B.E.; Hamano, Y.; Liu, C.; Goto, S. Changes in visual interaction: Viewing a Japanese garden directly, through glass or as a projected image. *J. Environ. Psychol.* **2018**, *60*, 116–121. [CrossRef]
54. Dixson, B.J.; Grimshaw, G.M.; Ormsby, D.K.; Dixson, A.F. Eye-tracking women's preferences for men's somatotypes. *Evol. Hum. Behav.* **2014**, *35*, 73–79. [CrossRef]
55. Leder, H.; Mitrovic, A.; Goller, J. How Beauty determines gaze! Facial attractiveness and gaze duration in images of real world scenes. *I-Perception* **2016**, *7*, 204166951666435. [CrossRef] [PubMed]
56. Chamberlain, L. Eye tracking methodology_theory and practice. *Qual. Mark. Res.* **2007**, *10*, 217–220.
57. Poole, A.; Ball, L.J. Eye Tracking in Human-Computer Interaction and Usability Research: Current Status and Future Prospects. In *Encyclopedia of Human-Computer Interaction*; Ghaoui, C., Ed.; Idea Group, Inc. Press: Hershey, PA, USA, 2006.
58. Wang, P.; Yang, W.; Wang, D.; He, Y. Insights into Public Visual Behaviors through Eye-Tracking Tests: A Study Based on National Park System Pilot Area Landscapes. *Land* **2021**, *10*, 497. [CrossRef]
59. Mohamed, A.A.; Kronenberg, J.; Łaszkiewicz, E.; Ali, F.A.; Mahmoud, S.; Abdelhameed, R. Parental perceived safety using PPGIS and photo survey across urban parks in cairo, egypt. *Leis. Sci.* **2023**, *45*, 1–24. [CrossRef]
60. Ode, Å.; Fry, G.; Tveit, M.S.; Messager, P.; Miller, D. Indicators of perceived naturalness as drivers of landscape preference. *J. Environ. Manag.* **2009**, *90*, 375–383. [CrossRef] [PubMed]
61. Dee, C. *Form and Fabric in Landscape Architecture: A Visual Introduction*; Spon Press: New York, NY, USA, 2003.
62. Zarghami, E.; Karimimoshaver, M.; Ghanbaran, A.; SaadatiVaghar, P. Assessing the oppressive impact of the form of tall buildings on citizens: Height, width, and height-to-width ratio. *Environ. Impact Assess. Rev.* **2019**, *79*, 106287. [CrossRef]
63. Hwang, T.; Yoshizawa, N.; Munakata, J.; Hirate, K. A study on the oppressive feeling caused by the buildings in urban space. *J. Environ. Eng. (Trans. AIJ)* **2007**, *616*, 25–30. (In Japanese) [CrossRef] [PubMed]
64. Alexander, C.; Ishikawa, S.; Silverstein, M.; Jacobson, M.; Fiksdahl-King, I.; Shlomo, A. *A Pattern Language*; Oxford University Press: Oxford, UK, 1977.
65. Tallar, R.Y.; Suen, J.P. Measuring the aesthetic value of multifunctional lakes using an enhanced visual quality method. *Water* **2017**, *9*, 233. [CrossRef]
66. Smith, D.G.; Davies-Colley, R.J. Perception of water clarity and colour in terms of suitability for recreational use. *J. Environ. Manag.* **1992**, *36*, 225–235. [CrossRef]
67. Palmer, S.E.; Schloss, K.B. An ecological valence theory of human color preference. *Proc. Natl. Acad. Sci. USA* **2010**, *107*, 8877–8882. [CrossRef]
68. Nordh, H.; Hartig, T.; Hagerhall, C.M.; Fry, G. Components of small urban parks that predict the possibility for restoration. *Urban For. Urban Green.* **2009**, *8*, 225–235. [CrossRef]
69. Jiang, B.; Chang, C.; Sullivan, W. A dose of nature: Tree cover, stress reduction, and gender differences. *Landsc. Urban Plan.* **2014**, *132*, 26–36. [CrossRef]
70. De la Fuente Suárez, L.A. Subjective experience and visual attention to a historic building: A real-world eye-tracking study. *Front. Archit. Res.* **2020**, *9*, 744–804. [CrossRef]
71. Luo, J.; Zhao, T.; Cao, L.; Biljecki, F. Water View Imagery: Perception and evaluation of urban waterscapes worldwide. *Ecol. Indic.* **2022**, *145*, 109615. [CrossRef]
72. Ode, A.; Hagerhall, C.M.; Sang, N. Sang Analysing visual landscape complexity: Theory and application. *Landsc. Res.* **2010**, *35*, 111–131. [CrossRef]

73. Ode, A.; Miller, D. Analysing the relationship between indicators of landscape complexity and preference. *Environ. Plan. B* **2011**, *38*, 24–40. [CrossRef]
74. Xu, W.; Zhao, J.; Huang, Y.; Hu, B. Design intensities in relation to visual aesthetic preference. *Urban For. Urban Green.* **2018**, *34*, 305–310. [CrossRef]
75. Deng, L.; Luo, H.; Ma, J.; Huang, Z.; Sun, L.; Jiang, M.; Zhu, C.; Li, X. Effects of integration between visual stimuli and auditory stimuli on restorative potential and aesthetic preference in urban green spaces. *Urban For. Urban Green.* **2020**, *53*, 126702. [CrossRef]
76. Gao, C.; Iqbal, J. An empirical study of Thailand cities' color landscapes. *Heliyon* **2023**, *9*, 17558. [CrossRef]
77. Lin, W.; Mu, Y.; Zhang, Z.; Wang, J.; Diao, X.; Lu, Z.; Guo, W.; Wang, Y.; Xu, B. Research on cognitive evaluation of forest color based on visual behavior experiments and landscape preference. *PLoS ONE* **2022**, *17*, 0276677. [CrossRef] [PubMed]
78. Zhang, R. Integrating ergonomics data and emotional scale to analyze people's emotional attachment to different landscape features in the Wudaokou Urban Park. *Front. Archit. Res.* **2023**, *12*, 175–187. [CrossRef]
79. Li, H.; Browning, M.; Rigolon, A.; Larson, L.; Taff, D.; Labib, S.; Benfield, J.; Yuan, S.; McAnirlin, O.; Hatami, N.; et al. Beyond "bluespace" and "greenspace": A narrative review of possible health benefits from exposure to other natural landscapes. *Sci. Total Environ.* **2023**, *856*, 159292. [CrossRef] [PubMed]
80. Ottosson, J.; Grahn, P. Nature archetypes—Concepts related to objects and phenomena in natural environments. A swedish case. *Front. Psychol.* **2021**, *11*, 612672. [CrossRef] [PubMed]
81. Marselle, M.; Hartig, T.; Cox, D.; de Bell, S.; Knapp, S.; Lindley, S.; Triguero-Mas, M.; Böhning-Gaese, K.; Braubach, M.; Cook, P.; et al. Pathways linking biodiversity to human health: A conceptual framework. *Environ. Int.* **2021**, *150*, 106420. [CrossRef] [PubMed]
82. Asgarzadeh, M.; Koga, T.; Yoshizawa, N.; Munakata, J.; Hirate, K. Investigating green urbanism; building oppressiveness. *J. Asian Archit. Build.* **2010**, *9*, 555–562. [CrossRef]
83. Wang, Y.; Wang, S.; Xu, M. The Function of Color and Structure Based on EEG Features in Landscape Recognition. *Int. J. Environ. Res. Public Health* **2021**, *18*, 4866. [CrossRef]
84. Duan, J.; Wang, Y.; Fan, C.; Xia, B.; De Groot, R. Perception of urban environmental risks and the effects of urban green infrastructures (UGIs) on human well-being in four public green spaces of Guangzhou, China. *Environ. Manag.* **2018**, *62*, 500–517. [CrossRef]
85. Wu, L.; Kim, S.K. Health outcomes of urban green space in China: Evidence from Beijing. *Sustain. Cities Soc.* **2021**, *65*, 102604. [CrossRef]
86. Berto, R.; Massaccesi, S.; Pasini, M. Do eye movements measured across high and low fascination photographs differ? Addressing Kaplan's fascination hypothesis. *J. Environ. Psychol.* **2008**, *28*, 185–191. [CrossRef]
87. Gungor, S.; Polat, A.T. Relationship between visual quality and landscape characteristics in urban parks. *J. Environ. Prot. Ecol.* **2018**, *19*, 939–948.
88. Qin, B.; Zhu, W.; Wang, J.; Peng, Y. Understanding the relationship between neighbourhood green space and mental wellbeing: A case study of Beijing, China. *Cities* **2021**, *109*, 103039. [CrossRef]

land

MDPI

Article

Ecological Functional Zoning in Urban Fringe Areas Based on the Trade-Offs Between Ecological–Social Values in Ecosystem Services: A Case Study of Jiangning District, Nanjing

Ning Xu [1,*] and Haoran Duan [2]

1 School of Architecture, Southeast University, Nanjing 210096, China
2 Binhu District Urban Management Bureau, Wuxi 214072, China; 19550010397@163.com
* Correspondence: 101011755@seu.edu.cn

Abstract: Amid the rapid socio-economic development of urban fringe areas, promoting the multi-functional supply of ecosystems and sustainable development is essential. Taking Jiangning District in Nanjing as a case study, this study explores the relationships and spatial clustering characteristics among various ecosystem service values in urban fringe areas, focusing on the trade-offs between ecological and social values. Ecological functional zones were delineated based on the ecosystem service clustering results and regional conjugation principles, followed by an analysis of the trade-offs and synergies among the values within each zone. The findings reveal the following: (1) trade-offs between ecological and social ecosystem service values are prevalent across the entire region, as well as within sub-regions in urban fringe areas; (2) Jiangning District can be divided into five key ecological functional zones—the Vibrant Industry-Urbanization Integration Zone, Important Habitat Conservation Zone, Livable Organic Renewal Zone, Characteristic Rural Landscape Development Zone, and Riparian Recreation and Ecological Conservation Zone. Each zone exhibits significant differences in the types and features of the services provided; and (3) understanding the relationships among ecological and social values within each zone may help to resolve trade-offs between them. This progressive trade-off analysis, from the regional to sub-regional level, enables more precise identification of ecosystem functions, providing reference for decision-making to enhance the overall regional value and guide sustainable planning and management practices in urban fringe areas.

Keywords: ecosystem service value trade-offs; ecological value; social value; urban fringe areas; ecological functional zoning

1. Introduction

Since the 1950s, rapid urban expansion has become an irreversible global trend. In this context, the emergence of urban fringe areas has become a defining feature in the evolution of urban spatial structures. Urban fringe areas are a crucial part of urban regional structures, serving as both a hub for the exchange of material and energy between urban and rural settings. It also serves as a zone where the natural environment integrates with human living spaces. As urbanization accelerates, these areas have experienced significant economic growth and faced challenges related to the degradation of both natural and cultural ecosystems. Urban fringe areas can be viewed as complex ecosystems that encompass the relationships among people, nature, and society [1]. The ecological and social values of ecosystem services reflect both the natural ecological characteristics and the socio-cultural attributes of these ecosystems. Their interactions manifest as either synergistic cooperation, conflicting trade-offs, or independent relationships. A synergistic or trade-off relationship occurs when an increase in the value of one ecosystem service corresponds to an increase or decrease in the value of another, respectively [2]. Due to limited resources and the existence of trade-off relationships, reliance on a single type of ecosystem service can hinder the maximization of overall ecological benefits. Compared to typical rural areas, urban fringe zones stand at the forefront of urbanization, balancing the needs of ecological protection and economic development for both the urban core and the fringe itself. As such, these areas face significant environmental pressures, and their ecosystems are characterized by high complexity, dynamic changes, and limited protection. The population and land uses are diverse, with complex demands from both urban and rural residents. Rapid changes in spatial forms and industrial structures make ecological spaces, settlement patterns, and traditional cultures highly susceptible to disruption. If the coordinative relationship between the ecological and social values of ecosystem services in urban fringe areas is disrupted, it can lead to the degradation and simplification of ecosystem functions. For example, the intense competition between large-scale crop production and natural or semi-natural land use often limits the enhancement of multiple ecosystem service values [3]. While programs such as returning farmland to forests may help to address this issue, farmland abandonment can sometimes result in loss of biodiversity, damage to cultural heritage, and even increase the risks of natural disasters and human–wildlife conflicts. Moreover, in suburban areas near densely populated regions, the rapid development of non-agricultural industries has led to fragmented agricultural and natural landscapes in urban fringe areas. This fragmentation, coupled with the diverse spatial demands of urban and rural residents for esthetics, recreation, and education [4], seriously threatens food security and habitat conservation. Given the complexities surrounding the interconnections of ecosystem service values in urban fringe areas, addressing key issues in ecological conservation is essential. This can be achieved through tailored spatial planning and ecological management strategies, with the goal of enhancing the comprehensive value of ecosystem services.

In existing studies on the trade-offs in ecosystem service values, carbon storage, food provision, biodiversity, water resource protection, and water conservation are the most-frequently mentioned factors. These studies place the greatest emphasis on regulating services, followed by provisioning and supporting services, with the least attention given to social values [5]. However, in recent years, an increasing number of scholars have begun to focus on the intangible values of ecosystem services, and incorporating socio-ecological variables into the assessment of trade-offs in ecosystem service values has become an emerging trend. For example, Alessa et al. [6] proposed the concept of and a mapping method for socio-ecological spatial hotspots, using landscape indices to explore the spatial structure characteristics of the hotspots. Bryan et al. [7] used the Analytic Hierarchy Process to determine the weights of ecological and social values and identified socio-ecological value hotspots through the local Moran's I index. Karimi et al. [8] evaluated social and ecological values through public participatory geographic information systems and software for species distribution and conservation priority, identifying socio-ecological value hotspots using multiple importance thresholds. In their subsequent study, they established the relationships between biodiversity, the social values of ecosystem services, and land management preferences. Bagstad et al. [9] used hotspot analysis tools to identify socio-ecological value hotspots and cold spots in the San Isabel National Forest, USA. Chi et al. [10] analyzed changes in ecosystem service values in karst areas through land use change assessments using quantitative methods, highlighting their role in understanding ecosystem service supply and demand, trade-offs, and implications for sustainable development and land resource planning. Lourdes et al. [11] examined the social values of ecosystem services in rapidly urbanizing Greater Kuala Lumpur, Malaysia, using the SolVES tool and a public participatory GIS survey to reveal distinct resident preferences for green versus gray development and identifying potential areas of land use conflict.

During the dynamic evolution of urban fringe areas, complex mechanisms of interaction occur among various ecosystem service values, presenting both trade-offs and synergies. Cueva et al. [12] quantified the provisioning, regulating, and supporting ecosystem services of urban and peri-urban forests, identifying a trade-off effect between supporting and provisioning services, as well as between supporting and regulating services. This suggests that planning for suburban forests should consider these trade-off patterns. Zhang et al. [13] examined the interactions and relationships among the ecological values of five ecosystem services—carbon storage, water yield, soil conservation, biodiversity protection, and food production—in urban–rural transitional areas of southwestern China. They noted that interactions among ecosystem services in urban fringe areas are stronger than those in urban and rural areas. Chen et al. [14] highlighted the complex relationships among ecosystem services in peri-urban areas, mainly characterized by agricultural landscapes, observing that provisioning services trade-off, to some extent, with regulating and cultural services.

Nevertheless, existing research on the interaction patterns of ecosystem service values in urban fringe areas has revealed notable gaps: regarding their research focus, many studies mainly examine relationships centered on ecological values, such as food supply and habitat quality [12]. As such, they often overlook social values that capture the non-material benefits of ecosystems, limiting the applicability of their methods and findings to the complex ecological and socio-cultural contexts of urban fringe areas. In terms of research content, the majority of studies are based on correlation analyses to quantify trade-offs between pairs of values [13], neglecting clustering analyses of multiple values. This limits the ability to identify dominant value types and regions with concentrated synergistic effects. As for outcomes, a few studies have delineated ecological functional zones based on the trade-off relationships between ecosystem service values [14]. However, they often fail to investigate the internal value relationships within these zones, leading to the oversight of critical ecological control points. In addition, compared to urban and rural areas, research on the application of ecosystem service values in spatial planning for urban fringe areas remains relatively lacking. Most practices are carried out based on different goal-oriented approaches, with spatial planning and management methods being determined through the definition of ecosystem types and value assessments. The focus is mainly on identifying key areas for ecological restoration [15], constructing habitat conservation patterns [16], and optimizing ecosystem service spaces based on public perceptions and preferences [17]. Moreover, its application in ecological functional zoning is particularly scarce, which, to some extent, limits the management level of regional ecosystems.

Ecological functional zoning emphasizes the spatial heterogeneity resulting from differences in environmental resources, ecological succession, and human interventions. This makes region-specific functional zoning a vital approach for managing and regulating ecological environments. Kareiva et al. [18] first introduced the concept of ecosystem service bundles, which later evolved into a significant research method guiding ecological functional zoning. Ecosystem service bundles are groups of ecosystem services that consistently occur across time and space [19], representing clusters of interacting ecosystem services within a given region. Identifying these bundles can help to reveal the dominant ecosystem service values in a region and their clustering patterns, offering insights that can help to prevent the negative consequences of poorly informed decision-making. This approach has broad prospects for application in ecological spatial zoning, landscape planning, management, optimizing ecosystem services, and other related fields. For example, Karimi et al. [20] used the K-means clustering algorithm to identify ecosystem service bundles in urban areas at a 2 m resolution. On the other hand, Cheng et al. [3] applied hierarchical clustering at the grid scale to identify four types of ecosystem service bundles and their key landscape features within a park. However, existing research on the trade-offs among ecosystem service bundles remains limited, which somewhat restricts their practical application in ecological functional zoning management. Moreover, most research on trade-offs among ecosystem service values for ecological functional zoning has focused on urban or rural areas, with relatively little attention paid to urban fringe areas.

Since the 1990s, China's cities have increasingly adopted a growth-oriented development model. Spatial planning has mainly focused on utilizing rural resources, supporting urban expansion, and promoting economic growth. In this "growth-oriented planning" [21] paradigm, urban fringe areas have become key zones for urban sprawl. However, with issues such as ecological degradation and resource depletion arising from uncontrolled urban sprawl and extensive rural development, the traditional growth model has become unsustainable. China's urban development has now entered an era of stock, and landscape planning urgently needs to shift toward the more efficient allocation and use of internal spatial resources. In response, urban fringe areas must adopt refined zoning and management strategies which are tailored to actual development needs. This approach should avoid the negative impacts of uniform development constraints and management approaches while preserving the natural and cultural beauty with minimal intervention. Furthermore, ecological landscape planning in these areas should not focus solely on environmental conservation; it must also adopt an integrated approach that thoroughly assesses multiple values. This includes ensuring the continued provision of ecological benefits at the natural level while minimizing social disruptions and addressing the legitimate demands of residents for non-material benefits. Therefore, to facilitate differentiated and refined ecosystem management in urban fringe areas, it is important to expand the scope of research on the trade-off relationships between ecosystem service values and strengthen the synchronized assessment of social values. Additionally, accurately identifying homogeneous regions with similar trade-off characteristics is essential for ecological functional zoning, as this enables a clearer analysis of the complex relationships among multiple values across different zones.

This study examines the Jiangning District in Nanjing, analyzing the trade-off relationships between the ecological and social values of ecosystem services. Using clustering methods, the functional and structural characteristics of various regions are identified and ecological functional zones are delineated. Additionally, the interaction patterns of values within each zone at a refined spatial scale are explored. The aim of this study is to provide scientific support for comprehensive value enhancement and sustainable planning and management practices of urban fringe areas oriented towards collaborative development. The structure of the remainder of this study is as follows: Section 2 provides an overview of the study area. Section 3 outlines the methodology and data sources used in the research. Section 4 presents the spatial patterns of both individual and comprehensive ecological–social values in the study area; the trade-offs between the ecological–social values of its ecosystem services are also analyzed. Section 5 presents the ecological functional zoning of Jiangning District based on the ecosystem service cluster identification results, followed by an in-depth analysis of the trade-offs within the district. Finally, Section 6 concludes the study with a summary of the main findings and their implications.

2. Overview of the Study Area

Jiangning District is located in southwestern Jiangsu Province, China (Figure 1). It boasts a favorable natural ecological environment along with distinct social, economic, and transportation positioning advantages. Since 2000, Jiangning has transformed from a suburban county into a vital hub for external communications in Nanjing. It serves as a pivotal growth pole in the city's strategic metropolitan expansion. As a result, it has emerged as a model for ecological civilization initiatives at both national and local levels, encompassing rural revitalization, technological innovation, comprehensive tourism, and historical resource preservation, making it a region of significant importance. By the end of 2023, Jiangning District comprised 10 sub-districts, 136 communities, and 71 administrative villages, spanning a total area of 1560.82 km^2 [22]. Of this area, 335.20 km^2 is arable land, 34.42 km^2 is garden land, 434.53 km^2 is forest land, 29.96 km^2 is grassland, and 2.79 km^2 is wetland. The remaining land includes 381.91 km^2 for urban, rural, industrial, and mining purposes; 86.00 km^2 for transportation; and 256.01 km^2 covered by water bodies and water conservancy infrastructure [23]. In terms of its natural environment, Jiangning District is characterized by low mountain ranges and hills to the north and south, with extensive plains in the central region. It is situated in the northern subtropical monsoon climate zone, characterized by abundant rainfall. The district boasts rich water resources, including the Yangtze River, Qinhuai River, and Tangshan Hot Springs. Economically, Jiangning District is the largest logistics hub for agricultural and side-line products in East China. The local economy and technology sector have grown rapidly, with the per capita income of residents increasing severalfold over the past decade. In addition, Jiangning District boasts a rich cultural and historical heritage, with more than 100 historical sites. Furthermore, its tourism resources are abundant, earning it more than 20 accolades, including recognition as a National Demonstration County for Leisure Agriculture and Rural Tourism, among other honors.

Jiangning District boasts a favorable natural ecological environment and distinctive social, economic, and transportation advantages. Since the 1990s, it has experienced rapid economic growth driven by significant industrialization and urbanization, with its built-up area and population growth among the highest in Nanjing's administrative districts. However, this rapid urbanization has led to "urban maladies," including the fragmentation of ecological spaces, landscape homogenization, and severe environmental pollution. Additionally, as public demands diversify, the conflict between land redevelopment in the stock development phase and ecological protection has intensified. This conflict hinders the healthy development of ecosystems and the living environment. In conclusion, this region serves as an exemplary case for examining the critical contradictions between environmental protection and social development in urban fringe areas. It also provides valuable insights for conducting research on ecological functional zoning, which can inform the management of ecosystems in other urban fringe regions.

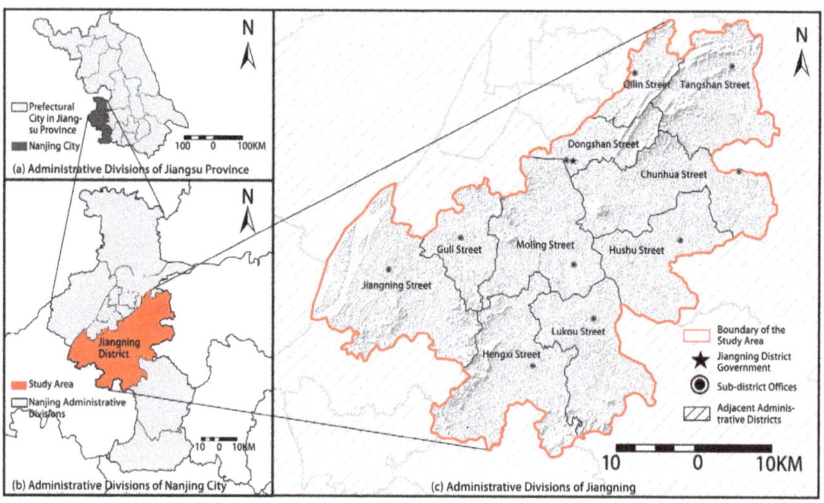

Figure 1. Location map of Jiangning District.

3. Research Methods and Data Sources

This research employs a three-step approach involving the assessment of ecosystem service values, the analysis of spatial trade-offs, and ecological functional zoning along with intra-zonal trade-off analysis (Figure 2). This workflow covers the complete process from database construction to spatial decision-making, establishing a comprehensive research methodology for ecological functional zoning. It transitions from a regional to a more localized level within urban fringe areas, facilitating ecological management and spatial optimization.

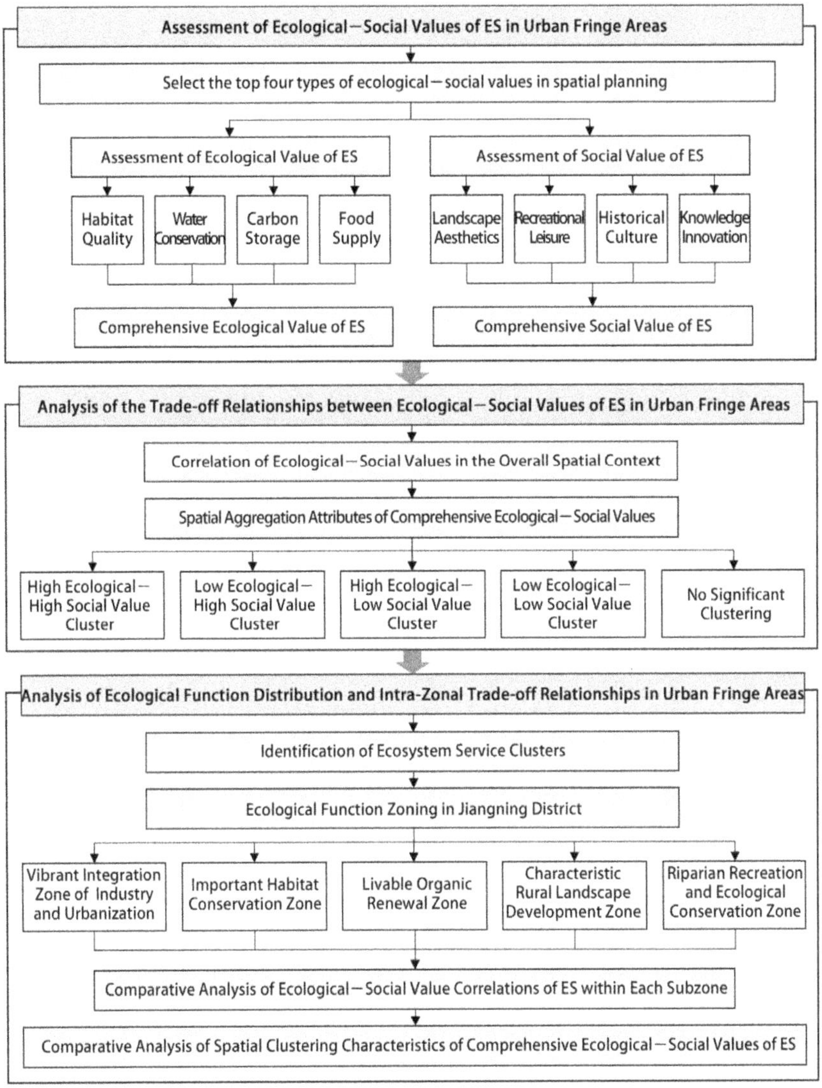

Figure 2. Framework of proposed research methods.

3.1. Selection of Ecosystem Service Value Types

Drawing on the classification frameworks of the United Nations Millennium Ecosystem Assessment (MA) and the widely recognized Common International Classification of Ecosystem Services (CICES) [24], this study systematically reviews 14 types of ecological and social values associated with ecosystem services. These values have been documented over the past decade in Jiangning District's planning initiatives for rural–urban integration, ecological construction, environmental protection, urban renewal, and comprehensive spatial land use planning. Given the key environmental challenges posed by urbanization, these values were classified into four levels: not mentioned (0 points), briefly mentioned (1 point), incorporated as planning principles or goals (2 points), or systematically applied through relevant theories or evaluations to guide spatial planning (3 points). These scores were then aggregated to derive quantitative insights into the extent to which each value has been applied in recent spatial planning efforts within Jiangning District (Figure 3). Among ecological values, the top four most frequently applied types were water conservation, habitat quality, food supply, and carbon storage. Among social values, the top four types were landscape esthetics, knowledge innovation, recreational leisure, and historical culture. These eight values have been widely applied in Jiangning District's recent spatial planning, serving as representative indicators. Therefore, these eight values were selected for use in this study.

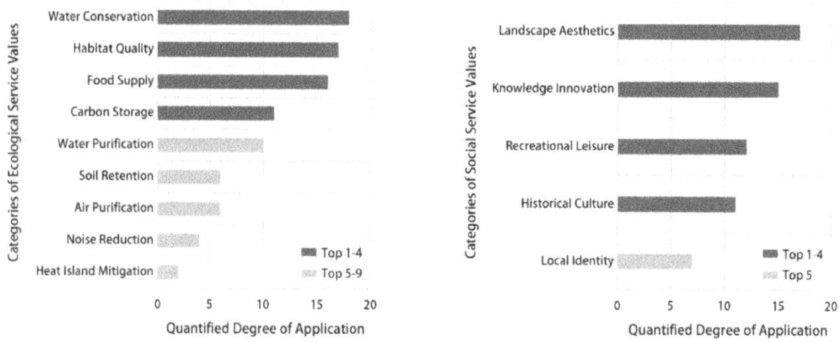

Figure 3. Quantitative analysis of the degree of ecosystem service value application in recent spatial planning efforts in Jiangning District.

3.2. *Ecosystem Service Ecological Value Assessment*

3.2.1. Methods for Assessing Ecological Values

The value of water conservation is reflected in its role in reducing surface runoff, recharging groundwater, mitigating seasonal flow fluctuations, and ensuring water quality. These efforts collectively support the continuous supply and maintenance of water resources. The evaluation of water conservation is carried out by calculating the amount of water conserved using the water balance equation [25]. In Formula (1), Q_x represents the water conservation amount (mm) for grid cell x, P_x denotes the average annual precipitation for grid cell x, R_x indicates the surface runoff coefficient for grid cell x, and ET_x denotes the average annual evapotranspiration.

$$Q_x = P_x - P_x \times R_x - ET_x \tag{1}$$

The value of habitat quality is used to quantify the extent to which ecosystems provide suitable habitats for natural biological communities, reflecting the survival conditions and reproductive capacity of species in specific habitats. Habitat quality is assessed using the habitat quality module of the InVEST model, which produces a dimensionless index ranging from 0 to 1, as shown in Formula (2) [26]. In this formula, Q_{xj} represents the habitat quality of grid cell x within land type j; D_{xj} denotes the level of disturbance affecting grid cell x in land type j, reflecting the degree of habitat degradation; K is the half-saturation constant; H_j refers to the habitat adaptability for land type j; and Z represents a conversion factor inherent to the model.

$$Q_{xj} = H_j\left[1 - \left(\frac{D_{xj}^Z}{D_{xj}^Z + K^Z}\right)\right] \tag{2}$$

The value of food supply refers to the production capacity and sustainability of various food crops intended for human consumption, which reflects the agricultural production function of ecosystems. The evaluation of food supply is based on the ratio of the $NDVI$ value for cultivated land grids to the total $NDVI$ value for all cultivated land [27]. In Formula (3), G_x denotes the food production (kg) for grid cell x, G_{sum} represents the total food production from the total area, $NDVI_x$ indicates the $NDVI$ value for cultivated land grid x, and $NDVI_{sum}$ refers to the aggregate $NDVI$ values for cultivated land within the study area.

$$G_x = \left(\frac{NDVI_x}{NDVI_{sum}}\right) \times G_{sum} \tag{3}$$

The value of carbon storage refers to the natural process through which carbon dioxide is converted into organic matter and stored within ecosystems. This process reflects the capacity to sequester carbon and reduce greenhouse gas emissions. Carbon storage is assessed through the carbon storage module of the InVEST model [28]. In Formula (4), C_i denotes the total carbon density for land type i, while C_{iabove} represents the above-ground carbon density, C_{ibelow} signifies the below-ground carbon density, C_{isoil} indicates the soil organic carbon density, and C_{idead} refers to the carbon density of dead organic matter (all measured in $t \cdot hm^{-2}$).

$$C_i = C_{iabove} + C_{ibelow} + C_{isoil} + C_{idead} \tag{4}$$

The overall ecological value of ecosystem services is assessed using a linear weighting approach, with the weights for the four ecological values determined through the Delphi method, as presented in Table 1.

Table 1. Weight of ecological values of ecosystem services in Jiangning District.

Weight of Ecological Value			
Habitat Quality	Water Conservation	Carbon Storage	Food Supply
0.260	0.246	0.252	0.242

3.2.2. Data Sources for Assessing Ecological Value

The necessary data for assessing the selected ecological value types are detailed in Table 2. All data were collected in 2023 and later resampled to obtain raster data with a consistent resolution of 30 m.

Table 2. Data sources required for assessing ecological values of ecosystem service in Jiangning District.

Data Components	Sources and Preprocessing
Biophysical coefficients	InVEST User Guide (https://www.sustainablehighways.org/ (accessed on 5 January 2023))
Average annual temperature, precipitation, and evaporation	Data obtained from the Resource and Environmental Science Data Center, Chinese Academy of Sciences (https://www.resdc.cn/ (accessed on 10 January 2023)), followed by spatial interpolation processing.
Carbon density	Carbon density dataset for terrestrial ecosystems in China (2010–2020) (https://nesdc.org.cn/ (accessed on 8 January 2023)).
Normalized Difference Vegetation Index	Landsat-8 remote sensing images downloaded from the Geospatial Data Cloud site (https://www.gscloud.cn/ (accessed on 8 January 2023)), processed using band calculations.
Food production	Jiangning District Statistical Yearbook 2022 (http://www.jiangning.gov.cn/sjfb/tjnj/ (accessed on 9 January 2023))

3.3. *Assessment of the Social Values of Ecosystem Services*

3.3.1. Methods for Assessing Social Values

The social values of ecosystem services encompass landscape esthetics, recreational leisure, historical culture, and knowledge innovation. Landscape esthetics not only directly reflect regional features but also serve as a vital medium for recognizing and expressing local values, highlighting the esthetic significance of ecosystems. Recreational leisure provides the public with various outdoor spaces for recreation and entertainment, emphasizing the importance of social interactions and leisure activities. Historical culture is evident in elements such as historical sites, the preservation of intangible cultural heritage, and the protection of cultural relics. These aspects demonstrate how ecosystems embody the lifestyles and cultural traditions of humanity across different historical periods. Knowledge innovation offers the public opportunities for scientific education, creative learning, and research, fostering both knowledge and innovation, thus reflecting the ecosystem's role in knowledge enrichment and intellectual development.

The spatial mapping of the social values of ecosystem services was conducted using MaxEnt 3.3.3. The environmental variable data and latitude–longitude coordinates of the social value points were input separately into the model's environmental variable and species distribution modules. A random selection of 75% of the data were used as training samples, while the remaining 25% were reserved for testing. The bootstrap repetition was set to 10, and the accuracy of the evaluation results was assessed using the area under the receiver operating characteristic curve (AUC).

The MaxEnt model, developed by Phillips et al., allows for the evaluation of the output value model to extrapolate the distributions of social values from known locations to unknown areas. The final value index, ranging from 0 to 10, along with its spatial distribution map, was generated using ArcGIS [29].

The overall social value was calculated using a linear weighting method, with the weights determined through the Delphi method, as shown in Table 3.

Table 3. Weight of ecosystem service ecological values in Jiangning District.

Weight of Social Values			
Landscape Esthetics	Recreational Leisure	Historical Culture	Knowledge Innovation
0.257	0.252	0.250	0.241

3.3.2. Data Sources for Assessing Social Values

The MaxEnt model requires both social value point data and environmental variable data. Points of Interest (POIs) in 2023 were selected as the social value data, and 69,326 POI entries from Jiangning District were obtained using the Gaode Map application interface. The POI data were then classified, filtered, and cleaned using a combination of tags and keywords (Table 4).

Table 4. Method for screening POI data in Jiangning District.

Types	Filtered by Labels	Filtered by Keywords	Examples
Landscape Aesthetics	Scenic Attractions, Leisure Activities, Natural Landmarks	Scenery, Forest, Grass, Flower, Apricot, Bamboo, Willow, Plum, Mountain, Rock, Peak, Valley, Cliff, Cave, Gorge, Island, Slope, Spring, Water, River, Sea, Lake, Stream, Bay, Ravine, Pond, Wetland, Village, Field, Garden, Courtyard, Bridge, Wheel, Square, Dock, etc.	Shike Lake Ecological Park, Verbena Flower Sea, Nanjing Niushou Mountain Cultural Tourism Zone—Yinlong Lake Square
Recreational Leisure	Leisure Activities, Sports and Fitness Services, Lifestyle Services	Recreation, Entertainment, Amusement, Activities, Sports, Fitness, Rehabilitation, Vacations, Picking, Fishing, Board Games, Hot Springs, Hotels, Cinemas, Clubs, Amusement Parks, Farms, Resorts, Rural Tourism, Inns, Towns, Squares, Centers, Bases, etc.	Tangshan No.1 Natural Hot Spring Resort, Huluba Tourism Fishing Center, Ginkgo Lake Golf Club
Historical Culture	Scenic Attractions, Educational and Cultural Sites, Leisure Activities	Pavilions, Towers, Platforms, Halls, Temples, Mansions, Buildings, Pagodas, Palaces, Tombs, Mausoleums, Steles, Ancestral Halls, Villages, Historical Residences, Historical Sites, Ruins, Fossils, Stone Inscriptions, Antiquities, History, Buddhism, Confucianism, Taoism, Ethics, Filial Piety, Culture, Memorials, etc.	Ancient Buildings of Yangliu Village, Daishan Memorial Hall, Huanglongxian Tea Culture Museum
Knowledge Innovation	Scenic Attractions, Educational and Cultural Sites, Lifestyle Services	Innovation, Intelligence, Technology, Popular Science, Science, Exhibitions, Presentations, Expansion, Team-Building, Training, Education, Creativity, Cultural Innovation, Experiments, Practice, Experiences, Exploration, Research, Development, Industries, Design, Bases, Parks, Zones, Ports, Academies, Institutes, Camps, etc.	Tangshan Fangshan National Geopark, Jiangning High-tech College Student Creative and Entrepreneurial Park, Xitian Ecological Civilization Education Base

This study identifies ten environmental variables that capture the key environmental characteristics of Jiangning District, encompassing both natural and anthropogenic factors. The natural factors include elevation, slope, Normalized Difference Vegetation Index, proximity to water bodies, and patch count, while the anthropogenic variables consist of land use type, distance to roads, distance to tourist attractions, distance to rural residential areas, and distance to government centers. All data, as shown in Table 5, were collected in 2023, with a raster data spatial resolution of 30 m.

Table 5. Data sources for environmental variables required for social value assessment.

Data Components	Sources and Preprocessing
Elevation and slope	Geospatial Data Cloud site (https://www.gscloud.cn/ (accessed on 8 January 2023))
Normalized Difference Vegetation Index	Landsat-8 remote sensing imagery obtained from Geospatial Data Cloud site (https://www.gscloud.cn/ (accessed on 8 January 2023)), processed through band calculations.
Distance to water bodies and roads	Vector data sourced from Open Street Map (https://www.openstreetmap.org/ (accessed on 11 January 2023)), cropped and analyzed to calculate distances using the Euclidean distance.
Land use types and distance to rural residential areas	Multi-temporal land use remote sensing monitoring data obtained from the Resource and Environmental Science Data Center, Chinese Academy of Sciences (https://www.resdc.cn/ (accessed on 10 January 2023)). Distances to rural residential areas are calculated using Euclidean distance.
Patch count	Land use data (https://www.gscloud.cn/ (accessed on 8 January 2023)) reclassified and analyzed using Fragstats V4.4.
Distance to tourist attractions	POI data downloaded from the Gaode Map API (https://lbs.amap.com/ (accessed on 11 January 2023)), filtered, and cleaned, with distances calculated using the Euclidean distance.
Distance to government centers	Administrative boundary and point data sourced from the Resource and Environmental Science Data Center, Chinese Academy of Sciences (https://www.resdc.cn/ (accessed on 10 January 2023)). Distances are calculated using the Euclidean distance.

3.4. *Analysis of the Ecological–Social Value Trade-Offs in Ecosystem Services*

3.4.1. Analysis of the Correlation Between Ecological–Social Values in Ecosystem Services

The Pearson correlation coefficient is commonly used to examine the paired relationships between ecosystem service values, providing a foundation for understanding the characteristics in a given region [30]. The correlation coefficient matrix was calculated and visualized using the Corrplot package in R. A negative correlation coefficient indicates an inverse relationship between two values, reflecting a trade-off, while a positive coefficient suggests a synergistic relationship.

3.4.2. Analysis of the Spatial Clustering Characteristics of Ecological–Social Values in Ecosystem Services

The bivariate local Moran's I statistic helps to uncover spatial correlations between two or more ecosystem service values [31]. The bivariate local spatial autocorrelation model in the GeoDa 1.22 was used to analyze the spatial clustering characteristics of trade-off relationships among comprehensive ecosystem service values. The bivariate local spatial autocorrelation clustering map classifies the study area into five types of spatial aggregation of ecological–social value: high–high, high–low, low–high, low–low, and no significant relationship.

3.5. *Methods for Ecosystem Functional Zoning*

3.5.1. Identification of Ecological Service Clusters

To identify ecosystem service clusters, this study evaluated the similarity between ecosystem service values through iterative calculations and relative distances. Units with high dissimilarity were grouped into distinct ecosystem service clusters, highlighting differences in both the quantity and type of ecosystem service values across regions. This analysis serves as a key foundation for ecological functional zoning [20]. The K-means clustering algorithm was employed, using the Factoextra package in R, in order to identify these clusters. By analyzing the significance of the classification results for regional ecological management, this study examined the trends in within-cluster sum of squares for cluster counts ranging from 2 to 10. The elbow method was used to determine the optimal number of ecosystem service clusters.

3.5.2. Ecological Functional Zoning and Analysis of Intra-Zone Trade-Offs

Based on the identification results of ecosystem service clusters, spatially adjacent areas with functional consistency or similarity were integrated, according to the principle of regional conjugation, to form coherent spatial units. Small, isolated, or peripheral areas were merged, minor adjustments were made, and smoothing of ecosystem service clusters was performed according to the specific characteristics of each patch. The clusters were then aggregated into ecological function zones based on their dominant value types, enhancing the overall coherence and manageability of the zoning. The overall characteristics of each ecological functional zone were assessed from three perspectives: spatial distribution, connection to land use, and value characteristics. The area proportions of different land use types within the ecological functional zones were calculated, and the cluster centers for ecosystem service values in each zone were identified. A rose diagram was then created, with distance values representing the average of each service value. This average was then compared to the standardized mean of all values in the study area, in order to derive the relative ecosystem service value statistics.

4. Assessments of Ecological–Social Values and Trade-Offs of Ecosystem Services in Jiangning District

4.1. *Spatial Patterns of Ecological–Social Values of Ecosystem Services in Jiangning District*

4.1.1. Spatial Patterns of Individual Ecological–Social Values in Jiangning District

By analyzing individual ecological values (Figure 4), it was found that water conservation is particularly high in the forests and certain agricultural areas of the southwestern, eastern, and southern parts of Jiangning District. Conversely, areas with extensive impervious surfaces in the northern low-altitude urban development exhibit poor water source conservation capacity. The spatial distribution of habitat quality closely mirrors that of carbon storage, with regions rated as good or higher primarily located in forested mountain areas far from high-density urban centers. In contrast, the eastern Yangtze River and the northern and southern ends of the Qinhuai River experience significant disturbances due to industrial development and tourism activities, resulting in lower values for both categories. The food supply follows a spatial distribution pattern, with higher values in the south and east and lower values in the north and west. Areas rated as "good" or above are mainly found in the southern and eastern regions, with nutrient-rich soils and extensive farmlands.

The AUC values for both the training and testing models for each social value exceeded 0.7, indicating the high reliability of the simulation results. When analyzing individual social values (Figure 5), the landscape esthetic value presented higher levels in the north and lower levels in the south, with both the eastern and western ends also exhibiting lower values. Areas rich in natural landscapes, scenic spots, and distinctive rural complexes tend to have higher values. The knowledge innovation value was clustered near research institutions, science exhibition venues, and modern industrial parks, followed by urban parks, rivers, and villages that are easily accessible and close to agricultural land. Recreational leisure value was mainly observed in densely populated urban areas and university towns, with additional concentrations around tourist attractions that leverage natural features and cultural heritage sites. Historical and cultural value displayed a spatial pattern of higher levels in the north and lower levels in the south, characterized by multi-point clustering, influenced by historical landscapes, local customs, and the distribution of industrial sectors.

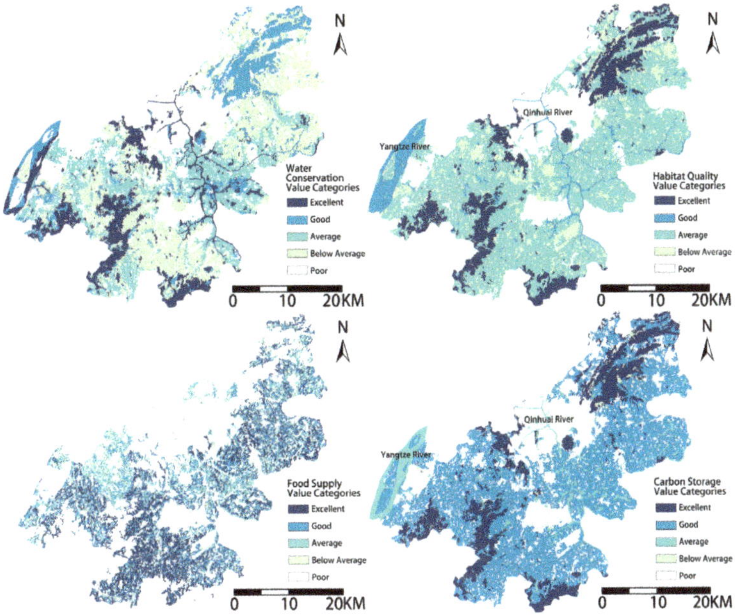

Figure 4. Spatial distribution of ecosystem service ecological value levels in Jiangning District.

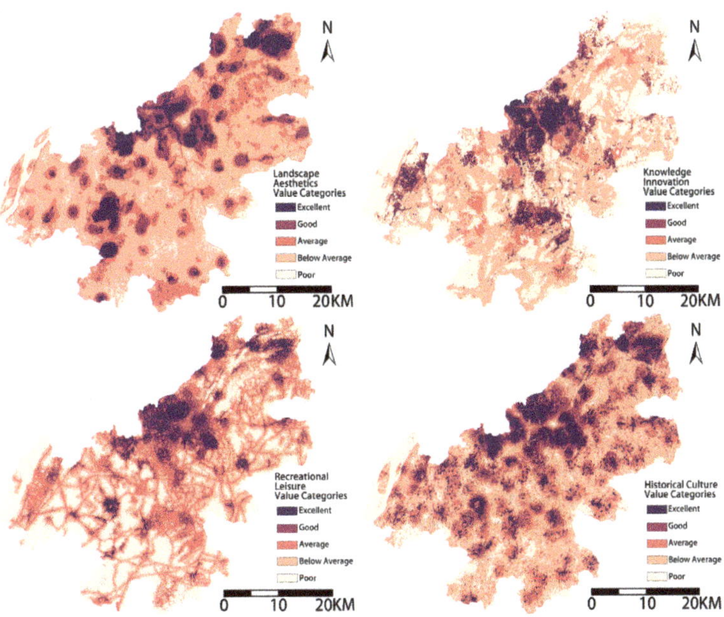

Figure 5. Spatial distribution pattern of social value levels of ecosystem services in Jiangning District.

4.1.2. Spatial Patterns of Comprehensive Ecological–Social Values in Jiangning District

The results indicated that Jiangning District has significant potential for comprehensive value. It is home to farmland, vegetable plots, and forests and grasslands that support the cultivation of various food crops, fruits, vegetables, timber, and livestock. These resources ensure a reliable supply of diverse food, freshwater, and raw materials. The high-altitude forest and grassland areas, which are widely distributed, continuous, and minimally disturbed by human activity, serve as significant sources of comprehensive ecological service value. They also play a significant role in maintaining natural ecological processes. Forest areas, which are less impacted by urbanization, maintain a relatively intact ecological structure and serve as significant carbon sink regions. They also play a significant role in maintaining natural ecological processes. Additionally, the distinctive landscape pattern of interspersed agriculture and forestry in the southern and eastern regions effectively enhances the overall ecological value (Figure 6). In contrast, the spatial distribution of social value related to comprehensive ecological services in Jiangning District shows higher values in the north and lower values in the south, with inland clusters. The main urban area in the north, along with the clustered new towns, is a key source of social value (Figure 7).

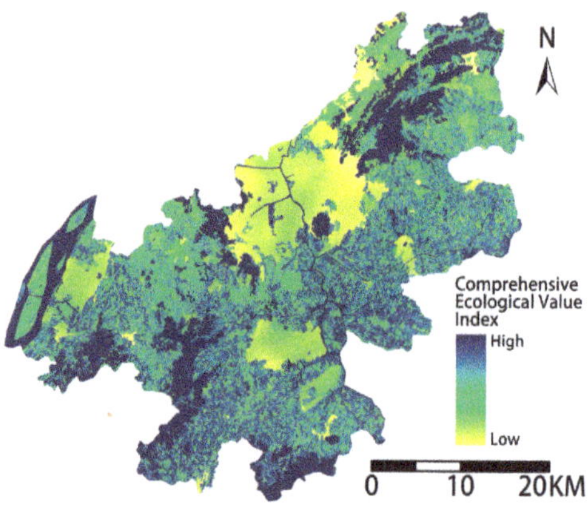

Figure 6. Spatial distribution pattern of comprehensive ecological value of ecosystem services in Jiangning District.

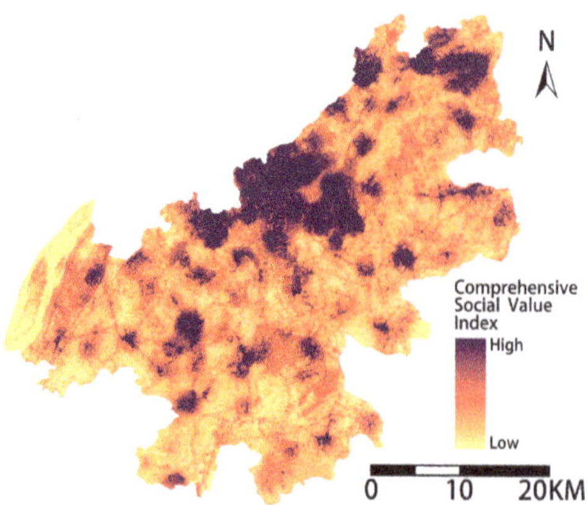

Figure 7. Spatial distribution of comprehensive social value of ecosystem services in Jiangning District.

The spatial distribution of both comprehensive ecological and social service values in Jiangning District reveals several challenges: First, the areas with high and medium–high comprehensive ecological value are relatively few. This highlights the need to address weaknesses in individual ecological components and improve the overall ecological network. Second, high social value zones are mainly clustered in distant locations, resulting in low connectivity, which limits the potential for high overall social value performance in the region. Third, potential conflicts may arise between the distributions of ecological and social values, requiring a more precise analysis of the trade-offs between the two.

4.2. Trade-Offs Between Ecological–Social Values of Ecosystem Services in Jiangning District

4.2.1. Correlation Between Ecological–Social Values in Jiangning District

Analyzing the correlations between ecosystem service values is crucial for understanding the characteristics of regional ecosystem services. Each pair of ecosystem service values in Jiangning District was tested for significance at the $p < 0.01$ level. In the correlation coefficient matrix, red indicates a negative correlation, while blue represents a positive correlation (Figure 8).

Negative correlations were identified between food supply and both habitat quality and carbon storage within the categories of ecological and social values. In contrast, strong positive correlations emerged between habitat quality and carbon storage, habitat quality and water conservation, and water conservation and carbon storage, highlighting the direct impacts of land use patterns on ecosystem services. Notably, the synergy between habitat quality and carbon storage was the strongest, consistent with the findings of Zhang et al. in their study on urban–rural transitional areas in southwestern China [13]. The explanation for this is that food supply often depends on agricultural expansion and intensive land use, which can degrade habitat integrity and negatively impact local biodiversity. Moreover, the lower vegetation density in farmland reduces its carbon sequestration capacity, leading to a decrease in overall carbon storage. There is a positive feedback loop among habitat quality, carbon storage, and water conservation values. Improvements in habitat quality foster biodiversity, which enhances ecosystem stability and increases both carbon storage and water conservation capacities. The ecosystem regulation ability of urban fringe areas is also closely linked to surface runoff absorption and their capacity to mitigate flooding and waterlogging disasters [32], underscoring the significance of water conservation functions. This emphasizes the need to accurately address the coordination of regional ecological values, including ecological integrity protection, carbon reduction, and resource supply.

From the perspective of social value groups, there was better compatibility among various values, which is in alignment with previous research [14]. Strong correlations were observed between recreational leisure and historical culture, landscape esthetics and historical culture, recreational leisure and knowledge innovation, and landscape esthetics and recreational leisure. This is because social values usually do not conflict in terms of resource demands. The needs of different social groups can be met simultaneously through multifunctional public spaces, reflecting the shared characteristics of such spaces and resources. However, the synergy between landscape esthetics and knowledge innovation was weaker, likely as knowledge innovation often depends on educational and cultural resources, while landscape esthetics focuses more on visual and sensory experiences. These two social values prioritize different aspects of demand. This suggests that, through effective spatial planning, various social values can generate mutually beneficial synergistic effects. These effects, in turn, promote the overall enhancement of regional social value.

Examining the trade-offs between ecological and social value groups revealed that only two pairs—landscape esthetic value with habitat quality and with carbon storage—showed positive correlations. The remaining pairs were negatively correlated. The correlation patterns between different values indicate the complexity and multi-dimensional needs characterizing regional development. In terms of correlation strength, two pairs exhibited strong correlations, seven pairs showed moderate correlations, six pairs had weak correlations, and one pair showed no significant correlation. Notably, the strongest negative correlation was between water conservation and knowledge innovation. Among the ecological values, water conservation exhibited the most significant trade-off with social values, while carbon storage showed the weakest negative correlation with social values. This may be linked to the widespread distribution of forests, grasslands, and farmlands with high carbon storage in Jiangning District. In addition, food supply was negatively correlated with all social values, supporting the conclusions of related studies [14]. This is because agricultural land competes with other land uses, limiting the availability of land for most other services [33]. Regarding social values, knowledge innovation had the most significant negative correlation with other ecological values, while landscape esthetic value demonstrated a lower level of trade-off with each ecological value, particularly with food supply. However, the interaction mechanisms between these values are quite complex; for example, research has indicated that the loss of traditional agricultural landscapes may diminish esthetic value and even weaken the emotional bonds that people have with these places [34].

Moreover, relevant research has shown that the value of ecosystem services in urban fringe areas is both broad and diverse [35]. These areas provide spaces that visually reflect natural processes, providing opportunities for people to learn about ecology and engage in environmental education [36]. Furthermore, traditional rural settlements and old industrial communities often have unique historical and cultural heritage. Local residents often form emotional and cultural attachments to urban fringe areas, imbuing them with significant spiritual value [37].

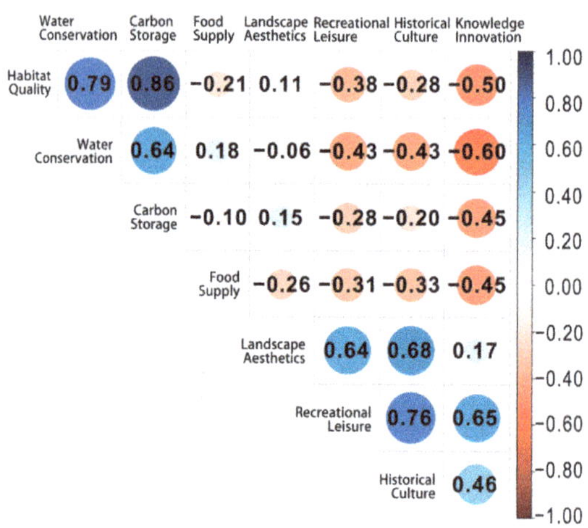

Figure 8. Correlation coefficient matrix of ecological–social values of ecosystem services in Jiangning District.

In general, there are notable trade-off relationships between ecological and social values in Jiangning District, although the compatibility within the groups of ecological and social values is relatively strong. This finding supports previous research [38], which has suggested that increases in most ecological values often lead to decreases in social values, and vice versa. Therefore, it is crucial to specifically identify, plan for, and manage the trade-off relationships between ecological and social values of ecosystem services in urban fringe areas from a spatial perspective.

4.2.2. Spatial Clustering Characteristics of Ecological–Social Value in Jiangning District

The LISA Clust Map (Figure 9), generated using a bivariate local autocorrelation model, revealed the spatial heterogeneity of trade-offs between ecological and social values within Jiangning District's comprehensive ecosystem services. Areas with lower compatibility between these values exhibited higher levels of statistical significance (Figure 10).

The high ecological, high social value clusters (7.23%) are mainly concentrated in regions such as Niushou Mountain, Tangshan, Ginkgo Lake, and Shitang Village. These areas are characterized by rich natural landscapes, including mountainous terrain, water bodies, and dense vegetation, convenient transportation and well-developed infrastructure. In contrast, the low ecological, low social value clusters (6.99%) are more scattered, typically situated around rural settlements and industrial or mining areas. The green spaces in these regions are vulnerable to high-intensity industrial or agricultural activities and have not been developed into tourist or leisure zones. The low ecological, high social value clusters (15.29%) are mainly found in highly urbanized zones, close to major tourist attractions or industrial parks. High-density development in these regions has severely degraded habitats, creating significant conflicts between ecological and social values. Meanwhile, the high ecological, low social value clusters (30.86%) are located along the eastern, western, and southern edges of Jiangning District. Although these areas are sparsely populated and maintain strong ecological integrity, they are constrained by challenging terrain and transportation conditions. Their potential non-material benefits remain largely untapped, providing opportunities to enhance their integrated ecological and social benefits.

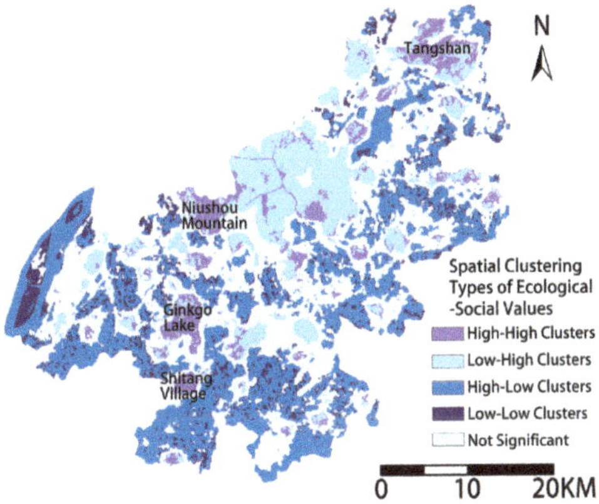

Figure 9. LISA clustering of ecological–social values in ecosystem services.

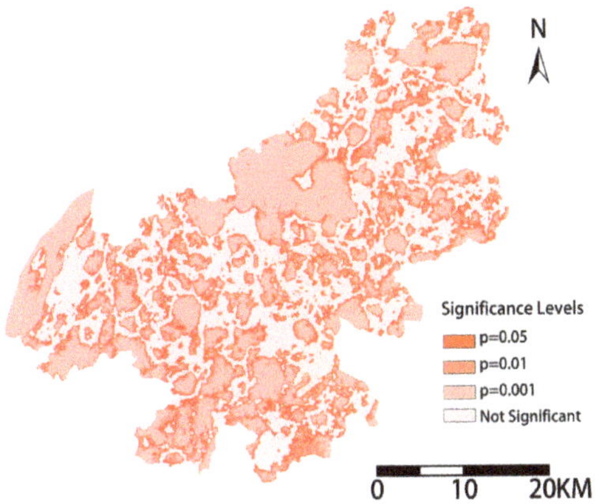

Figure 10. LISA significance of ecological–social values in ecosystem services.

In conclusion, trade-off relationships between low–high and high–low clusters predominate in Jiangning District. Their spatial distribution is characterized by significant fragmentation, with a relatively even spread across various land use types. This results in more complex interactions between humans and the environment, underscoring the urgent need to optimize the ecological functional space in urban fringe areas at a more detailed scale.

5. Ecological Functional Zoning and Intra-Zone Trade-Offs in Jiangning District

5.1. Identification Results of Ecosystem Service Clusters

The identified ecosystem service clusters revealed the differences in the quantity and types of ecological and social values of ecosystem services, reflecting the supply conditions of ecosystem services in different regions, as well as how spatial planning and land uses influence the value levels of these services. These clusters can serve as a primary basis for ecological functional zoning. An analysis of changes in the cluster variance explanation curve revealed a clear inflection point in the sum of squared errors (Figure 11), which indicated that five was the optimal number of ecosystem service clusters.

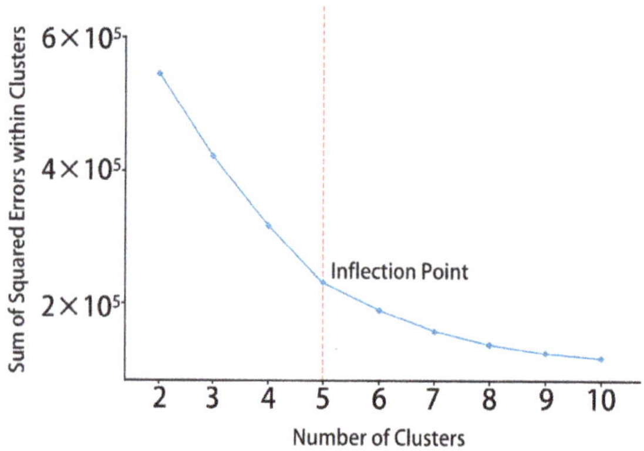

Figure 11. Sum of squared errors within cluster statistics.

The five types of ecosystem service clusters were named clusters A through E, and the area proportion of each cluster was calculated by street (Figure 12). Cluster A accounts for 23.35% of Jiangning District's area, ranking second in area proportion, and is primarily distributed in Lukou Street and Jiangning Street, followed by Moling Street, Qilin Street, and Chunhua Street, with the lowest distribution in Dongshan Street. Cluster B accounts for 18.04% of Jiangning District's area, concentrated in Tangshan Street, Hengxi Street, and Jiangning Street, with minimal distribution in Hushu Street. Cluster C covers 13.17% of the district, with the highest proportion in Moling Street, followed by Chunhua Street and Dongshan Street, while in the other seven streets, its distribution is less than 10%. Cluster D, covering 34.55%, has the largest area but is nearly absent in Qilin Street and Dongshan Street. Cluster E, covering 10.89%, is the smallest in area but represents nearly 45% of the area in Jiangning Street, with much lower distributions across the other nine streets. Overall, the clusters ranked in area, from largest to smallest, as follows: D > A > B > C > E. Clusters E, C, and B exhibit clear spatial differentiation across streets, Cluster A shows relatively minor spatial variation, and Cluster D has the broadest distribution.

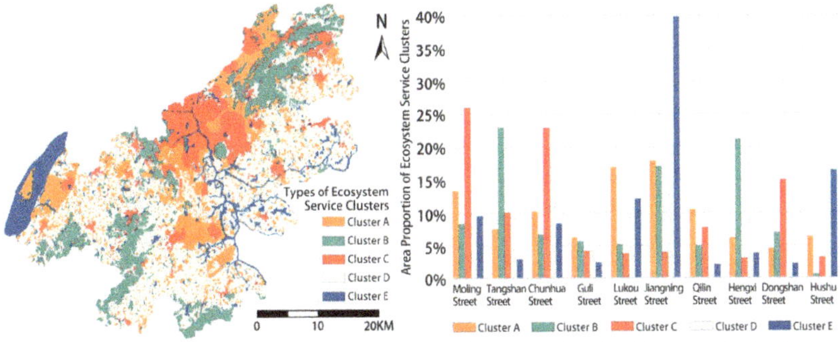

(a) LISA Clustering of Ecological - Social Values in the Riparian Recreation and Ecological Conservation Zone

(b) Area Proportion of Each Ecosystem Service Cluster by Administrative Street

Figure 12. Spatial distribution pattern of ecosystem service clusters in Jiangning District and area proportion by street.

5.2. Results of Ecological Functional Zoning in Jiangning District

Due to the overly complex spatial nesting relationships between the ecosystem service clusters, they are not suitable to be directly used as ecological functional zoning results. In adherence to the principle of regional conjugation and considering the spatial distribution, land use characteristics, and dominant value types of each ecological function zone, five main ecological function zones were delineated: the Vibrant Integration Zone of Industry and Urbanization, the Important Habitat Conservation Zone, the Livable Organic Renewal Zone, the Characteristic Rural Landscape Development Zone, and the Riparian Recreation and Ecological Conservation Zone (Figure 13). These zones exhibit distinct spatial differentiation patterns. From the perspective of "land sharing" versus "land sparing" [39], the Important Habitat Conservation Zone, Characteristic Rural Landscape Development Zone, and Riparian Recreation and Ecological Conservation Zone should generally follow a conservation-oriented landscape model based on the dominant land uses. Conversely, areas such as the Vibrant Integration Zone of Production and Urbanization, the Livable Organic Renewal Zone, and other regions involving tourism, mixed land uses, forest recreation, and similar activities are characterized by a sharing-based spatial configuration that integrates various land uses.

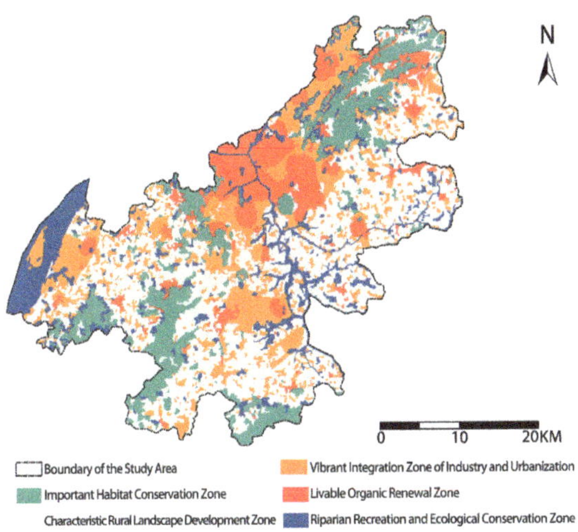

Figure 13. Ecological functional zoning in Jiangning District.

5.2.1. Vibrant Integration Zone of Industry and Urbanization

The Vibrant Integration Zone of Industry and Urbanization is mainly located in the Binjiang area, Lukou New Town, and the towns of Moling, Guli, and Qilin (Figure 14A). In this zone, social values slightly outweigh ecological values, with a notable emphasis on historical culture, recreational leisure, and water conservation. In contrast, values such as carbon storage, knowledge innovation, and landscape esthetics are relatively lower (Figure 14B). This area plays a key role in driving the development of the robust industrial sector in Jiangning, underscoring the need for enhanced knowledge innovation. The limited distribution of forests, grasslands, and water bodies hinders the effectiveness of regulatory services. Furthermore, inefficient agricultural land outside of the designated permanent farmland offers significant potential for conversion and re-greening. This can be achieved through moderate development and the addition of green spaces, enhancing various values such as carbon storage and landscape esthetics.

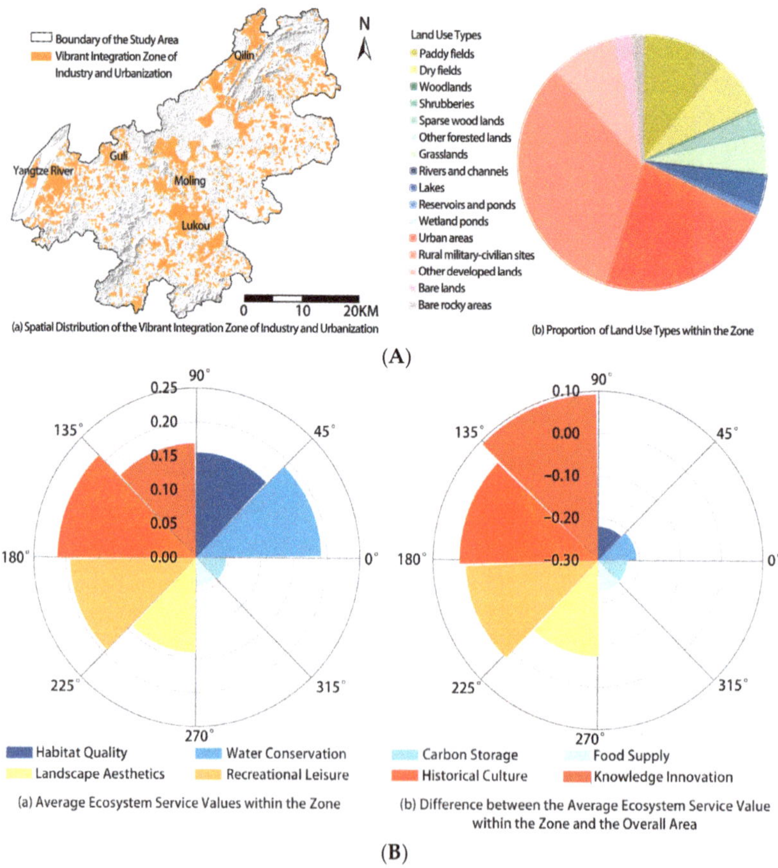

Figure 14. Spatial pattern of the Vibrant Integration Zone of Industry and Urbanization and the distribution of ecosystem service values. (**A**) Spatial distribution of the Vibrant Integration Zone of Industry and Urbanization and land use distribution. (**B**) Average ecosystem service values within the zone and the differences from overall values.

5.2.2. Important Habitat Conservation Zone

The Important Habitat Conservation Zone mainly covers large mountainous and hilly regions, including Qinglong Mountain, Niushou Mountain, and Yuntai Mountain, which are home to diverse habitats and rich vegetation communities (Figure 15A). This area excels in carbon storage, habitat quality, and water conservation, while food supply and knowledge innovation are its weakest value categories (Figure 15B). The zone is the ecological and landscape backbone not only of Jiangning District but of the entire city of Nanjing, underscoring the need to protect carbon sinks and wildlife habitats. It plays a vital role in safeguarding biodiversity and supporting carbon sequestration. However, the presence of numerous fragmented green patches emphasizes the need to balance ecological protection with urban development by establishing an interconnected and multi-functional ecological security network.

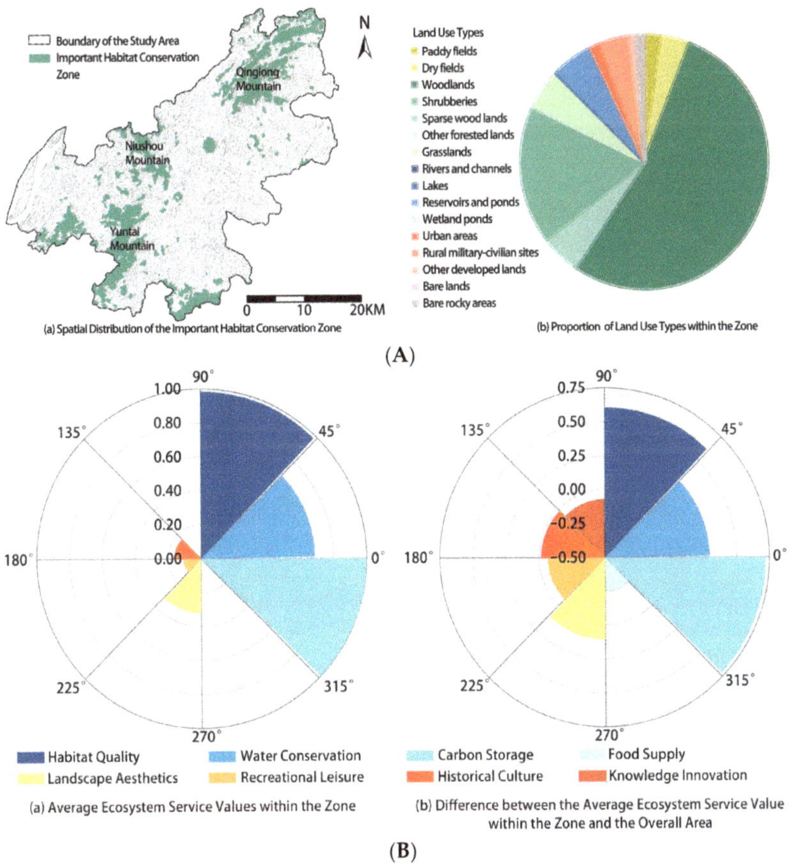

Figure 15. Spatial pattern of the Important Habitat Conservation Zone and the distribution of ecosystem service values. (**A**) Spatial distribution of the Important Habitat Conservation Zone and land use distribution. (**B**) Average ecosystem service values within the zone and the differences from overall values.

5.2.3. Livable Organic Renewal Zone

The Livable Organic Renewal Zone is mainly concentrated in densely populated areas, including Dongshan Sub-city, Baijia Lake Central Area, Tangshan New City, and Chunhua New Town (Figure 16A). This zone holds notable social value, particularly in terms of recreational leisure and landscape esthetics. However, its ecological value is significantly lower than that of other zones (Figure 16B). The main contributing factors include the limited availability of large urban parks and green corridors, as well as the lack of systematic protection for fragmented and degraded ecological areas. As an essential space for the daily lives of Jiangning District residents and a flourishing tourism industry, locations with strong geographical advantages—such as waterfront areas, industrial zones, and areas with distinctive landscapes—hold significant potential for redevelopment. There is an urgent need to enhance land use efficiency in these areas, addressing diverse social demands while simultaneously improving overall ecological quality.

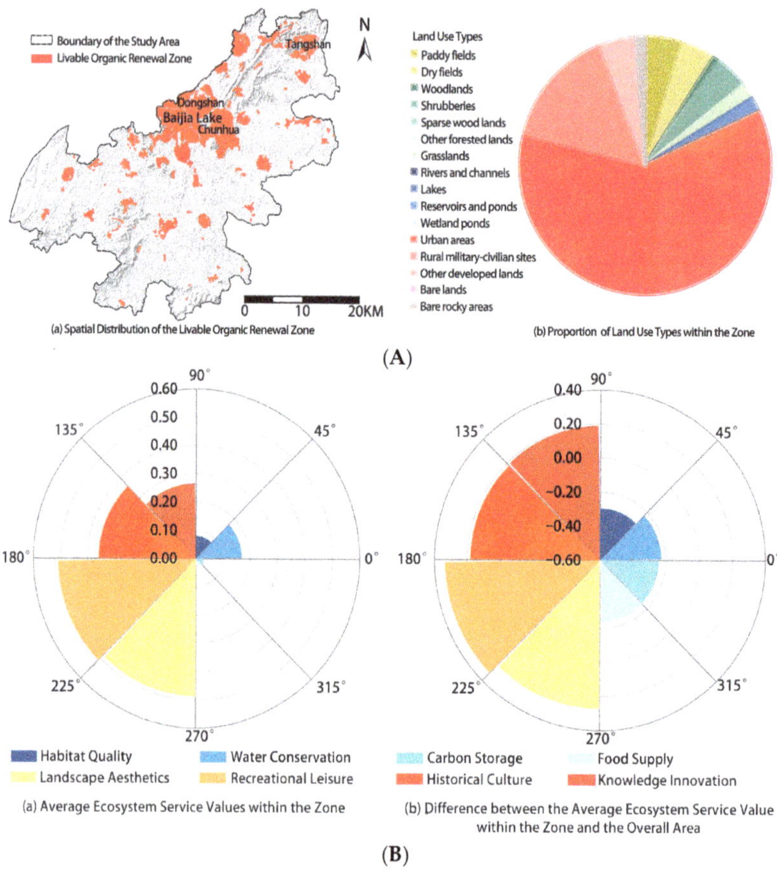

Figure 16. Spatial pattern of the Livable Organic Renewal Zone and the distribution of ecosystem service values. (**A**) Spatial distribution of the Livable Organic Renewal Zone and land use distribution. (**B**) Average ecosystem service values within the zone and the differences from overall values.

5.2.4. Characteristic Rural Landscape Development Zone

The Characteristic Rural Landscape Development Zone spans the eastern, southern, and western parts of Jiangning District, encompassing paddy fields, dry lands, and rural settlements (Figure 17A). While this zone excels in grain production and offers certain water conservation benefits, its other ecosystem services are notably lower (Figure 17B). This underscores how the fragmented and relatively uniform land use patterns in the area constrain the overall potential of ecosystem services. Previous studies have shown that protecting agricultural areas surrounding urban regions is crucial for ensuring food security and promoting urban sustainability [40]. As such, these areas should also be considered in strategic landscape planning. Despite the presence of extensive farmland, countryside parks, and historic towns and villages, the zone struggles with weak social value. Issues such as the uniformity of agricultural landscapes, limited opportunities for rural knowledge engagement, and inadequate protection of cultural heritage are prevalent. Moving forward, it is important to enhance this zone's development by securing food supply and water regulation while also gaining a deeper understanding of natural and public resources. Tailored development strategies can then address the diverse cultural and social needs of local residents.

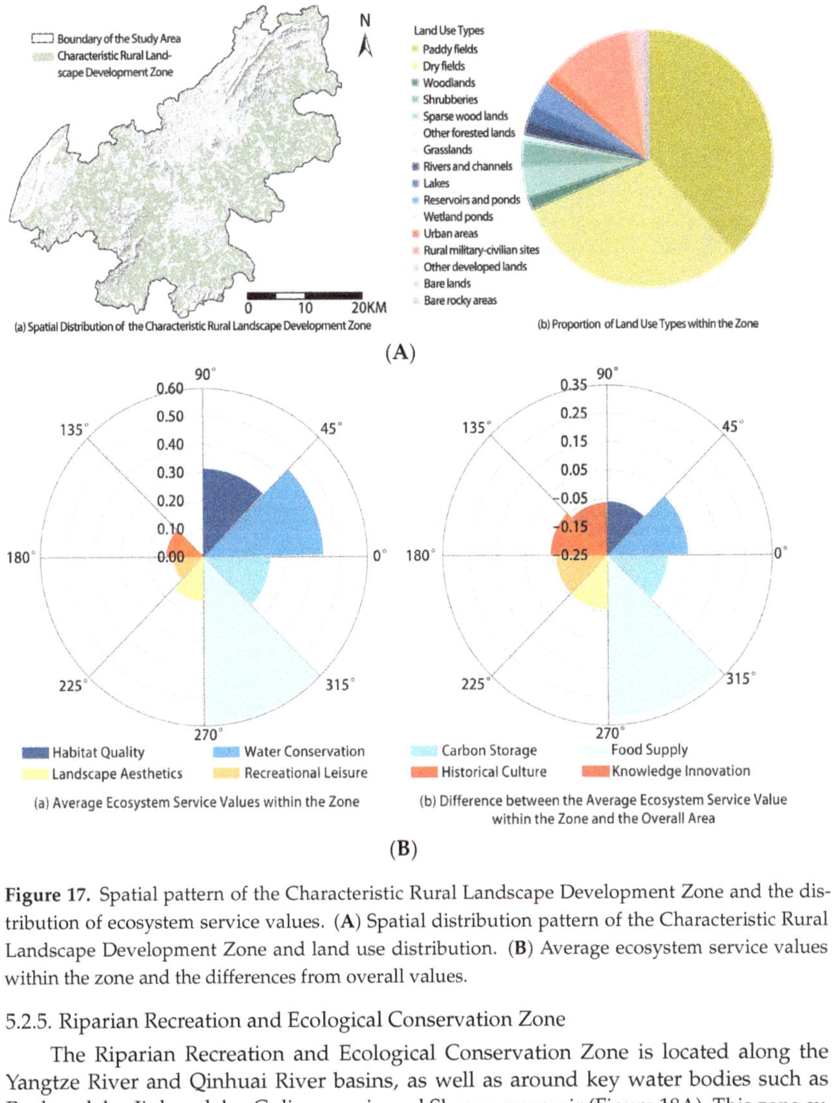

Figure 17. Spatial pattern of the Characteristic Rural Landscape Development Zone and the distribution of ecosystem service values. (**A**) Spatial distribution pattern of the Characteristic Rural Landscape Development Zone and land use distribution. (**B**) Average ecosystem service values within the zone and the differences from overall values.

5.2.5. Riparian Recreation and Ecological Conservation Zone

The Riparian Recreation and Ecological Conservation Zone is located along the Yangtze River and Qinhuai River basins, as well as around key water bodies such as Panlong lake, Jiulong lake, Guli reservoir, and Shecun reservoir (Figure 18A). This zone excels in water conservation and habitat quality but has significant potential for improvement in terms of carbon storage, recreational spaces, and cultural heritage values (Figure 18B). This area plays a crucial role in shaping a cityscape where parks seamlessly blend with water features, while also protecting water-related cultural heritage. However, the current water system suffers from poor connectivity, fragmented green spaces around large lakes and reservoirs, inadequate accessibility, and unevenly distributed waterfront recreational facilities. There is an urgent need to develop an interconnected water network that integrates various basins and supports the healthy evolution of aquatic ecosystems. By implementing flexible management and lenient interventions in waterfront areas, water-related recreational functions can be expanded, fostering the synergy of ecological and social values.

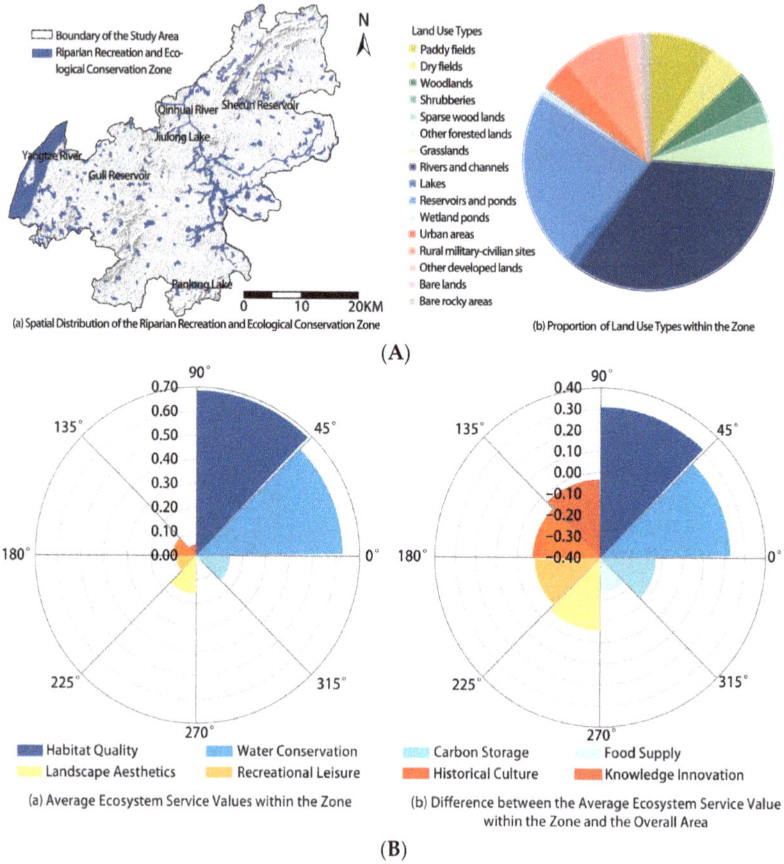

Figure 18. Spatial pattern of the Riparian Recreation And Ecological Conservation Zone and the distribution of ecosystem service values. (**A**) Spatial distribution pattern of the Riparian Recreation And Ecological Conservation Zone and land use distribution. (**B**) Average ecosystem service values within the zone and the differences from overall values.

5.3. *Trade-Offs Within Ecological Function Zones in Jiangning District*

5.3.1. Comparative Analysis of the Correlation Between Ecological–Social Values of Ecosystem Services Across Zones

The statistical results—in terms of Pearson correlation coefficients across different zones (Figure 19)—show that, within each ecological function zone, ecological and social values generally exhibit a synergistic relationship. However, the relationships between ecological and social values tend to reflect trade-offs. In areas dominated by large expanses of construction land or farmland, the trade-off between ecological and social values is particularly pronounced. Intensive industrial and agricultural practices often neglect landscape diversity, undermining the intrinsic health of ecosystems. Moreover, synergistic effects are more common among social values than among ecological values.

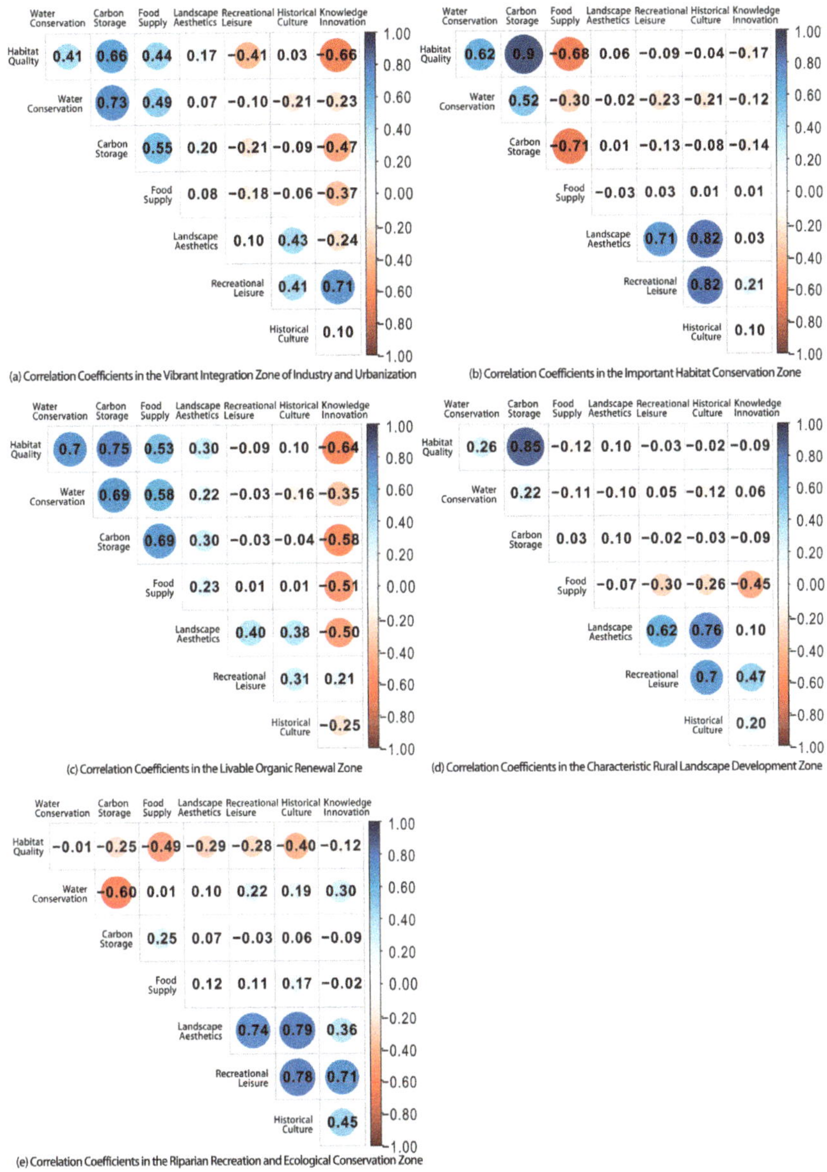

Figure 19. Correlation coefficient matrices of ecological–social values of ecosystem services in the ecological functional zones of Jiangning District.

5.3.2. Comparative Analysis of the Spatial Clustering Characteristics of Comprehensive Ecological–Social Values of Ecosystem Services Across Zones

Based on zoning data, the dominant spatial clusters representing the ecological and social values of ecosystem services within each ecological function zone were identified (Figure 20). Most zones are characterized by clusters that indicate a trade-off between ecological and social values. In the Important Habitat Conservation Zone, the Riparian Recreation and Ecological Conservation Zone, and the Characteristic Rural Landscape Development Zone, clusters with high ecological but low social values are most common.

These areas should prioritize preserving and enhancing spaces of high ecological value while also exploring opportunities to improve non-material benefits, such as natural education and eco-tourism. In contrast, the Livable Organic Renewal Zone is dominated by clusters with low ecological and high social values. Strategies such as controlling visitor numbers and increasing the connectivity of green spaces should be implemented to mitigate the negative environmental impacts of human activities.

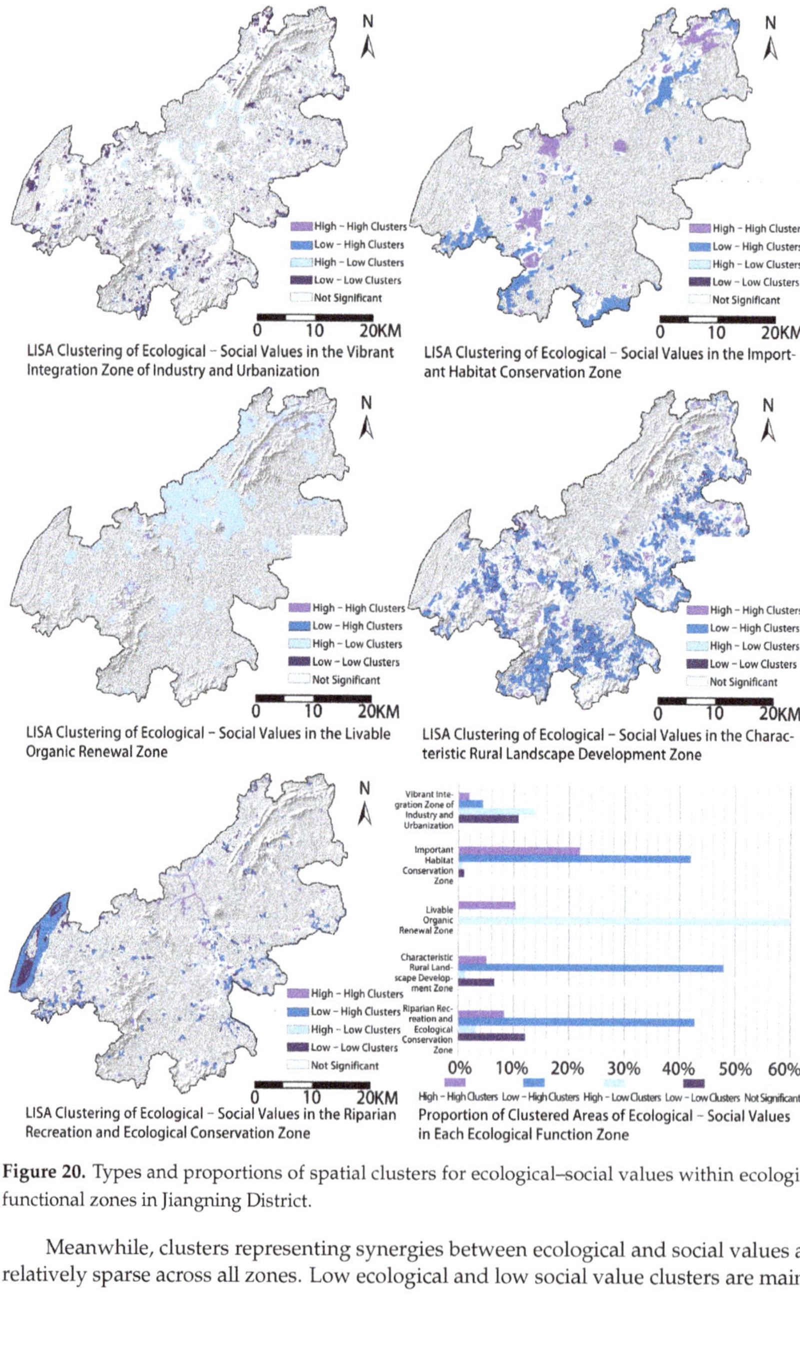

Figure 20. Types and proportions of spatial clusters for ecological–social values within ecological functional zones in Jiangning District.

Meanwhile, clusters representing synergies between ecological and social values are relatively sparse across all zones. Low ecological and low social value clusters are mainly found in areas with poor landscape quality and late-stage ecological restoration efforts, particularly within the Vibrant Integration Zone of Industry and Urbanization and the Riparian Recreation and Ecological Conservation Zone. In these areas, prioritizing the restoration of ecological value is crucial to fostering positive synergies across the region. Conversely, high ecological and high social value clusters are mostly located within the Important Habitat Conservation Zone, where access to both major roads and water bodies is often excellent. This suggests that proximity to infrastructure, along with the integrity and connectivity of ecological structures, are key factors in maximizing ecological–social benefits. Strengthening the management and protection of these green spaces is essential for maintaining this ecological–social synergy.

6. Spatial Optimization Strategy for Ecological Functional Zones in Jiangning District Based on Trade-Off Relationships

This study takes into account the critical environmental issues that need to be addressed during the process of urbanization in Jiangning District, as well as the interactions between ecosystem services and ecological–social values, both in a region-wide manner and within each ecological functional zone. It is proposed that the spatial optimization objective for the ecological functional zones in Jiangning District involves reducing trade-offs and promoting synergistic benefits between ecosystem services and ecological–social values.

In the Vibrant Integration Zone of Production and Urbanization, numerous trade-off relationships exist between ecosystem services and ecological–social values. These are often found in regions with a large spatial concentration characterized by low ecological value and high social value. The direction of spatial optimization involves promoting the green development of industrial spaces while enhancing the synergistic effects of multi-functional spaces. This includes building ecological industrial parks to strengthen the integration of production and environmental sustainability, introducing low-impact development technologies to reduce human interference, and rehabilitating urban public spaces to foster the synergy of social values.

In the Important Habitat Conservation Zone, the main trade-offs occur both within the ecological value group and between ecological and social values; meanwhile, the social values generally align well with each other. In terms of spatial clustering, there is a noticeable pattern of high ecological and low social value clusters. The direction for spatial optimization in this area involves protecting natural resources and ecological structures while enhancing natural esthetics. This includes strictly prohibiting reckless farmland reclamation to safeguard core ecological functions, resolving conflicts between structural elements to maintain ecosystem stability, and utilizing forest landscape resources to create spaces that provide esthetic experiences.

In the Livable Organic Renewal Zone, the main trade-offs occur between ecosystem services and ecological and social values, followed by trade-offs within the social value group. Ecological values tend to be highly compatible, with most areas exhibiting a low ecological and high social value clustering. The direction for spatial optimization in this area involves enhancing the utilization efficiency of existing spaces and improving ecological and environmental quality. This involves expanding green spaces in the old city from various perspectives to increase the supply of ecological values, restoring the ecological network system to ensure the stability of the ecological structure, and promoting the development of multi-functional spaces to diversify the use of existing land.

In the Characteristic Rural Landscape Development Zone, there are notable trade-offs between food supply, water conservation, and social values, with potential conflicts arising among ecological values; however, the compatibility within the social value group is relatively strong. In addition, the spatial clustering mainly exhibits high ecological and low social value areas. The spatial optimization strategy for this zone should focus on promoting the sustainable development of ecological agriculture while preserving the unique local cultural heritage. This involves adjusting diverse planting structures to enhance compound environmental benefits, optimizing the agricultural development area to control the intensity of land development, and revitalizing rural areas with distinctive characteristics, all while advancing cultural development in a targeted manner.

In the Riparian Recreation and Ecological Conservation Zone, trade-offs mainly occur between ecological and social values and within the ecological value group. Water conservation and social values, as well as the social value group, exhibit a high degree of compatibility. Spatially, this is mainly reflected in a high ecological and low social value cluster. The direction for spatial optimization in this area involves protecting the intrinsic health of the water ecosystem while cautiously expanding leisure industries. This involves ecological restoration of nearshore areas to preserve the water system's intrinsic health, adjusting and optimizing structural components to minimize the impacts of human activities, and flexibly managing waterfront spaces to support moderate tourism and sightseeing development.

7. Conclusions

Urban fringe areas play a critical role in coordinating land resources and fostering urban–rural integration, acting as sensitive zones where natural environments and human societies often intersect. Addressing pressing issues such as the decline in ecosystem functions, the rising demand for high-quality urban–rural development, and the increasing contradictions in ecosystem service values, this study investigated a method for delineating ecological function zones, focusing on the trade-offs between ecological–social values in ecosystem services. Moreover, it clarified how these values interact as we move from a broader spatial context to more localized zones.

The findings indicated that the trade-off relationships between ecological and social values are widespread in urban fringe areas, being observable at both regional and zonal levels. Based on the ecosystem service clustering results, Jiangning District was categorized into five ecological function zones: the Vibrant Integration Zone of Industry and Urbanization, Important Habitat Conservation Zone, Livable Organic Renewal Zone, Characteristic Rural Landscape Development Zone, and Riparian Recreation and Ecological Conservation Zone. Each of these zones exhibited significant differences in terms of the types and characteristics of the service values that they provide. Resolving the trade-offs between ecological and social values requires a deeper understanding of the varying relationships within the zones, necessitating the development of tailored and refined ecological management strategies to foster mutual benefits between ecosystems and living environments.

A comprehensive understanding of the complex interactions between ecosystem service values and social–ecological contexts is crucial for achieving multi-functional development goals in urban fringe areas. This study goes beyond a single perspective on ecological function zoning in these areas, clarifying the trade-offs between ecological and social values of ecosystem services across the overall spatial landscape of urban fringe areas. Based on the identification of ecosystem service clusters, ecological function zones were delineated, and an analysis of the trade-offs within each zone and their spatial differentiation characteristics was conducted. A systematic, progressive trade-off analysis framework was constructed, advancing from whole-area spatial assessment to sub-regional zones. This analytical framework is versatile and applicable to zoning management practices in other urban fringe areas with diverse ecological characteristics and complex urban–rural functions, particularly when aiming to mitigate trade-offs and enhance synergies. However, adjustments to the parameters and zoning types may be necessary to address the specific ecological, social, and management needs of each area.

Future research should further refine the study of ecological function zoning in urban fringe areas by optimizing the ecosystem service value assessment models, analyzing factors that influence trade-off relationships, and exploring the value perceptions and preferences of different stakeholders. This will promote ecological balance and enhance collaborative value in urban fringe ecosystems. In particular, incorporating the perspectives of diverse stakeholders—such as the government, the public, and businesses—through interviews, surveys, and consultative mechanisms will allow for an in-depth analysis of their positions, needs, and decision-making influence in the ecological–social value trade-offs of ecosystem services. This enhances the practical applicability of ecological zoning, thereby increasing its external validity.

Author Contributions: Conceptualization, N.X. and H.D.; methodology, N.X. and H.D.; software, H.D.; validation, H.D.; formal analysis, H.D. and N.X.; investigation, H.D.; resources, N.X.; data curation, H.D.; writing—original draft preparation, H.D.; writing—review and editing, N.X.; visualization, H.D.; supervision, N.X.; project administration, N.X.; funding acquisition, N.X. All authors have read and agreed to the published version of the manuscript.

Funding: This research was funded by the National Natural Science Foundation of China (52378046) and the National Natural Science Foundation of China (51978147).

Data Availability Statement: The data presented in this study are available from the corresponding author on reasonable request. The data are not publicly available due to some of them being used in other studies that have not yet been publicly published.

Conflicts of Interest: The authors declare no conflicts of interest.

References

1. Cui, L.; Wang, J.; Sun, L.; Lv, C. Construction and optimization of green space ecological networks in urban fringe areas: A case study with the urban fringe area of Tongzhou district in Beijing. *J. Clean. Prod.* **2020**, *276*, 124266. [CrossRef]
2. Cord, A.F.; Bartkowski, B.; Beckmann, M.; Dittrich, A.; Hermans-Neumann, K.; Kaim, A.; Lienhoop, N.; Locher-Krause, K.; Priess, J.; Schröter-Schlaack, C.; et al. Towards systematic analyses of ecosystem service trade-offs and synergies: Main concepts, methods and the road ahead. *Ecosyst. Serv.* **2017**, *28*, 264–272. [CrossRef]
3. Cheng, X.; Damme, S.V.; Li, L.; Uyttenhove, P. Cultural ecosystem services in an urban park: Understanding bundles, trade-offs, and synergies. *Landsc. Ecol.* **2022**, *37*, 1693–1705. [CrossRef]
4. Huang, S.; Wang, Y.; Liu, R.; Jiang, Y.; Qie, L.; Pu, L. Identification of land use function bundles and their spatiotemporal trade-offs/synergies: A case study in jiangsu coast, China. *Land* **2022**, *11*, 286. [CrossRef]
5. Aryal, K.; Maraseni, T.; Apan, A. How much do we know about trade-offs in ecosystem services? A systematic review of empirical research observations. *Sci. Total Environ.* **2022**, *806*, 151229. [CrossRef]
6. Alessa, L.; Kliskey, A.; Brown, G. Social–ecological hotspots mapping: A spatial approach for identifying coupled social–ecological space. *Landsc. Urban Plan.* **2008**, *85*, 27–39. [CrossRef]
7. Bryan, B.A.; Raymond, C.M.; Crossman, N.D.; King, D. Comparing spatially explicit ecological and social values for natural areas to identify effective conservation strategie. *Conserv. Biol.* **2011**, *25*, 172–181. [CrossRef] [PubMed]
8. Karimi, A.; Brown, G.; Hockings, M. Methods and participatory approaches for identifying social-ecological hotspots. *Appl. Geogr.* **2015**, *63*, 9–20. [CrossRef]
9. Bagstad, K.J.; Reed, J.M.; Semmens, D.J.; Sherrouse, B.C.; Troy, A. Linking biophysical models and public preferences for ecosystem service assessments: A case study for the Southern Rocky Mountains. *Reg. Environ. Change* **2015**, *16*, 2005–2018. [CrossRef]
10. Chi, Y.; He, C. Impact of land use change on the spatial and temporal evolution of ecosystem service values in south china karst areas. *Forests* **2023**, *14*, 893. [CrossRef]
11. Lourdes, K.T.; Gibbins, C.N.; Sherrouse, B.C.; Semmens, D.J.; Hamel, P.; Sanusi, R.; Azhar, B.; Diffendorfer, J.; Lechner, A.M. Mapping development preferences on the perceived value of ecosystem services and land use conflict and compatibility in Greater Kuala Lumpur. *Urban For. Urban Green.* **2024**, *92*, 128183. [CrossRef]
12. Cueva, J.; Yakouchenkova, I.A.; Froehich, K.; Dermann, A.F.; Dermann, F.; Koehler, M.; Grossmann, J.; Meier, W.; Bauhus, J.; Schroder, D.; et al. Synergies and trade-offs in ecosystem services from urban and peri-urban forests and their implication to sustainable city design and planning. *Sustain. Cities Soc.* **2022**, *82*, 103903. [CrossRef]
13. Zhang, H.; Deng, W.; Zhang, S.; Peng, L.; Liu, Y. Impacts of urbanization on ecosystem services in the Chengdu-Chongqing Urban Agglomeration: Changes and trade-offs. *Ecol. Indic.* **2022**, *139*, 108920. [CrossRef]
14. Chen, S.; Chen, H.; Yang, R.; Ye, Y. Linking social-ecological management and ecosystem service bundles: Lessons from a peri-urban agriculture landscape. *Land Use Policy* **2023**, *131*, 106697. [CrossRef]
15. Fang, G.; Sun, X.; Sun, R.; Liu, Q.; Tao, Y.; Yang, P.; Tang, H. Advancing the optimization of urban–rural ecosystem service supply-demand mismatches and trade-offs. *Landsc. Ecol.* **2024**, *39*, 32. [CrossRef]
16. Duvernoy, I.; Zambon, I.; Sateriano, A.; Salvati, L. Pictures from the other side of the fringe: Urban growth and peri-urban agriculture in a post-industrial city (Toulouse, France). *J. Rural Stud.* **2018**, *57*, 25–35. [CrossRef]
17. Sun, X.; Liu, H.; Liao, C.; Nong, H.; Yang, P. Understanding recreational ecosystem service supply-demand mismatch and social groups' preferences: Implications for urban–rural planning. *Landsc. Urban Plan.* **2024**, *241*, 104903. [CrossRef]
18. Kareiva, P.; Watts, S.; McDonald, R.; Boucher, T. Domesticated nature: Shaping landscapes and ecosystems for human welfare. *Science* **2007**, *316*, 1866–1869. [CrossRef] [PubMed]

19. Raudsepp-Hearne, C.; Peterson, G.D.; Bennett, E.M. Ecosystem service bundles for analyzing tradeoffs in diverse landscapes. *Proc. Natl. Acad. Sci. USA* **2010**, *107*, 5242–5247. [CrossRef]
20. Karimi, J.D.; Corstanje, R.; Harris, J.A. Bundling ecosystem services at a high resolution in the UK: Trade-offs and synergies in urban landscapes. *Landsc. Ecol.* **2021**, *36*, 1817–1835. [CrossRef]
21. Wu, F. *Planning for Growth: Urban and Regional Planning in China*; Routledge: London, UK, 2015.
22. National Bureau of Statistics. Statistical Zoning Code for 2023. 2024. Available online: https://www.stats.gov.cn/sj/ (accessed on 13 September 2023).
23. The People's Government of Jiangning District Nanjing. Main Data Bulletin of the Third National Land Survey in Jiangning District. Available online: www.jiangning.gov.cn/jnqrmzf/202209/t20220927_3709893.html (accessed on 11 January 2023).
24. Haines-Young, R.; Potschin, M. Common International Classification of Ecosystem Services (CICES): Consultation on Version 4, August-December 2012. 2013. EEA Framework Contract No EEA/IEA/09/003. Available online: https://cices.eu/content/uploads/sites/8/2012/07/CICES-V43_Revised-Final_Report_29012013.pdf (accessed on 13 September 2023).
25. Wang, S.; Zhang, B.; Wang, S.; Xie, G. Dynamic changes in water conservation in the Beijing–Tianjin sandstorm source control project area: A case study of Xilin Gol League in China. *J. Clean. Prod.* **2021**, *293*, 126054. [CrossRef]
26. Wang, X.; Peng, J.; Luo, Y.; Qiu, S.; Dong, J.; Zhang, Z.; Vercruysse, K.; Grabowski, R.C.; Meersmans, J. Exploring social-ecological impacts on trade-offs and synergies among ecosystem services. *Ecol. Econ.* **2022**, *197*, 107438. [CrossRef]
27. Cui, F.; Tang, H.; Zhang, Q.; Wang, B.; Dai, L. Integrating ecosystem services supply and demand into optimized management at different scales: A case study in Hulunbuir, China. *Ecosyst. Serv.* **2019**, *39*, 100984. [CrossRef]
28. González-García, A.; Arias, M.; García-Tiscar, S.; Alcorlo, P.; Santos-Martín, F. National blue carbon assessment in Spain using InVEST: Current state and future perspectives. *Ecosyst. Serv.* **2022**, *53*, 101397. [CrossRef]
29. He, S.; Su, Y.; Shahtahmassebi, A.R.; Huang, L.; Zhou, M.; Gan, M.; Deng, J.; Zhao, G.; Wang, K. Assessing and mapping cultural ecosystem services supply, demand and flow of farmlands in the Hangzhou metropolitan area, China. *Sci. Total Environ.* **2019**, *692*, 756–768. [CrossRef]
30. Gong, J.; Jin, T.; Liu, D.; Zhu, Y.; Yan, L. Are ecosystem service bundles useful for mountainous landscape function zoning and management? A case study of Bailongjiang watershed in western China. *Ecol. Indic.* **2022**, *134*, 108495. [CrossRef]
31. Lyu, Y.; Wang, M.; Zou, Y.; Wu, C. Mapping trade-offs among urban fringe land use functions to accurately support spatial planning. *Sci. Total Environ.* **2022**, *802*, 149915. [CrossRef]
32. Soto-Montes-de-Oca, G.; Cruz-Bello, G.M.; Bark, R.H. Enhancing megacities' resilience to flood hazard through peri-urban nature-based solutions: Evidence from Mexico City. *Cities* **2023**, *143*, 104571. [CrossRef]
33. Turner, K.G.; Odgaard, M.V.; Bøcher, P.K.; Dalgaard, T.; Svenning, J. Bundling ecosystem services in Denmark: Trade-offs and synergies in a cultural landscape. *Landsc. Urban Plan.* **2014**, *125*, 89–104. [CrossRef]
34. Sasaki, K.; Hotes, S.; Ichinose, T.; Wolters, V. Hotspots of agricultural ecosystem services and farmland biodiversity overlap with areas at risk of land abandonment in Japan. *Land* **2021**, *10*, 1031. [CrossRef]
35. O'Brien, L.; De Vreese, R.; Kern, M.; Sievänen, T.; Stojanova, B.; Atmiş, E. Cultural ecosystem benefits of urban and peri-urban green infrastructure across different European countries. *Urban For. Urban Green.* **2017**, *24*, 236–248. [CrossRef]
36. Kovács, B.; Uchiyama, Y.; Miyake, Y.; Quevedo, J.M.D.; Kohsaka, R. Capturing landscape values in peri-urban Satoyama forests: Diversity of visitors' perceptions and implications for future value assessments. *Trees For. People* **2022**, *10*, 100339. [CrossRef]
37. Cheung, L.T.; Hui, D.L. Influence of residents' place attachment on heritage forest conservation awareness in a peri-urban area of Guangzhou, China. *Urban For. Urban Green.* **2018**, *33*, 37–45. [CrossRef]
38. Karimi, A.; Yazdandad, H.; Fagerholm, N. Evaluating social perceptions of ecosystem services, biodiversity, and land management: Trade-offs, synergies and implications for landscape planning and management. *Ecosyst. Serv.* **2020**, *45*, 101188. [CrossRef]
39. Lin, B.B.; Fuller, R.A. FORUM: Sharing or sparing? How should we grow the world's cities. *J. Appl. Ecol.* **2013**, *50*, 1161–1168. [CrossRef]
40. Camps-Calvet, M.; Langemeyer, J.; Calvet-Mir, L.; Gómez-Baggethun, E. Ecosystem services provided by urban gardens in Barcelona, Spain: Insights for policy and planning. *Environ. Sci. Technol.* **2016**, *62*, 14–23. [CrossRef]

land

MDPI

Article

Nature-Based Solutions Planning for Urban Microclimate Improvement and Health: An Integrated Ecological and Economic Approach

Teodoro Semeraro [1,*], Antonio Calisi [2], Jian Hang [3], Rohinton Emmanuel [4] and Riccardo Buccolieri [5]

1 Research Institute on Terrestrial Ecosystems (IRET-URT Lecce), National Research Council of Italy (CNR), Campus Ecotekne, 73100 Lecce, Italy
2 Department of Science and Technological Innovation (DISIT), University of Eastern Piedmont, Viale Michel 11, 15121 Alessandria, Italy; antonio.calisi@uniupo.it
3 School of Atmosphere Sciences, Sun Yat-sen University, Zhuhai 519082, China; hangj3@mail.sysu.edu.cn
4 The Research Centre for Built Environment Asset Management (BEAM), Glasgow Caledonian University, Cowcaddens Rd, Glasgow G4 0BA, UK; rohinton.emmanuel@gcu.ac.uk
5 Department of Biological and Environmental Sciences and Technologies, University of Salento, S.P. 6 Lecce-Monteroni, 73100 Lecce, Italy; riccardo.buccolieri@unisalento.it
* Correspondence: teodoro.semeraro@cnr.it

Abstract: Nature-based Solutions (NbSs) play a pivotal role in mitigating the impact of microclimates on human well-being. The effectiveness of NbSs is contingent upon the synergy between natural capital, defined by the ecological structure and functions of the ecosystem, and human-derived capital, encompassing the economic investments required for implementation. This study introduces a decision-making framework designed to evaluate the impact of NbSs and advocate for optimal solutions for human health at the local scale, amalgamating ecological and economic assessments. Physiological Equivalent Temperature (PET) was chosen as a key urban parameter to assess the efficacy of NbSs in mitigating urban microclimates and enhancing human health. The PET analysis was conducted using ENVI-met 5.0.3 software across diverse urban scenarios in Gallipoli city, Italy. Integrated with a cost–benefit analysis of NbSs considering various investment scenarios, the study aimed to identify the most effective solution. Results indicated positive effects of NbSs in open spaces and around building blocks where the PET levels remained below 30 °C. Conversely, scenarios without NbSs exhibited PETs exceeding 40 °C, with peaks of 50 °C, posing potential risks to human health. Considering the social and economic benefits associated with PET mitigation, the cost–benefit analysis suggests that implementing NbSs using a mix of young and mature plants in the initial phase is advantageous compared to using only young plants. Thus, in establishing NbSs, it is crucial to consider not only the quantity of vegetation but also the strategic timing of implementation. In conclusion, our work offers an innovative framework that combines ecological and economic perspectives, providing valuable insights for decision-makers in urban planning and promoting the practical application of NbSs for enhanced human well-being.

Keywords: ecosystem services; nature-based solutions; ecological urban planning; cost–benefit analysis; physiological equivalent temperature; social benefits

1. Introduction

Urban areas, marked by the Urban Heat Island (UHI) effect, face pressing environmental challenges requiring innovative solutions. Urban expansion and industrialization amplify the UHI effect by transforming natural or agricultural land into dense networks of structures [1–3]. Glasgow, with its industrial legacy, exemplifies these challenges, experiencing a year-round UHI of over 3 °C. Similar challenges are observed globally in cities like Tokyo, each grappling with unique UHI dynamics [4].

The multifaceted impact of the Urban Heat Island (UHI) effect includes increased heat-absorbing surfaces, heightened anthropogenic heat production, altered air circulation patterns, and diminished evapotranspiration [5–8]. Research suggests that even a modest increase in Green Infrastructure could potentially eliminate the UHI effect. Additionally, elevated temperatures in urban areas increase pollution levels, affecting air quality [9–12] and human health. Impermeable surfaces further worsen these challenges by reducing water infiltration capacity and increasing flood risk [1,13]. This interplay highlights the need for comprehensive research to support resilient urban planning. In this context, natural capital—encompassing vegetation, green spaces, and ecosystems—is critical. Nature-based Solutions (NbSs) provide a holistic framework for mitigating microclimatic impacts on human well-being [14]. Their effectiveness depends on the integration of natural and human-derived capital [15]. Incorporating NbSs into urban planning enhances urban resilience and fosters sustainability [16,17]. The lack of vegetation intensifies UHI effects, increasing thermal stress and its associated health risks. Addressing urban heat, especially in the context of climate change and more frequent heat waves, underscores the importance of prioritizing microclimate mitigation in urban planning [18,19].

In the context of global warming, urban forests represent a crucial NbS for enhancing environmental and human well-being in cities. They play a significant role in microclimate regulation and UHI reduction by providing shade, cooling the surrounding area through evapotranspiration, and lowering the demand for air conditioning. This cooling effect is particularly valuable in increasingly hot urban environments [16,20,21]. To ensure sustainability, urban forests must be integrated into urban planning frameworks as a key component of climate action, land-use planning, and public health strategies. Their valuation should comprehensively consider social, economic, and environmental dimensions [21,22]. The Physiological Equivalent Temperature (PET) serves as a useful index for assessing the effects of NbSs' impacts [23,24]. The PET comfort index evaluates the human heat balance with the environment, considering meteorological and thermo-physiological parameters. The assessment of PET for NbS planning needs to consider vegetation structure, human preferences, and human-derived capital, balancing cost investments with human benefits [15,25].

The effectiveness of NbSs is heavily influenced by the quality of their implementation, particularly regarding the ecological structures and functions that define the intended natural capital. The costs associated with NbSs represent a substantial investment in human-derived capital, encompassing financial resources and other components necessary for their implementation, rather than focusing solely on ecological aspects [14,15]. Human-derived capital plays a crucial role in determining the feasibility and success of NbS projects, as it establishes the conditions under which these solutions can be effectively applied and sustained. Understanding the role of human-derived capital in enhancing natural capital is crucial for urban decision-making, as it helps determine the optimal balance between natural capital and technological alternatives that influence investment cost [16]. This manuscript establishes a novel decision-making framework for NbS planning, assessing microclimate ecosystem services' impact on human health while incorporating cost–benefit analyses. The proposed framework adopts a systems approach to sustainable urban development, integrating natural and human-derived capital to inform decisions that promote long-term resilience and well-being in cities.

2. Materials and Methods

2.1. Study Area

The study area is a peri-urban region characterized by agroecosystems, consisting of arable land interspersed with sprawling residential areas. These residential zones are in line with the urban development plan outlined by the municipality of Gallipoli for new residential expansion (Figure 1). Gallipoli is a medium-sized town located on the western coast of the Salentine Peninsula, in the southern part of the Apulia Region. As of 2023, the town has a population of approximately 20,000 residents (https://demo.istat.it/

app/?i=RIC&l=it, accessed on 1 December 2023). Gallipoli is a key center in the region, known for its unique cultural heritage and coastal charm. The cityscape features traditional Mediterranean architecture, with two- to three-story buildings and winding narrow streets. The city's location is of particular interest for urban microclimate research because it is situated in the Mediterranean region, which has been recognized as a climate change "hot spot". This designation signifies that the impact of global warming, which showed an increasing trend over time, especially during the summer months, was recognized for the Mediterranean region, along with documented changes in precipitation patterns over the 20th-century [26]. According to the Köppen–Geiger climate classification, which is widely used to categorize global climates, Gallipoli falls into the Warm Mediterranean Climate 'Csa' class. This classification indicates a temperate–hot climate, with the average temperature of the coldest month ranging between 6 °C and 10 °C, the annual average temperature ranging between 14.5 °C and 17 °C, and having four months with an average temperature exceeding 20 °C. The unique climatic and geographical features of Gallipoli contribute to its distinctive atmosphere and make it a noteworthy area for further exploration and study.

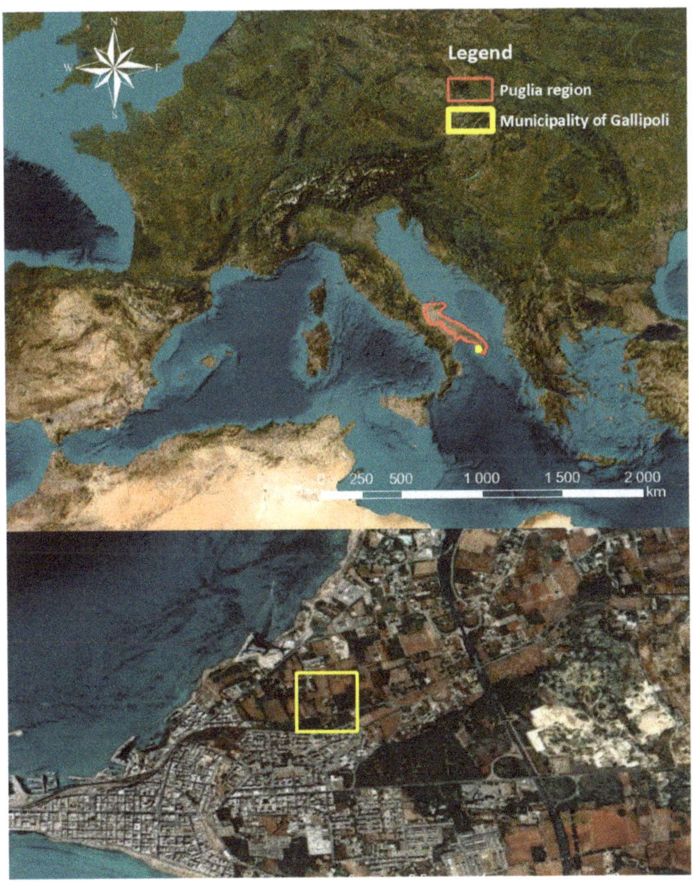

Figure 1. Study area of interest in Gallipoli municipality (base map from Google Earth).

2.2. *Applied Decision-Making Framework*

The decision-making framework evaluates the efficacy of ecosystem structure and function, defining the planned natural capital in NbSs, to achieve a favorable impact on priority ecosystem services. This entails comprehending how these natural elements contribute to lowering air temperature and PET in the urban setting through ENVI-met simulations across diverse scenarios [24,27]. Variations in the structure and function of natural capital, aimed at enhancing ecosystem service flows, significantly influence the investment costs associated with implementing NbSs. Therefore, different combinations of natural and human capital can result in different levels of ecosystem service flows and hence different human and economic benefits. For this reason, a cost–benefit analysis was conducted to appraise the direct social benefits of the different scenarios of the proposed NbSs. Various potential investment costs for its implementation were considered, aiding in selecting the optimal realization solution (Figure 2).

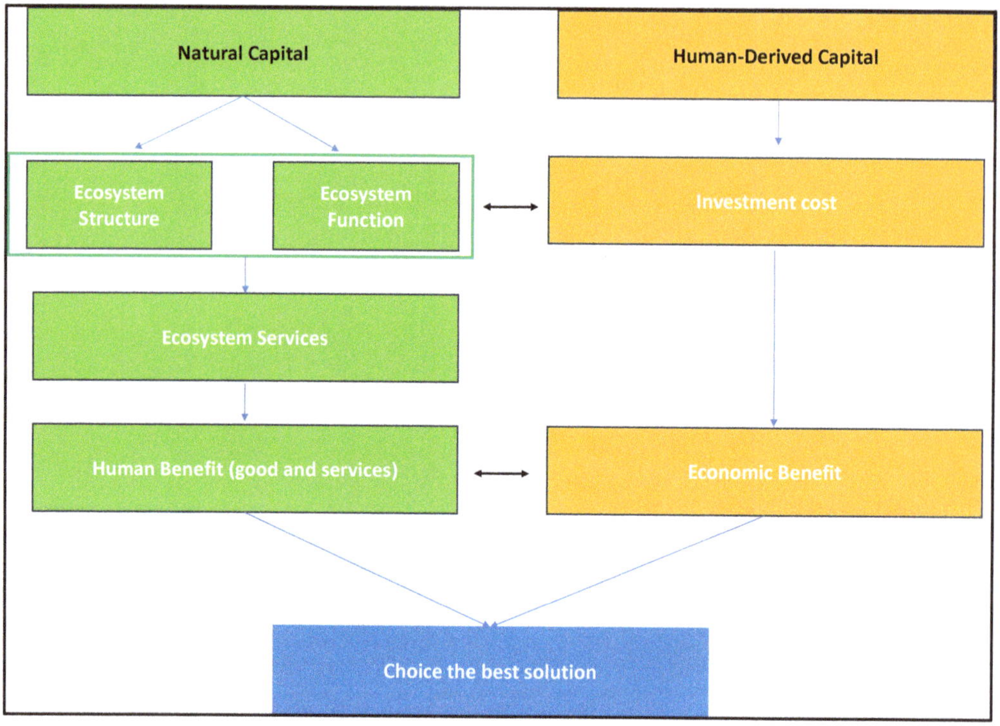

Figure 2. Framework applied to support decision-making processes.

2.2.1. ENVI-Met Analysis and Data Input

The PET serves as a crucial metric for assessing individuals' thermal comfort in outdoor settings [28–33]. The PET scale (Table 1) provides a classification of thermal comfort based on temperature, with the ranges listed below [28]. In the context of this study, a PET below 40 °C is considered acceptable for human health, as higher values may lead to increased thermal stress and health risks.

Temperature and PET simulations were conducted for different urban scenarios (Figure 3):

- First scenario (*Scenario zero*): This represents the current state scenario. It is characterized by the original land cover of the study area before urbanization, characterized mainly by arable land with a few buildings, including old rural structures mixed with new houses.
- Second scenario (*Intermedium scenario*): This represents the urban planned scenario without NbSs. It is characterized by the planned urbanization processes, featuring potential buildings, parking areas, roads, and arable land. Notably, no vegetation,

such as a green park, is included in this scenario. This setup allows for a comparison of urbanization effects on temperature and PET with both Scenario zero and the third scenario, which anticipates the implementation of NbS.

- Third scenario (*NbS scenario*): *This represents* the urban planned scenario with the planned urban forest NbS. It shares the same land cover attributes as the second scenario in terms of buildings, roads, and parking areas. However, it introduces the establishment of an urban forest as an NbS in place of arable land. The urban forest is strategically designed to address microclimate issues in the urban garden, along walkways, and around buildings, thereby enhancing thermal comfort.

Table 1. Ranges of Physiologically Equivalent Temperature (PET in °C) for different grades of thermal perception and physiological stress [28].

PET (°C)	Thermal Perception	Grade of Physiological Stress
<4	Very cold	Extreme cold stress
4–8	Cold	Strong cold stress
8–13	Cool	Moderate cold stress
13–18	Slightly cool	Slight cold stress
18–23	Comfortable	No thermal stress
23–29	Slightly warm	Slight heat stress
29–35	Warm	Moderate heat stress
35–41	Hot	Strong heat stress
>41	Very hot	Extreme heat stress

A thorough examination of microclimate and PET was conducted using ENVI-met software, a powerful tool for predicting intricate interactions between plants and atmospheric conditions in complex urban settings [24,27]. This software incorporates a 3D vegetation model that adapts to different biophysical structures in green spaces. As a result, it proves invaluable in assessing dynamic interactions between various types of vegetation and the surrounding environment, including built-up areas [16,28,29,34,35]. The suitability of ENVI-met for microclimate simulations in the Salento area has been validated through several local studies, exemplified by Gatto et al. [36]. Table 2 details the parameters utilized to configure the model for simulation.

Table 2. Initial and boundary conditions used in ENVI-met simulations.

Parameter	Definition	Value
Simulation time	Start date	31 July 2021
	Start of simulation (h)	00:00
	Total simulation time	19 h (5 h spin-up + 11 h)
Computational domain and grid	Grid cells (x, y, z)	170 × 170 × 25
	$\delta x \times \delta y \times \delta z$	2 m × 2 m × 2 m (equidistant: 5 cells close to the ground)
	Nesting grids	7
	Boundary conditions	Cyclic

Figure 3. Urban scenarios analyzed: (**A**) *Scenario zero*, (**B**) *Intermedium scenario*, and (**C**) *NbS scenario*.

The meteorological data incorporated into the model included air temperature, relative humidity, wind speed, wind direction, and precipitation. This information was sourced from the ARPA (Agenzia Regionale per la Prevenzione e la Protezione Ambientale)—Puglia station, measured at a height of 10 m above ground [37]. Specifically, values were selected from 1 July 2023 to 31 August 2023. The average values for each meteorological variable were calculated at every half-hour interval to establish the daily average profile, as shown in Figure 4.

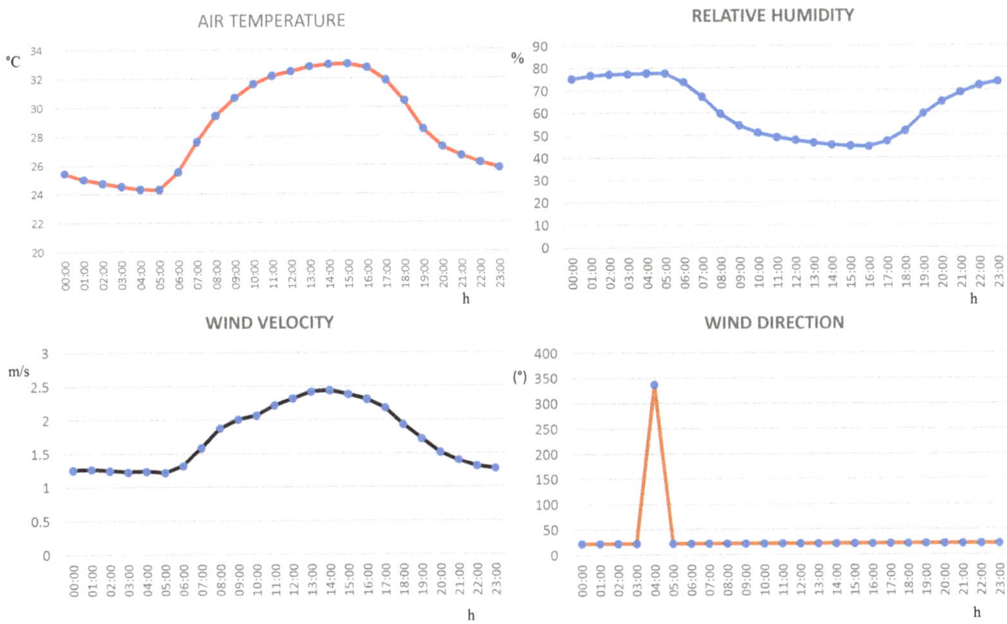

Figure 4. Daily profiles of meteorological parameters (Air Temperature, Relative Humidity, Wind Velocity and Wind Direction) employed as input in ENVI-met simulations.

2.2.2. Cost–Benefit Analysis

We conducted an analysis to assess the optimal method of implementing the NbS, focusing on planned social benefits through a cost–benefit analysis (CBA) centered on microclimate ecosystem services.

CBA is a systematic process employed to evaluate the potential social benefits and costs associated with a project, program, or policy [38–40]. It serves as a structured framework supporting decision-makers in assessing the effectiveness and desirability of various project options by quantifying and comparing their costs and benefits [38,39]. In this study, cost–benefit analysis (CBA) was used to evaluate the cost-effectiveness of three different implementation types for the same planned urban forest NbS. Given that urban forests provide a wide range of ecosystem services, the analysis focused on their impact on climate change mitigation and thermal comfort, as measured by the PET index, highlighting the direct human benefits derived from these services. These were defined as priority ecosystem services, while the energy saved from building cooling was considered an indirect benefit.

Three different implementation modalities of the urban forest were considered:

- In the first implementation modality of the NbS (Type A), the urban forest is planned to be established with plants 6–7 m in height, capable of developing into a mature forest in a short time. This type of urban forest is expected to support microclimate regulation and improve human health within a few years across the entire study area.
- In the second implementation modality of the NbS (Type B), the urban forest is planned to be established with plants 6–7 m in height, which require more time to grow and develop into a mature forest (approximately 20 years). As a result, this urban forest will not be able to support microclimate regulation or improve human health within the first few years.
- In the third implementation modality of the NbS (Type C), the urban forest combines elements of the first two modalities. It uses 6–7 m trees for 3/4 of the total tree population to mitigate microclimate effects around buildings, while incorporating

younger trees in the park area. This urban forest will be able to support microclimate regulation and improve human health within a few years, but only in the immediate vicinity of the buildings.

By assessing three different types of implementation costs, we calculated the Net Present Value (NPV) and benefit–cost ratio (BCR) for each. Specifically, the NPV represents the difference between the total Present Value of Future Benefits (PVFB) and the Present Value of Future Costs (PVFC) for each alternative considered [38,39]:

$$\text{NPV} = \sum \text{PVFB} - \sum \text{PVFC}$$

BCR represents the ratio between the total Present Value of Future Benefits (PVFB) and the total Present Value of Future Costs (PVFC), providing an indication of the feasibility and convenience of implementing a project [33,35,36].

$$BCR = \frac{\sum \text{PVFB}}{\sum \text{PVFC}}$$

The Present Value of Future Benefits (PVFB) is a method of assessing the benefits expected from a project or investment in today's terms and comparing them with its costs. Calculating the PVFB helps decision-makers understand whether the benefits justify the initial and ongoing costs of the project. The Present Value of Future Costs (PVFC), on the other hand, is the present value of the costs that will be incurred on the project over time, adjusted for the time value of money. The PVFB and PVFC were calculated using the Actualized Coefficient (AC), which is discount factor for restating future amounts to current values. Their values were calculated according to the following equations [38,40,41]:

$$\text{Present Value of Future Benefits} = \sum_{i=0}^{n} benefits * \text{AC}$$

$$\text{Present Value of Future Costs} = \sum_{i=0}^{n} Cost * \text{AC}$$

where $\text{AC} = \frac{1}{(1+r)^n}$, where r is discounting rate capital, representing the cost of money or capital, which was considered equal to 5% in relation to actual capital cost, and n represent years of analysis. In this study, n was considered for tree growth time windows of 5 years, 10 years, and 20 years [40].

The total cost comprises both initial costs (capital expenditures) and ongoing operational costs. In this instance, the total cost was estimated by considering the practices necessary to realize the urban forest, as detailed in Table 3. Therefore, an estimated calculation metric was made for the realization of the urban forest, considering the plants and materials needed for its realization. For each type of implementation considered, the price difference of the vegetation was determined by the size of the plants planned in the initial phase. Prices for each item were extrapolated from the price list commonly used for public projects in the Apulian region and informed by market analysis. Operational costs were omitted as they are likely to be similar across the three scenarios and, therefore, do not significantly impact the analysis. Similarly, the analysis systematically identifies and quantifies all benefits, both direct and indirect, arising from each alternative. In this study, the benefits considered for each scenario remain consistent, although the time required to achieve them varies. This accounts for the varying durations associated with growing urban forests using different practices (types A, B, and C). Primarily, the structural aspects of the plants change for each scenario, while the type remains consistent throughout the planting actions.

Table 3. Types of interventions considered for cost estimation of NBS in the cost–benefit analysis, aligned with the Puglia regional price list [42].

Code Price	Type of Intervention
OF 01.11	Tillage of the soil to a depth of no less than 60 cm on agricultural land
OF 01.03	Clearing of scrub and herbaceous vegetation on unwooded land, carried out by hand or using mechanical equipment (back-brush cutters), including collection, transport, and destruction of the resulting material
OF 01.24	Hole opening with a mechanical tool
OF 01.27	Planting of resinous and broad-leaved plants, including backfilling and compaction of the soil adjacent to the roots of the plants, and any other work necessary to ensure that the work is carried out in a workmanlike manner (excluding the delivery of the plant)
OF 01.32	Providing wooden stakes on site, including loading/unloading, transport and any other costs
OF 01.34	Supply and installation of a protective net cylinder for seedlings (tree shelter) to protect the seedling from ungulates
OF 01.30	Supply of broadleaf seedlings
Not available	Supply of adult broadleaf seedlings
OF 01.21	Localized mulching with discs or squares of biodegradable lignocellulosic material, minimum size 40 × 40 cm, including delivery, installation, and anchoring with stakes
OF 03.05	Crop care
OF 03.07	Emergency irrigation, including water supply at any distance and in any quantity, distribution of water by any means and in any way for each operation and seedling (20 L) (for the first three years after planting)

3. Results

3.1. ENVI-Met Analysis

The analysis of air temperature reveals a lower average in the urban scenario with the NbS (*NbS scenario*), as opposed to the higher average temperature observed in the urban scenario without the NbS (*Intermedium scenario*) and the current state scenario (*Scenario zero*) (Figure 5). Urbanization can have a detrimental effect on air temperature, resulting in elevated values compared to the area's natural state. However, this adverse impact can be mitigated through the implementation of NbSs, particularly those that incorporate forested elements. In this case study, NbSs play a crucial role in reducing the effects of urbanization, leading to a notable decrease in the average air temperature by approximately 3 °C during the hottest period of the day, specifically from 12:00 to 16:00. The analysis of variance revealed significant differences (Table 4), indicating that NbSs, especially those with forested components, contribute substantially to cooling effects and help counteract the temperature rise associated with urban development.

The maximum average air temperature was consistently recorded across all urban scenarios at 14:00, and Figure 6 illustrates the spatial distribution of air temperature between the *NbS scenario* and *Intermedium scenario* (the urban scenario without NbS). Specifically, it is noticeable that the air temperature around the residential building area can be reduced to below 30 °C thanks to the NbS, in comparison to the urban scenarios without the NbS. Mainly, the air temperature in the residential area can be reduced by more than 5 °C.

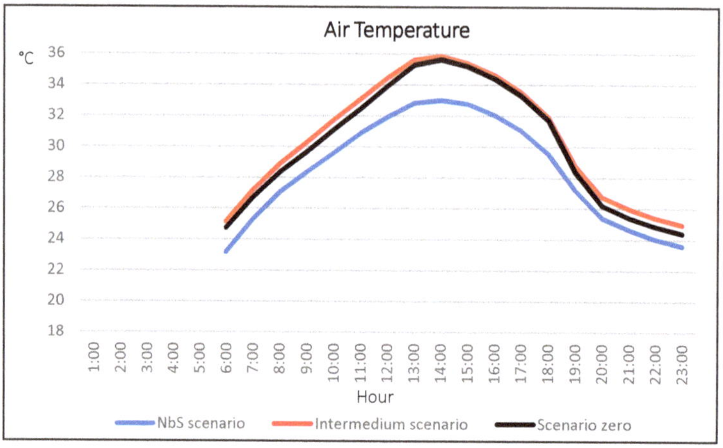

Figure 5. Daily profiles of air temperature obtained from ENVI-met simulations. The value represents the average air temperature of the study area.

Table 4. Analysis of variance between daily profiles of air temperature: (A) *NbS scenario* vs. *Intermedium scenario* and (B) *Intermedium scenario* vs. *Scenario zero*.

(A)																		
NbS scenario* vs. *Intermedium scenario																		
h	6:00	7:00	8:00	9:00	10:00	11:00	12:00	13:00	14:00	15:00	16:00	17:00	18:00	19:00	20:00	21:00	22:00	23:00
F Test	16.5	13.8	15.2	24.6	9.7	75.8	17.8	24.8	42.4	51.6	60.9	15.6	9.1	7.8	15.9	82.0	72.9	9.6
N	29,476																	
p Value	$p < 0.01$																	
(B)																		
Intermedium scenario* vs. *Scenario zero																		
h	6:00	7:00	8:00	9:00	10:00	11:00	12:00	13:00	14:00	15:00	16:00	17:00	18:00	19:00	20:00	21:00	22:00	23:00
F Test	37.1	78.6	23.3	40.0	15.7	50.6	10.7	16.1	28.2	33.6	35.5	32.3	29.3	23.5	66.3	11.4	170.0	66.6
N	29,476																	
p Value	$p < 0.01$																	

The positive impact of NbSs on the urban microclimate can have significant benefits for human health. The results of the PET analysis demonstrate the favorable influence of NbSs on enhancing livability in urban areas. This is reflected in the reduction in the spatially averaged (over the study area) PET value throughout the day when compared to an urban scenario without vegetation. The most notable change occurs at 17:00, where a difference of 11 °C in PET is observed. It is important to note that, in the urban setting with NbS, the highest spatially averaged PET value occurs at 13:00, reaching an average of 41.5 °C (Figure 7). However, during other times of the day, the spatially averaged PET values remain below 40 °C, which is considered a critical threshold for human health. In contrast, in the *Intermedium scenario*, spatially averaged PET values exceed 40 °C from 10:00 to 17:00, with the maximum recorded value reaching 50.6 °C. The analysis of variance showed significant differences (Table 5).

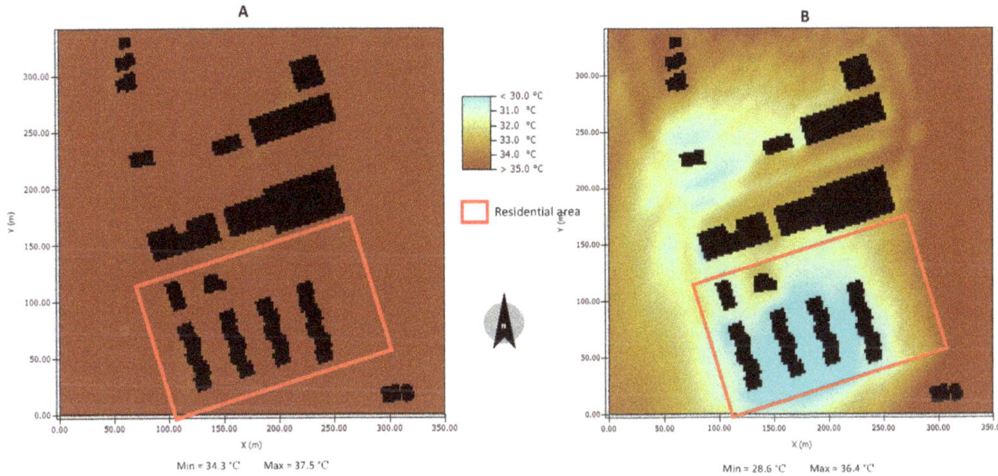

Figure 6. Spatial distribution of air temperature at z = 1.4 m at 14:00: (**A**) *Intermedium scenario* (urban scenario without NbS) and (**B**) *NbS scenario* (urban scenario with NbS).

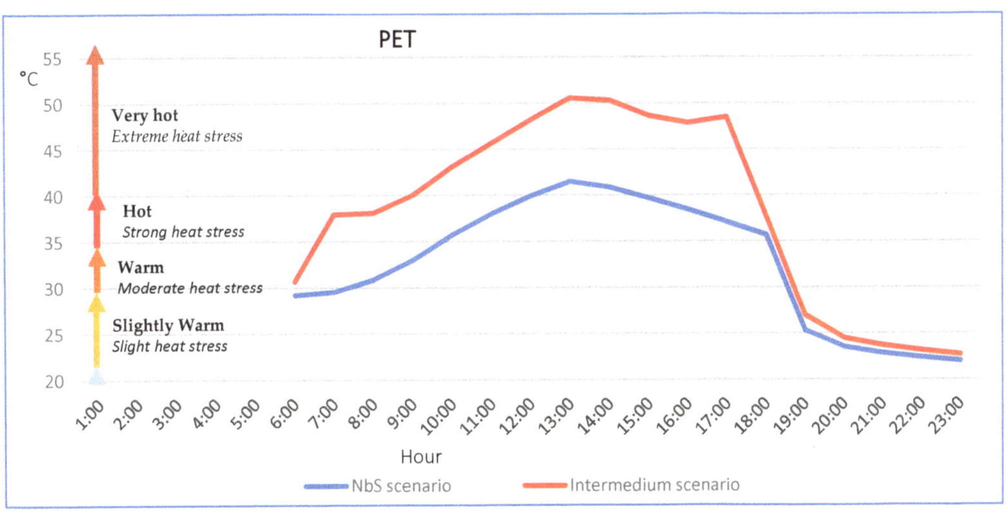

Figure 7. Daily profiles of spatially averaged PET values obtained from ENVI-met simulations, with indication of the thermal perception (bold) and the degree of physiological stress (italic) reported in Table 1.

Table 5. Analysis of variance between daily profiles of PET: *NBS scenario* (urban scenario with NbS) vs. *Intermedium scenario* (urban scenario without NbS).

	NbS scenario* vs. *Intermedium scenario																	
h	6:00	7:00	8:00	9:00	10:00	11:00	12:00	13:00	14:00	15:00	16:00	17:00	18:00	19:00	20:00	21:00	22:00	23:00
F Test	1.8	1.4	0.7	1.3	1.4	1.4	1.4	1.2	1.3	1.5	1.8	1.7	1.6	1.8	2.1	1.8	1.6	1.7
N	29,476																	
p Value	$p < 0.01$																	

Figure 8 illustrates the spatial distribution of PET between the *NbS scenario* (the urban scenario with NbS) and the *Intermedium scenario* (the urban scenario without NbS). Specifically, it is noticeable that the PET around the residential building area can be reduced to below 30 °C thanks to the NbS, in comparison to the *Intermedium scenario*, which recorded values above 48 °C.

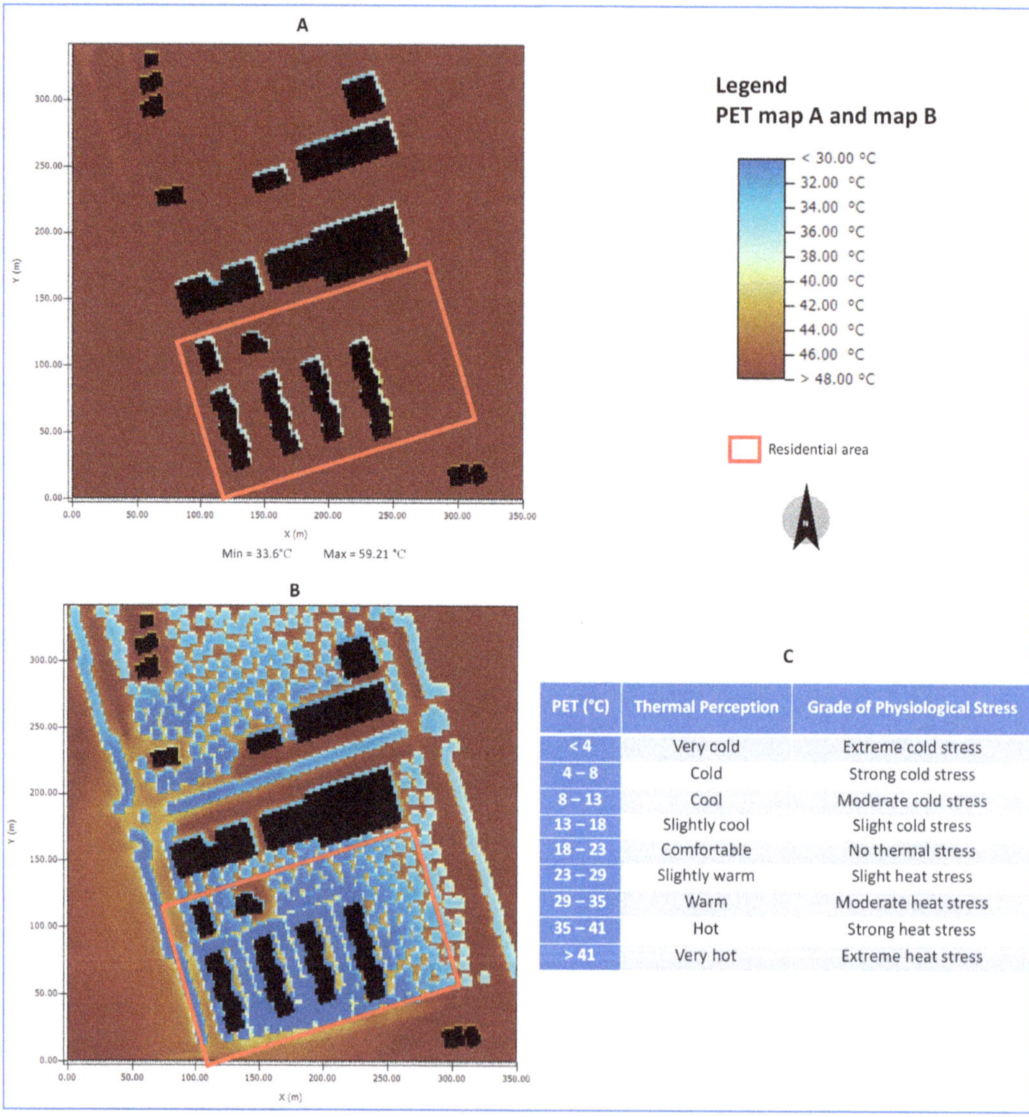

Figure 8. Spatial distribution of PET at z = 1.4 m at 13:00: (**A**) *Intermedium scenario* (urban scenario without NbS) and (**B**) *NbS scenario* (urban scenario with NbS). (**C**) Thermal Perception and Grade of Physiological Stress concerning the PET value (see Section 2.2.1).

NbSs can help reduce PET across the entire urban area, with varying impacts depending on the relationship between the urbanized zones and vegetation improvements (Figure 9). The most significant reduction in PET is observed in the residential areas. This decrease can positively influence the livability of these spaces, including outdoor areas like balconies, and potentially affect indoor environments as well. Importantly, this reduction is linked to the lowered air temperature resulting from the attenuation of solar irradiance throughout the day, as well as the wind intensity generated by the vegetation surrounding the buildings.

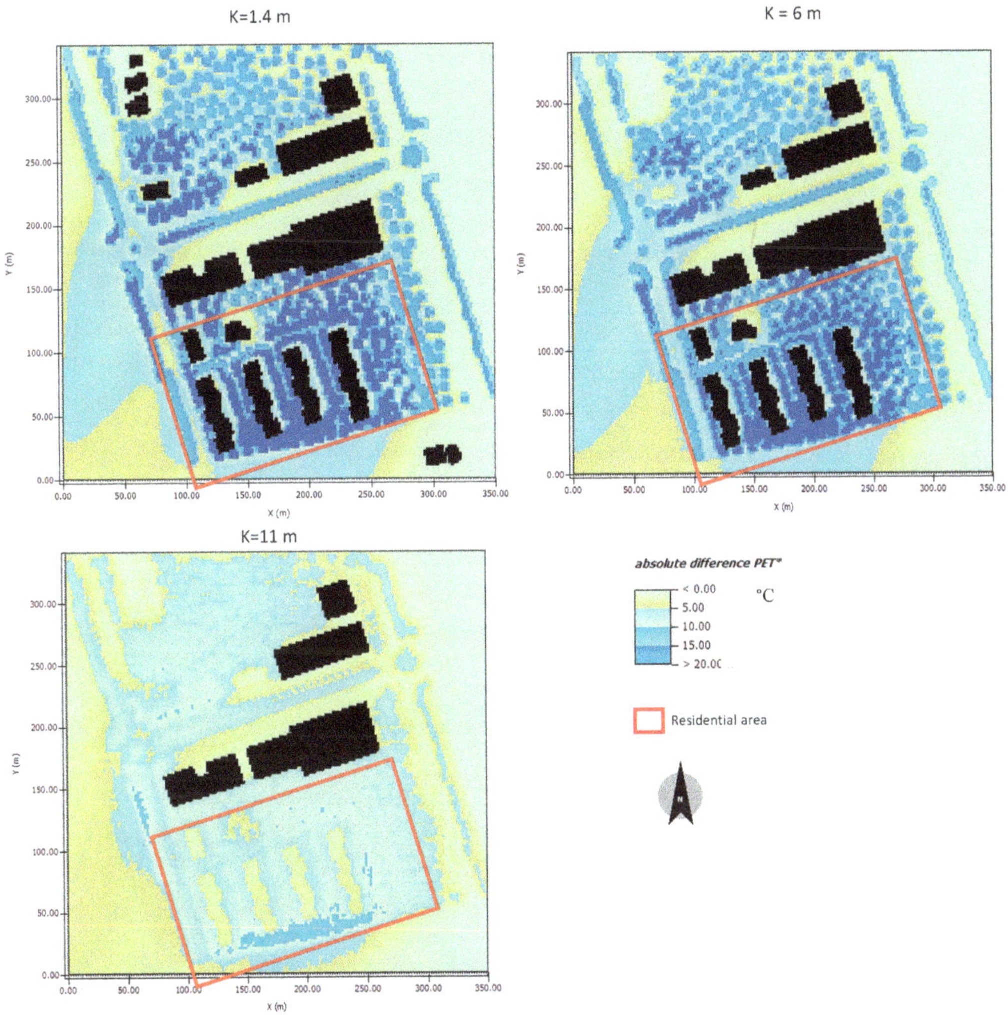

Figure 9. The difference in PET between the *NbS scenario* (urban scenario with NbS) and the *Intermedium scenario* (urban scenario without NbS) at 13:00 analyzed at different heights (z = 1.4 m, 5 m, 11 m).

These findings emphasize the effectiveness of NbSs at mitigating extreme temperatures and enhancing overall conditions in urban areas, especially during peak heat hours. Notably, the decrease in PET around the four residential buildings can contribute to a reduction in the human perception of heat and thermal stress.

This reduction in PET is crucial, as it not only improves comfort but also has a positive impact on human health by reducing heat-related risks. Additionally, by moderating outdoor temperatures, NbS can help lower energy consumption for cooling, potentially reducing the need for air conditioning. This dual benefit—improving public health and reducing energy consumption—emphasizes the importance of integrating NbSs into urban planning strategies, particularly in densely populated and heat-prone areas.

3.2. Cost–Benefit Analysis Results

Considering the characteristics of the study area, the priority ecosystem service considered when establishing the NbS is microclimate regulation to mitigate the human body's sensitivity to high temperatures, thereby reducing thermal discomfort in hot conditions, as measured by the PET. It can have a significant impact on human health when the PET exceeds 40 °C, leading to issues such as dehydration, cardiovascular problems, increased respiratory difficulties, and hyperthermia, which can escalate to heatstroke, potentially fatal if not promptly treated [14]. Furthermore, extended exposure to extreme temperatures can result in psychological effects, including heightened irritability and a reduced quality of sleep [14,43–47]. Therefore, the direct social and economic benefits that could be generated by an urban forest through microclimate regulation include the reduction in hospital costs for treating individuals affected by high PET and a decrease in building cooling expenses to lower air temperature and PET in the building. Specifically, trees in the residential area can reduce air temperature and direct solar radiation on the building during the hottest times. In this study, it was estimated that the total annual benefits could amount to EUR 67,500 for the forest capable of lowering the temperature and PET (Table 6).

Table 6. Estimation of economic benefits provided by NbSs. Benefit 1 is based on the daily public costs saved by hospitals in caring for single individuals in Italy [48]. Benefit 2 is estimated by considering the energy saved due to reduced cooling needs for buildings as a result of lower air temperatures and PET.

1	Benefit	**Hospital admission/day (EUR)**	**N days**	**N People**	**Total**
		500	3	9	13,500
2	Benefit	**Energy saved/day/study area (EUR)**	**N Days**		**Total**
		1200	45		54,000
				Total (1 + 2)	67,500

The practical framework for analyzing the impact of NbSs on social capital omust consider various scenarios in terms of the financial capital required to implement them and their capacity to provide ecosystem services [14,15]. Even when starting with the same planned NbS, different social and environmental impacts can be expected due to variations in the initial practices used for implementation. These differences may necessitate varying capital investments and yield diverse benefits over time.

In this study, investment cost scenarios for the realization of an urban forest have been examined (Figure 10):

- In the first NbS implementation type of (A), the investment cost for its realization was estimated at EUR 456,655. The ability of the forest to serve as a mature ecosystem for the provision of ecosystem services could begin five years after its establishment. Here, social benefits can be accounted for at 50% in the first five years and 100% thereafter.
- In the second NbS implementation type (B), the investment cost for its realization was estimated at EUR 50,029, and it may take up to 15 years before the forest generates direct benefits. Its full potential as an ecosystem for providing services will be reached 20 years after its establishment. Here, social benefits can be accounted for at 50% from 15 to 19 years and 100% from 20 years.
- In the third NbS implementation type (C), investment cost for its realization was estimated EUR 299,529. In this case, social benefits can be accounted for at 50% from 0 to 5 years, at 90% from 5 to 10 years, and then at 100%.

Year	AC	Type A) PVFC = 456 655 €				Type B) PVFC = 50 029 €				Type C) PVFC= 299 529€			
		Value Benefit	**PVFB**	**NPV**	**BCR**	Value Benefit	**PVFB**	**NPV**	**BCR**	**Value Benefit**	**PVFB**	**NPV**	**BCR**
0	0.00	0 €								0 €			
1	1.05	45,000 €								32,143 €			
2	1.10	42,857 €	220,434 €	-236,221 €	0.48		0	0	0	30,612 €	167,275 €	-132,254 €	0.56
3	1.16	40,816 €								29,155 €			
4	1.22	38,873 €								27,766 €			
5	1.28	52,888 €								47,599 €			
6	1.34	50,370 €								45,333 €			
7	1.41	47,971 €								43,174 €			
8	1.48	45,687 €	449,412 €	-7243	0.98		0	0	0	41,118 €	377,499 €	77,969 €	1.26
9	1.55	43,511 €								39,160 €			
10	1.63	41,439 €								41,439 €			
11	1.71	39,466 €								39,466 €			
12	1.80	37,587 €								37,587 €			
13	1.89	35,797 €								35,797 €			
14	1.98	34,092 €								34,092 €			
15	2.08	32,469 €	769,394 €	312,739 €	1.68	16,234 €	99,240 €	49,211 €	1.98	32,469 €	699,177 €	399,648 €	2.33
16	2.18	30,923 €				15,461 €				30,923 €			
17	2.29	29,450 €				14,725 €				29,450 €			
18	2.41	28,048 €				14,024 €				28,048 €			
19	2.53	26,712 €				13,356 €				26,712 €			
20	2.65	25,440 €				25,440 €				27,136 €			

Figure 10. The cost–benefit analysis carried out for three scenarios: (**A**) urban forest with mature trees; (**B**) urban forest with young trees, (**C**) urban forest with a mix of mature trees in residential areas and young trees elsewhere.

The CBA indicates that the first scenario is less advantageous than the other two, primarily in terms of their benefit–cost ratios (BCRs). While the first scenario may grant more benefits within the initial 5- and 10-year timeframes, these benefits are insufficient to offset the investment costs. The second scenario is unable to produce benefits before 15 years, as it takes time for the trees to reach the required height to reduce microclimate temperature by 3 °C and bring the PET below 40 °C in the residential area. In contrast, the third scenario is more advantageous compared to the other two because it can generate significant benefits within the first 5 years, effectively covering the initial investment costs. Therefore, it can be considered the most suitable solution, as it combines the natural capital required to provide priority ecosystem services with human-derived capital, considering implementation costs and direct social benefits over time.

The analysis was tested with different discount rates (r) to assess their potential influence on the results. The rates varied from 0.03 to 0.05 and 0.07, yet no significant differences in the main outcomes were observed. Specifically, the benefit–cost ratio (BCR) values consistently favor scenario C across all discount rates tested. For example, at a 7% discount rate, the BCR values at 20 years are A = 1.42, B = 1.42, and C = 1.94. At a 3% discount rate, the corresponding values are A = 2.03, B = 2.79, and C = 2.85. Additionally, the estimation of energy savings for cooling the building is considered a critical factor. Even when reducing the daily values to EUR 400 (daily economic value of the energy saved in cooling the buildings planned to be built in the area studied) for the total area, scenario C consistently demonstrates higher BCR values compared to the other two scenarios. At 20 years, the BCR values are A = 0.79, B = 0.93, and C = 1.09. Therefore, the results exhibit independence from the discount rate, indicating robustness in the findings.

4. Conclusions

NbSs provide a comprehensive framework for addressing the challenges of urbanization, including land use and land cover changes, while promoting sustainable urban development. This work shows the value of integrating ecological and economic perspectives to evaluate and implement NbSs that can mitigate urban microclimates and improve public health. Our findings demonstrate that urban forests, as a type of NbS, effectively reduce average temperatures by approximately 3 °C in peri-urban areas and lower the PET around residential buildings to below 40 °C, a critical threshold for human health [7,8]. These outcomes are particularly relevant in the context of global warming and urban heat islands, where high temperatures pose significant risks to population well-being. The results emphasize the dual benefits of NbS: enhancing thermal comfort and reducing energy consumption for building cooling.

The methodology developed here is not limited to local applications but offers potential for broader use, particularly in the Strategic Environmental Assessment (SEA) process. SEA, a mandatory practice across EU countries, evaluates the environmental impacts of urban plans and programs, making it an ideal framework to apply our approach. By assessing different urban scenarios, this study provides a replicable model for integrating environmental, social and economic considerations into urban planning decisions [49]. While the methodology focuses on comparative rather than absolute evaluations, this approach enables planners to understand the relative effectiveness of NbSs in specific contexts.

The integration of ecological and economic factors represents a key strength of this study, offering a holistic perspective on the planning and implementation of NbSs. However, we are aware that the accuracy of the results depends on the capacity of the models to simulate air temperature and PET in urban settings with specific vegetation configurations. Future research should prioritize improving model accuracy and testing the methodology across different urban contexts to enhance its applicability and reliability.

Additionally, this study reveals the importance of considering the temporal dynamics of NbS implementation. For example, urban forests that incorporate mature and younger trees can provide long-term benefits. This highlights the necessity of strategic timing and resource allocation when implementing NbSs to optimize their effectiveness.

From a policy perspective, this study underlines the need to focus not only on the quantity and type of NbSs but also on their real-world effectiveness in delivering ecosystem services. Urban policies should integrate the following:

- Natural capital, representing the ecological structures and functions of NbSs, which are essential for ensuring the provision of ecosystem services aligned with priority objectives [15,16];
- Human-derived capital, reflecting the economic investments and operational efforts required for NbS implementation. This integration ensures that NbSs are socially, economically, and environmentally sustainable [15].

This study also highlights the importance of considering cost-effectiveness in NbS planning. The results of the cost–benefit analysis reveal that implementation strategies combining mature and younger trees offer the highest BCRs. Policymakers must therefore develop strategies that balance ecological impacts with financial feasibility to maximize long-term benefits.

In conclusion, this research highlights the critical role of NbSs in addressing urban challenges associated with global warming. The key insights include the following:

- The necessity of defining priority ecosystem services, such as microclimate regulation, to guide NbS implementation [14,16];
- The importance of aligning natural and human-derived capital to ensure effective NbS delivery [15,16];
- The value of a comparative evaluation framework that enables urban planners to assess the relative benefits of different NbS scenarios.

By focusing on the real-world implementation and long-term sustainability of NbSs, this work offers actionable insights for policymakers and urban planners trying to create more resilient and livable cities.

Author Contributions: Conceptualization, T.S. and R.B.; methodology, T.S.; software, T.S., R.B. and R.E.; validation, T.S., A.C., J.H., R.B. and R.E.; formal analysis, T.S.; investigation, T.S.; resources, T.S.

and R.B.; data curation, T.S.; writing—original draft preparation, T.S.; writing—review and editing, R.B., A.C., J.H. and R.E.; visualization, T.S.; supervision, R.B.; project administration, T.S.; funding acquisition, T.S. All authors have read and agreed to the published version of the manuscript.

Funding: This research received no external funding.

Data Availability Statement: The data will be made available upon a specific request.

Acknowledgments: T.S. thanks ITINERIS Project funded by EU—Next Generation EU Mission 4 "Education and Research"—Component 2: "From research to business"—Investment 3.1: "Fund for the realisation of an integrated system of research and innovation infrastructures"—Project IR0000032—ITINERIS—Italian Integrated Environmental Research Infrastructures System—CUP B53C22002150006. He acknowledges the Research Infrastructures participating in the ITINERIS project with their Italian nodes: ACTRIS, ANAEE, ATLaS, CeTRA, DANUBIUS, DISSCO, e-LTER, ECORD, EMPHASIS, EMSO, EUFAR, Euro-Argo, EuroFleets, Geoscience, IBISBA, ICOS, JERICO, LIFEWATCH, LNS, N/R Laura Bassi, SIOS, SMINO.

Conflicts of Interest: The authors declare no conflicts of interest.

References

1. Emmanuel, R.; Krüger, E.L. Urban Heat Island Intensity in Sydney, Australia. *Procedia Environ. Sci.* **2012**, *14*, 85–96.
2. Yadav, N.; Rajendra, K.; Awasthi, A.; Singh, C.; Bhushan, B. Systematic exploration of heat wave impact on mortality and urban heat island: A review from 2000 to 2022. *Urban Clim.* **2023**, *51*, 101622. [CrossRef]
3. Rajagopal, P.; Priya, R.S.; Senthil, R. A review of recent developments in the impact of environmental measures on urban heat island. *Sustain. Cities Soc.* **2023**, *88*, 104279. [CrossRef]
4. Emmanuel, R.; Krüger, E. Urban Heat Islands and its impact on climate change resilience in a shrinking city: The case of Glasgow, UK. *Build. Environ.* **2012**, *53*, 137–149. [CrossRef]
5. You, M.; Huang, J.; Guan, C.H. Are New Towns Prone to Urban Heat Island Effect? Implications for Planning Form and Function. *Sustain. Cities Soc.* **2023**, *99*, 104939. [CrossRef]
6. He, B.; Zhao, Z.; Shen, L.; Wang, H.; Li, L. An approach to examining performances of cool/hot sources in mitigating/enhancing land surface temperature under different temperature backgrounds based on Landsat 8 image. *Sustain. Cities Soc.* **2019**, *44*, 416–427. [CrossRef]
7. Qi, J.; Ding, L.; Lim, S. A Decision-Making Framework to Support Urban Heat Mitigation by Local Governments. *Resour. Conserv. Recycl.* **2022**, *184*, 106420. [CrossRef]
8. Emmanuel, R.; Loconsole, A. Green infrastructure as an adaptation approach to tackle urban overheating in the Glasgow Clyde Valley Region. *Landsc. Urban Plan.* **2015**, *138*, 71–86. [CrossRef]
9. Wang, Y.; Guo, Z.; Han, J. The relationship between urban heat island and air pollutants and their influencing factors in the Yangtze River Delta, China. *Ecol. Indic.* **2021**, *129*, 107976. [CrossRef]
10. Swamy, G.S.N.V.K.S.N.; Nagendra, S.M.; Sch, U. Impact of urban heat island on meteorology and air quality at microenvironments. *J. Air Waste Manag. Assoc.* **2020**, *70*, 876–891. [CrossRef]
11. Ernst, M.; Le Mentec, S.; Louvrier, M.; Loubet, B.; Personne, E.; Stella, P. Impact of urban greening on microclimate and air quality in the urban canopy layer: Identification of knowledge gaps and challenges. *Front. Environ. Sci.* **2022**, *10*, 924742. [CrossRef]
12. Chaudhary, M.T. Urban Air Pollution, Urban Heat Island and Human Health: A Review of the Literature. *Sustainability* **2022**, *14*, 9234. [CrossRef]
13. Somers, K.A.; Bernhardt, E.S.; McGlynn, B.L.; Urban, D.L. Downstream dissipation of storm flow heat pulses: A case study and its landscape-level implications. *J. Am. Water Resour. Assoc.* **2016**, *52*, 281–297. [CrossRef]
14. Semeraro, T.; Gatto, E.; De Bellis, L.; Luvisi, A.; Emmanuel, R.; Buccolieri, R. A decision-making framework for promoting the optimum design and planning of Nature-based Solutions at the local scale. *Urban For. Urban Green.* **2023**, *84*, 127945. [CrossRef]
15. Tan, P.Y.; Zhang, J.; Masoudi, M.; Alemu, J.B.; Edwards, P.J.; Grêt-Regamey, A.; Richards, D.R.; Saunders, J.; Song, X.P.; Wong, L.W. A conceptual framework to untangle the concept of urban ecosystem services. *Landsc. Urban Plan.* **2020**, *200*, 103837. [CrossRef] [PubMed]
16. Semeraro, T.; Scarano, A.; Pandey, R. Ecosystem Services Analysis and Design through Nature-Based Solutions in Urban Planning at a Neighbourhood Scale. *Urban Sci.* **2022**, *6*, 23. [CrossRef]
17. Semeraro, T.; Radicchio, B.; Medagli, P.; Arzeni, S.; Turco, A.; Geneletti, D. Integration of Ecosystem Services in Strategic Environmental Assessment of a Peri-Urban Development Plan. *Sustainability* **2021**, *13*, 122. [CrossRef]
18. Akbari, H.; Kolokotsa, D. Three decades of urban heat islands and mitigation technologies research. *Energy Build.* **2016**, *133*, 834–842. [CrossRef]
19. Wong, N.H.; Tan, C.L.; Kolokotsa, D.D.; Takebayashi, H. Greenery as a mitigation and adaptation strategy to urban heat. *Nat. Rev. Earth Environ.* **2021**, *2*, 166–181. [CrossRef]
20. Semeraro, T.; Scarano, A.; Buccolieri, R.; Santino, A.; Eeva, A. Planning of Urban Green Spaces: An Ecological Perspective on Human Benefits. *Land* **2021**, *10*, 105. [CrossRef]

21. Zhao, J.; Davies, C.; Veal, C.; Xu, C.; Zhang, X.; Yu, F. Review on the Application of Nature-Based Solutions in Urban Forest Planning and Sustainable Management. *Forests* **2024**, *15*, 727. [CrossRef]
22. Battisti, L.; Giacco, G.; Moraca, M.; Pettenati, G.; Dansero, E.; Larcher, F. Spatializing Urban Forests as Nature-based Solutions: A methodological proposal. *Cities* **2024**, *144*, 104629. [CrossRef]
23. Zhang, S.; Zhang, X.; Niu, D.; Fang, Z.; Chang, H.; Lin, Z. Physiological equivalent temperature-based and universal thermal climate index-based adaptive-rational outdoor thermal comfort models. *Build. Environ.* **2023**, *228*, 109900. [CrossRef]
24. Höppe, P. The physiological equivalent temperature—A universal index for the biometeorological assessment of the thermal environment. *Int. J. Biometeorol.* **1999**, *43*, 71–75. [CrossRef] [PubMed]
25. Jones, L.; Norton, L.; Austin, Z.; Browne, A.L.; Donovan, D.; Emmett, B.A.; Grabowski, Z.J.; Howard, D.C.; Jones, J.P.G.; Kenter, J.O.; et al. Stocks and flows of natural and human-derived capital in ecosystem services. *Land Use Policy* **2016**, *52*, 151–162. [CrossRef]
26. Lionello, P.; Abrantes, F.; Congedi, L.; Dulac, F.; Gacic, M.; Gomis, D.; Goodess, C.; Hoff, H.; Kutiel, H.; Luterbacher, J.; et al. Introduction: Mediterranean climate: Background information. In *The Climate of the Mediterranean Region: From the Past to the Future*; Lionello, P., Ed.; Elsevier: Amsterdam, The Netherlands, 2012.
27. Bruse, M.; Fleer, H. Simulating surface–plant–air interactions insideurban environments with a three dimensional numerical model. *Environ. Softw.* **1998**, *13*, 373–384. [CrossRef]
28. Matzarakis, A.; Mayer, H.; Iziomon, M. Applications of a universal thermal index: Physiological equivalent temperature. *Int. J. Biometeorol.* **1999**, *43*, 76–84. [CrossRef]
29. Binarti, F.; Koerniawan, M.D.; Triyadi, S.; Matzarakis, A. The predicted effectiveness of thermal condition mitigation strategies for a climate-resilient archaeological park. *Sustain. Cities Soc.* **2022**, *76*, 103457. [CrossRef]
30. Binarti, F.; Triyadi, S.; Koerniawan, M.D.; Pranowo, P.; Matzarakis, A. Climate characteristics and the adaptation level to formulate mitigation strategies for a climate-resilient archaeological park. *Urban Clim.* **2021**, *36*, 100811. [CrossRef]
31. Cai, Y.; Li, C.; Ye, L.; Xiao, L.; Gao, X.; Mo, L.; Du, H.; Zhou, Y.; Zhou, G. Effect of the roadside tree canopy structure and the surrounding on the daytime urban air temperature in summer. *Agric. For. Meteorol.* **2022**, *316*, 108850. [CrossRef]
32. Manavvi, S.; Rajasekar, E. Evaluating outdoor thermal comfort in urban open spaces in a humid subtropical climate: Chandigarh, India. *Build. Environ.* **2022**, *209*, 108659.
33. Paramita, B.; Kusuma, H.E.; Matzarakis, A. Urban performance based on biometeorology index in high-density, hot, and humid cities. *Sustain. Cities Soc.* **2022**, *80*, 103767. [CrossRef]
34. Roth, M.; Lim, V.H. Evaluation of canopy-layer air and mean radiant temperature simulations by a microclimate model over a tropical residential neighbourhood. *Build. Environ.* **2017**, *112*, 177–189. [CrossRef]
35. Tsoka, S.; Tsikaloudaki, A.; Theodosiou, T. Analyzing the ENVI-met microclimate model's performance and assessing cool materials and urban vegetation applications–a review. *Sustain. Cities Soc.* **2018**, *43*, 55–76. [CrossRef]
36. Gatto, E.; Buccolieri, R.; Aarrevaara, E.; Ippolito, F.; Emmanuel, R.; Perronace, L.; Santiago, J.L. Impact of urban vegetation on outdoor thermal comfort: Comparison between a Mediterranean City (Lecce, Italy) and a Northern European City (Lahti, Finland). *Forests* **2020**, *11*, 228. [CrossRef]
37. Agenzia Regionale per la Prevenzione e la Protezione Ambientale. 2023. Available online: https://www.arpa.puglia.it/ (accessed on 1 December 2023).
38. Raihan, A.; Said, M.N.M. Cost–Benefit Analysis of Climate Change Mitigation Measures in the Forestry Sector of Peninsular Malaysia. *Earth Syst. Environ.* **2022**, *6*, 405–419. [CrossRef]
39. Biancardo, S.A.; Gesualdi, M.; Savastano, D.; Intignano, M.; Henke, I.; Pagliara, F. An innovative framework for integrating Cost-Benefit Analysis (CBA) within Building Information Modeling (BIM). *Socio-Econ. Plan. Sci.* **2023**, *85*, 101495. [CrossRef]
40. European Commission. *Guide to Cost-Benefit Analysis of Investment Projects. Economic Appraisal Tool for Cohesion Policy 2014–2020*; European Commission: Brussels, Belgium, 2014; ISBN 978-92-79-34796-2.
41. Roberts, R.J.; Naimy, V. Strategic Adoption of Genetically Modified Crops in L ebanon: A Comprehensive Cost–Benefit Analysis and Implementation Framework. *Sustainability* **2024**, *16*, 8350. [CrossRef]
42. Regione Puglia. Prezzario Regionale Puglia 2023. Dipartimento Bilancio, Affari Generali e Infrastrutture, Sezione Opere Pubbliche e Infrastrutture. 2023. Available online: https://regione.puglia.it/ (accessed on 1 December 2023).
43. Nastos, P.T.; Matzarakis, A. The effect of air temperature and human thermal indices on mortality in Athens, Greece. *Theor. Appl. Climatol.* **2012**, *108*, 591–599. [CrossRef]
44. Sharafkhani, R.; Khanjani, N.; Bakhtiari, B.; Jahani, Y.; Tabrizi, J.S. Physiological equivalent temperature index and mortality in Tabriz (The northwest of Iran). *J. Therm. Biol.* **2018**, *71*, 195–201. [CrossRef]
45. Dastoorpoor, M.; Khodadadi, N.; Masoumi, K.; Khanjani, N.; Idani, E.; Borsi, S.H.; Goudarzi, G.; Raji, H.; Sharafkhani, R. Physiological equivalent temperature (PET) and nonaccidental, cardiovascular and respiratory disease mortality in Ahvaz, Iran. *Environ. Geochem. Health* **2021**, *44*, 2767–2782. [CrossRef] [PubMed]
46. National Institute of Environmental Health Sciences. Temperature-Related Death and Illness. 2023. Available online: https://www.niehs.nih.gov/research/programs/climatechange/health_impacts/heat (accessed on 16 December 2023).
47. World Health Organization. Heat and Health. 2023. Available online: https://www.who.int/news-room/fact-sheets/detail/climate-change-heat-and-health (accessed on 13 December 2023).

48. Il Sole 24 Ore Journal. l Sole 24 Ore Ministero Della Salute. 2020. Available online: https://www.ilsole24ore.com/art/dal-check-up-ricovero-quanto-costa-curarsi-senza-sistema-sanitario-nazionale-AFrRBWQ?refresh_ce=1 (accessed on 1 November 2023).
49. EU. Strategic Environmental Assessment. 2024. Available online: https://environment.ec.europa.eu/law-and-governance/environmental-assessments/strategic-environmental-assessment_en (accessed on 30 November 2024).

MDPI AG
Grosspeteranlage 5
4052 Basel
Switzerland
Tel.: +41 61 683 77 34

Land Editorial Office
E-mail: land@mdpi.com
www.mdpi.com/journal/land

www.ingramcontent.com/pod-product-compliance
Lightning Source LLC
Chambersburg PA
CBHW040743250926
51850CB00007BB/1390
* 9 7 8 3 7 2 5 8 5 4 4 9 3 *